BASIC BUILDING
& CONSTRUCTION SKILLS

BASIC BUILDING & CONSTRUCTION SKILLS

Adrian Laws

CONSTRUCTION + PLUMBING Skills Series

7e

TAFE NSW

Basic Building and Construction Skills
7th Edition
Adrian Laws

Portfolio manager: Sophie Kaliniecki
Product manager: Sandy Jayadev
Content developer: Margie Asmus
Project editor: Raymond Williams
Cover and Text Design: Cengage Creative Studio
Cover Illustration: Studio Fable/Antonia Pesenti
Art direction: Danielle Maccarone
Editor: Sylvia Marson
Proofreader: James Anderson
Permissions/Photo researcher: Lumina Datamatics
Indexer: KnowledgeWorks Global Ltd
Typeset by KnowledgeWorks Global Ltd

Any URLs contained in this publication were checked for currency during the production process. Note, however, that the publisher cannot vouch for the ongoing currency of URLs.

Sixth edition published in 2020

For product information and technology assistance,
in Australia call 1300 790 853;
in New Zealand call 0800 449 725

For permission to use material from this text or product, please email aust.permissions@cengage.com

National Library of Australia Cataloguing-in-Publication Data
ISBN: 9780170463126
A catalogue record for this book is available from the National Library of Australia

Cengage Learning Australia
Level 5, 80 Dorcas Street
Southbank VIC 3006 Australia

For learning solutions, visit cengage.com.au

Printed in China by 1010 Printing International Limited.
1 2 3 4 5 6 7 28 27 26 25 24

BRIEF CONTENTS

CONTENTS

Part 1 Basic Workplace Skills 1

Guide to the text

As you read this text you will find a number of features in every chapter to enhance your study of Basic Building and Constructions Skills and help you understand how the theory is applied in the real world.

PART-OPENING FEATURES

Part openers introduce each of the chapters within the part and give an overview of how the chapters in the text relate to each other

PART 1
BASIC WORKPLACE SKILLS

CHAPTER-OPENING FEATURES

A list of key competencies give you a clear sense of what topics and competencies each chapter will cover. It will indicate what you should be able to do after reading the chapter within that part, and give an overview of how the chapters in the text relate to each other.

1 APPLY WHS REQUIREMENTS, POLICIES AND PROCEDURES IN THE CONSTRUCTION INDUSTRY

This chapter covers the following topics from the competency CPCCCA3008 Apply WHS requirements, policies and procedures in the construction industry:
* identify and assess risks
* identify hazardous materials and other hazards on work sites
* plan and prepare for safe work practices
* apply safe work practices
* follow emergency procedures.

This unit specifies the outcomes required to carry out work health and safety (WHS) requirements through safe work practices in all on- or off-site construction workplaces.

It requires the performance of work in a safe manner through awareness of risks and work requirements, and the planning and performance of safe work practices with concern for personal safety and the safety of others.

The unit covers fundamental WHS requirements necessary to undertake work tasks within any sector in the construction industry. It includes the identification of hazardous materials, including asbestos, and compliance with legislated work safety practices. It does not cover removal of asbestos, which is a licensed activity.

It applies to workers in the construction industry.

Learning Tasks encourage you to practically apply the knowledge and skills that you have just read about.

ICONS

GREEN TIP

Chemicals are normally very concentrated and can cause a lot of damage to the environment. Make sure you work with chemicals according to the SDSs that are available from the chemical's manufacturer.

Green Tip boxes highlight material that relates to environmentally-sustainable workplace practices.

Acute hazards are sometimes given priority when working because they cause a result quickly. However, chronic hazards can result in very debilitating health complications. The urgency of the job is not worth getting injured for.

Caution boxes highlight material relating to workplace health and safety.

 COMPLETE WORKSHEET 4

Worksheet icons indicate when it is appropriate to stop reading and complete a worksheet at the end of the chapter.

END-OF-CHAPTER FEATURES

At the end of each chapter you will find several tools to help you to review, practise and extend your knowledge of the key learning objectives.

Chapter summaries highlight the important concepts covered in each chapter as well as link back to the key competencies.

SUMMARY

In Chapter 1 you have learnt how to carry out work health and safety (WHS) requirements through safe work practices in all onsite or offsite construction workplaces.

You have learnt how to perform work in a safe manner through awareness of risks and work requirements, and the planning and performance of safe work practices with concern for personal safety and the safety of others.

You have also learnt about fundamental WHS requirements necessary to undertake work tasks within any sector in the construction industry. You also learned

about the identification of hazardous materials, including asbestos, and compliance with legislated work safety practices. We did not cover removal of asbestos, which is a licensed activity. You have also learnt to:

- identify and assess risks
- identify hazardous materials and other hazards on work sites
- plan and prepare for safe work practices
- apply safe work practices
- follow emergency procedures.

PART 1

1

The **References and Further Reading** sections provide you with a list of each chapter's references, as well as links to important text and web-based resources.

REFERENCES AND FURTHER READING

Acknowledgement
Reproduction of the following Resource List references from DET, TAFE NSW C&T Division (Karl Dunkel, Program Manager, Housing and Furniture) and the Product Advisory Committee is acknowledged and appreciated.

Texts
Comcare government website, search for General construction induction training card information,

Web-based resources
Regulations/Codes/Laws
http://www.austlii.edu.au search for Australian laws
http://www.safeworkaustralia.gov.au, search for Model WHS Regulation 2011
https://www.safeworkaustralia.gov.au/doc/model-code-practice-construction-work, Safe Work Australia – Model Code of Practice – Construction Work
https://www.worksafe.vic.gov.au

The **Get It Right** photo case study shows an incorrect technique or skill and encourages you to identify the correct method and provide reasoning.

GET IT RIGHT

In the photo below, the person is using the drop saw unsafely.

Identify the unsafe practices and provide reasoning for your answer.

Source: Richard Moran

GET IT RIGHT

1

Worksheets help assess your understanding of the theory and concepts in each chapter.

WORKSHEET 4

Student name: _____

Enrolment year: _____

Class code: _____

Competency name/Number: _____

To be completed by teachers
Student competent ☐
Student not yet competent ☐

WORKSHEETS

1

Task: Read through the sections beginning at *Apply safe work practices* then complete the following questions.

1   List two (2) reasons why a safety helmet must be worn on a construction site:

   1 _____

   2 _____

2   When wearing a safety helmet, is it acceptable to wear a helmet without the helmet harness fitted? Circle the correct answer.

   YES   NO

3   How is the back of the neck protected from sunburn when wearing a safety helmet?

   _____

   _____

Guide to the online resources

MINDTAP

Premium online teaching and learning tools are available on the *MindTap* platform – the personalised eLearning solution.

MindTap is a flexible and easy-to-use platform that helps build student confidence and gives you a clear picture of their progress. We partner with you to ease the transition to digital – we're with you every step of the way.

MindTap for **Certificate III in Carpentry** is full of innovative resources to support critical thinking, and help your students move from memorisation to mastery! Includes:

- Basic Building and Construction Skills, 7e eBook
- Site Establishment, Footings and Framework, 5e eBook
- Advanced Building and Joinery Skills, 4e eBook
- Construction Skills, 4e eBook.
- Instructional Videos
- Worksheets
- Revision Quizzes
- And more!

MindTap is a premium purchasable eLearning tool. Contact your Cengage learning consultant to find out how *MindTap* can transform your course.

INSTRUCTOR RESOURCE PACK

Premium resources that provide additional instructor support are available for this text, including Mapping Grid, Worksheets, Testbank, PowerPoints, Solutions Manual and more. These resources save you time and are a convenient way to add more depth to your classes, covering additional content and with an exclusive selection of engaging features aligned with the text.

SOLUTIONS MANUAL

The **Solutions Manual** includes solutions to Learning Tasks, end-of-chapter Worksheets and Get It Right case studies.

POWERPOINT™ PRESENTATIONS

Use the chapter-by-chapter **PowerPoint slides** to enhance your lecture presentations and handouts by reinforcing the key principles of your subject.

ARTWORK FROM THE TEXT

Add the digital files of graphs, pictures and flow charts into your course management system, use them in student handouts, or copy them into your lecture presentations.

WORD-BASED TEST BANK

This bank of questions has been developed in conjunction with the text for creating quizzes, tests and exams for your students. Deliver these through your LMS and in your classroom.

COMPETENCY MAPPING GRID

The downloadable Competency Mapping Grid demonstrates how the text aligns to the Certificate III in Carpentry.

The Instructor Resource Pack is included for institutional adoptions of this text when certain conditions are met. The pack is available to purchase for course-level adoptions of the text or as a standalone resource. Contact your Cengage learning consultant for more information.

FOR THE STUDENT

MINDTAP

MindTap is the next-level online learning tool that helps you get better grades!

MindTap gives you the resources you need to study – all in one place and available when you need them. In the MindTap Reader, you can make notes, highlight text and even find a definition directly from the page.

If your instructor has chosen *MindTap* for your subject this semester, log in to *MindTap* to:

- Get better grades
- Save time and get organised
- Connect with your instructor and peers
- Study when and where you want, online and mobile
- Complete assessment tasks as set by your instructor.

When your instructor creates a course using *MindTap*, they will let you know your course link so you can access the content. Please purchase MindTap only when directed by your instructor. Course length is set by your instructor.

FOREWORD

This seventh edition of *Basic Building and Construction Skills* has been updated to support the latest endorsed Construction, Plumbing and Services Training Package (9.0). As with previous editions, this seventh edition continues to address the first eight units of competence contained within the Carpentry, Shopfitting and Construction Carpentry qualifications.

Each chapter is based on the relevant competencies providing building trade apprentices, trainees and school students undertaking VET courses with the materials required to develop a sound understanding of the skills and knowledge required by each unit of competence.

The main aim of this text is to bestow a firm foundation for lifelong learning and will prepare new construction workers for a long and rewarding career in the building industry. It is anticipated that the outcomes of this learning will lead to the development of qualified tradespersons who are able to work safely, efficiently and prolifically in the building and construction industry for many years to come.

I would like to thank Richard Moran for his contribution of time and expertise in ensuring that the content of this publication is current and relevant to the building industry as it stands today and for the future of the industry.

Shayne Fagan
Head of Skills Team
Innovative Manufacturing, Robotic and Science
Western Sydney Region
TAFE New South Wales

PREFACE

The building and construction industry within Australia provides employment across a wide range of vocations and services, as well as the infrastructure that enables all sectors of our community to function. The industry can be divided into three main sectors: residential, non-residential and engineering (civil) construction. It is one of the most diverse single industries and contributes a large percentage to Australia's annual GDP, helping to make this country competitive on the global stage.

For Australian building and construction enterprises to remain competitive within both the global and national marketplace, Australian building and construction companies need well-trained team members. This means that the vocational education and training (VET) sector involved in providing high quality and effective training must respond quickly and efficiently to the ever-changing needs of these enterprises.

This text is a starting point, designed to meet the needs of the latest national training package by providing information and activities that reflect basic vocational and employability skills. The knowledge and skills derived from this text will provide a strong foundation for future learning, and will prepare new construction workers for a long and rewarding career in the industry.

I wish to thank Richard Moran for this excellent resource and all the teachers of building and construction who have contributed their time, knowledge and expertise over many years to ensure this text remains current, and that it provides the essential underpinning information to enhance the knowledge and skill of workers around the country in such an important industry.

I have adjusted and updated the content of the work Richard Moran has built so that this latest edition aligns directly to the latest training package. The elements and performance criteria have been used as the headings and content has been added as needed to match the training package. Don't forget to look for the additional resources that are available to supplement this textbook.

Adrian Laws
Teacher – Carpentry and Joinery, Building and Construction
TAFE New South Wales

ACKNOWLEDGEMENTS

Cengage would like to thank the many reviewers whose incisive feedback helped shape this new edition:

- Ben Wright – TasTAFE
- Brendan Kelly – North Metropolitan TAFE
- Clint Davey – TAFE NSW
- Josef Fritzer – TAFE NSW
- Oliver Hawkins – Box Hill Institute
- Paul Johnson – TAFE NSW
- Shane Wright – Box Hill Institute.

Adrian Laws would like to thank Richard Moran for the excellent resource and work that he did in the sixth edition.

Cengage would like to extend special thanks to Edward Hawkins for his numerous contributed photos, as well as to Shayne Fagan and Paul Kubisch from South Western Sydney Institute for their numerous contributions to this exciting new edition.

Every attempt has been made to trace and acknowledge copyright holders. Where the attempt has been unsuccessful, the publishers welcome information that would redress the situation.

COLOUR PALETTE FOR TECHNICAL DRAWINGS

Colour name	Colour	Material
Light Chrome Yellow		Cut end of sawn timber
Chrome Yellow		Timber (rough sawn), Timber stud
Cadmium Orange		Granite, Natural stones
Yellow Ochre		Fill sand, Brass, Particle board, Highly moisture resistant particle board (Particle board HMR), Timber boards
Burnt Sienna		Timber – Dressed All Round (DAR), Plywood
Vermilion Red		Copper pipe
Indian Red		Silicone sealant
Light Red		Brickwork
Cadmium Red		Roof tiles
Crimson Lake		Wall and floor tiles
Very Light Mauve		Plaster, Closed cell foam
Mauve		Marble, Fibrous plasters
Very Light Violet Cake		Fibreglass
Violet Cake		Plastic
Cerulean Blue		Insulation
Cobalt Blue		Glass, Water, Liquids
Paynes Grey		Hard plaster, Plaster board
Prussian Blue		Metal, Steel, Galvanised iron, Lead flashing
Lime Green		Fibrous cement sheets
Terra Verte		Cement render, Mortar
Olive Green		Concrete block
Emerald Green		Terrazzo and artificial stones
Hookers Green Light		Grass
Hookers Green Deep		Concrete
Raw Umber		Fill
Sepia		Earth
Vandyke Brown		Rock, Cut stone and masonry, Hardboard
Very Light Raw Umber		Medium Density Fibreboard (MDF), Veneered MDF
Very Light Van Dyke Brown		Timber mouldings
Light Shaded Grey		Aluminium
Neutral Tint		Bituminous products, Chrome plate, Alcore
Shaded Grey		Tungsten, Tool steel, High-speed steel
Black		Polyurethane, Rubber, Carpet
White		PVC pipe, Electrical wire, Vapour barrier, Waterproof membrane

LIST OF TABLES AND FIGURES

PART 1

BASIC WORKPLACE SKILLS

1

APPLY WHS REQUIREMENTS, POLICIES AND PROCEDURES IN THE CONSTRUCTION INDUSTRY

This chapter covers the following topics from the competency CPCCCA3008 Apply WHS requirements, policies and procedures in the construction industry:

- identify and assess risks
- identify hazardous materials and other hazards on work sites
- plan and prepare for safe work practices
- apply safe work practices
- follow emergency procedures.

This unit specifies the outcomes required to carry out work health and safety (WHS) requirements through safe work practices in all on- or off-site construction workplaces.

It requires the performance of work in a safe manner through awareness of risks and work requirements, and the planning and performance of safe work practices with concern for personal safety and the safety of others.

The unit covers fundamental WHS requirements necessary to undertake work tasks within any sector in the construction industry. It includes the identification of hazardous materials, including asbestos, and compliance with legislated work safety practices. It does not cover removal of asbestos, which is a licensed activity.

It applies to workers in the construction industry.

1. Identify and assess risks

The *Work-related traumatic injury fatalities, Australia 2016* report from Safe Work Australia (2017) shows that from 2003 to 2016, 3414 Australians died while working. This means approximately 243 people died each year from work-related **accidents**. The construction industry is the third-highest killer after agriculture and logistics. Construction accounted for 24 worker fatalities in 2021, after transport and agriculture. (https://www.safeworkaustralia.gov.au/doc/key-work-health-and-safety-statistics-australia-2022) (Figure 1.1).

The prevention of accidents in industry is not only the concern of experts, but of all workers, employers and **persons conducting a business or undertaking (PCBUs)**. Employees must learn how to work without hurting themselves or endangering their colleagues.

1.1 Identify, assess and report hazards in the work area to designated personnel

WHS legislation

To fully understand any piece of **legislation**, it is important to appreciate the origins and reasons for the law first being introduced. There were five main reasons for the development of **work health and safety (WHS)** laws, and these are outlined in Table 1.1.

In Australia, prior to the implementation of the model *Work Health and Safety Act 2011*, each state and territory had separate responsibility for making and enforcing laws concerning WHS. By and large, each principal WHS Act sets out the requirements for ensuring that workplaces are safe and healthy. These requirements spell out the duties of different groups of people who play a role in workplace health and safety (see Table 1.2).

- South Australia was the first state to introduce legislation. In 1972 it introduced the *Industrial Safety and Welfare Act*. In 1986 it enacted the *Occupational Health, Safety and Welfare Act*, following the general form of Victorian state legislation. Finally, on 1 January 2013, South Australia adopted the model *Work Health and Safety Act 2012.*
- In New South Wales, the *Occupational Health and Safety Act* was proclaimed in 1983. It was enacted following the Williams Inquiry into health and safety practices in the workplace. In 1987 major changes were made to that Act. In 2001, that legislation gave way to the *Occupational Health and Safety Act 2000, No 40*. Later, New South Wales enacted the *Work Health and Safety Act 2011*.

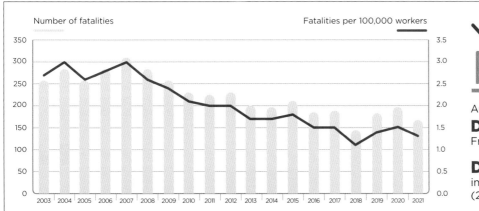

Australia's fatality rate
Decreased 57%
From the peak in **2007**

Decreased 35%
in the past **10 years**
(2012)

FIGURE 1.1 Trends in worker fatalities, 2003 to 2021 Australia

TABLE 1.1 Five reasons for the development of WHS laws

Self-regulation not working	Allowing organisations to regulate their own WHS programs was not working; at one stage, over 500 people nationally were dying each year due to work-related accidents, injuries and diseases.
National and overseas legal developments	There were efforts internationally as well as nationally to produce and then update existing WHS legislation, to bring it into line with the twentieth-century working environment.
Many workers not covered	When WHS legislation began, only approximately one-third of the workforce was covered. Today **all** workers are covered.
Too much legislation	Although only one-third of the workforce was covered by any WHS legislation at all, there were up to 26 different Acts in one state alone relating to occupational/work health and safety. Enforcement procedures for these Acts were a legal nightmare.
Cost of compensation	Every year billions of dollars were spent on workplace compensation and rehabilitation for those who had been injured at work. WHS legislation was introduced in conjunction with worker's compensation to provide for injured workers who were unable to work.

- Victoria introduced legislation in 1985 that became the model for the rest of the country at the time. Far-reaching social and industrial concepts were incorporated into the legislation. Victoria currently uses the *Occupational Health and Safety Act 2004*.
- Western Australia introduced the *Occupational Health, Safety and Welfare Act* in 1984. The current Act for Western Australia will be listed on the government website. At time of this publication it was *Work Health and Safety Act 2020*.
- Tasmania first introduced legislation in 1977. The current Act is the *Work Health and Safety Act 2012*.
- Queensland first introduced legislation in 1989. The current Act is the *Work Health and Safety Act 2011*.
- Northern Territory first introduced legislation in 1989. The current Act is the *Work Health and Safety (National Uniform Legislation) Act 2011* [in force 1 May 2016].
- ACT first introduced the *Occupational Health and Safety Act 1989*. The current Act is the *Work Health and Safety Act 2011*.

General construction induction training

The Work Health and Safety Regulations 2022 (Commonwealth) requires workers to complete general construction induction training before they can carry out construction work.

General construction induction training provides basic knowledge of construction work, the WHS laws that apply, common hazards likely to be encountered in construction work and how the associated risks can be controlled.

On completion of a WHS general construction induction training session, a worker will be issued with a statement that outlines the training they have received, identifies the training body, identifies the training assessor and the date of the assessment.

Although introducing a nationally consistent construction induction card has been discussed, each jurisdiction or state/territory differs and provides their own particular card. All cards must show the cardholder's name, the date training was completed, the number of the registered training organisation (RTO) providing the training, the jurisdiction in which it was issued and a unique identifying number.

As each state/territory provides their own card (see Figures 1.2–1.5), there has been a mutual agreement between the applicable state and territory regulating authorities to accept WHS induction cards from other states and territories if the training meets existing standards for currency.

The card should be always carried onsite and produced on demand for inspection.

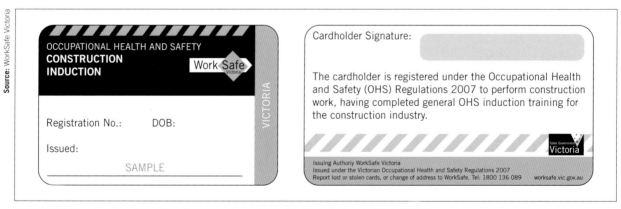

FIGURE 1.2 Construction Induction Card sample as issued in Victoria (post 1 July 2008)

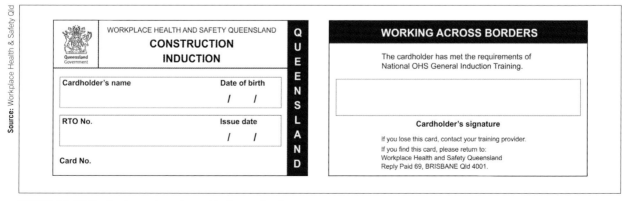

FIGURE 1.3 White Card sample as issued in Queensland

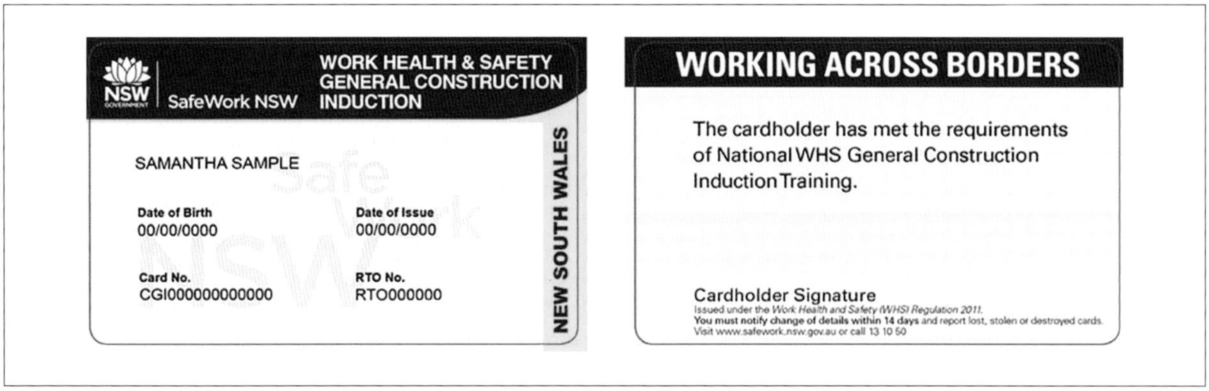

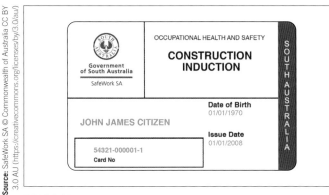

FIGURE 1.4 White Card sample as issued in New South Wales

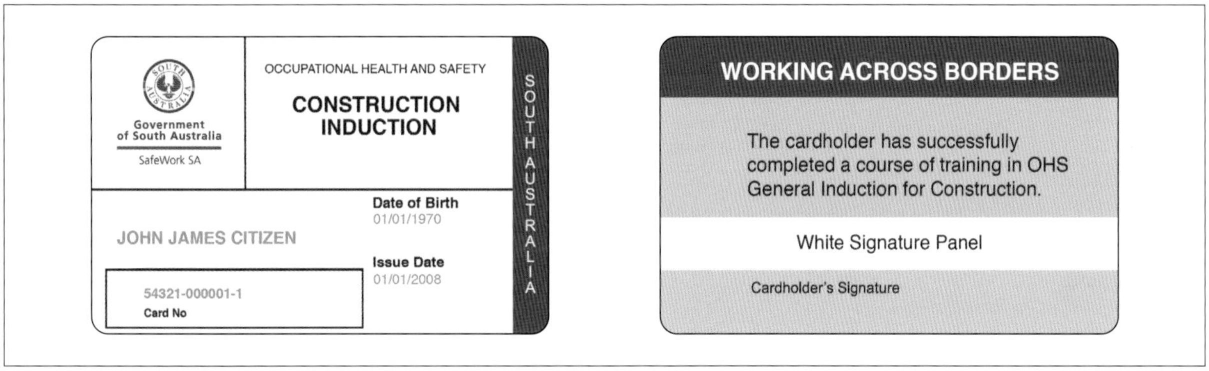

FIGURE 1.5 White Card sample as issued in South Australia

Note: A statement of WHS general construction induction training may cease to be valid when a person has not carried out construction work for a specific consecutive period; for example, two years. This means that if an individual has not carried out construction work for the period stipulated, they must undergo general construction induction training again.

Hazards in the work area

In the building industry, electricity, falls, collapsing trenches and melanoma often kill. Chemicals, corrosives, noise and dust inhalation can result in blindness, deafness, burns and injuries to lungs. Back problems or other serious strains or sprains can slow workers down and put them out of action for weeks or even permanently.

It is therefore important to be able to identify workplace *hazards* and implement a process of identifying, assessing, and controlling *risks* as per **Figure 1.6**.

A **hazard** is anything in the workplace that has the *potential* to injure, harm, damage, kill or destroy people, property or plant and equipment. Note the important word is *potential*; this means that a hazard is not only something that has caused damage or injury but includes anything that *can* possibly do damage or injury.

Hazards can include objects in the workplace, such as machinery or chemicals; they also include work processes where manual handling, excessive noise and fatigue may cause long-term damage. Hazards are divided into two categories – acute and chronic.

FIGURE 1.6 The risk management process

Acute hazards

An acute hazard is one where *short-term exposure* to the hazard will cause an injury or sickness (e.g. being burnt in an explosive fire).

Chronic hazards

A chronic hazard is one where *long-term exposure* to the hazard will cause an injury or sickness (e.g. melanoma from extended exposure to the sun, or slow poisoning from chemicals building up in the body's system over a long period of time).

 Acute hazards are sometimes given priority when working because they cause a result quickly. However, chronic hazards can result in very debilitating health complications. The urgency of the job is not worth getting injured for.

Risk

Risk is the *likelihood or probability* of a hazard causing damage or injury. The level of risk will depend on factors such as *how often* the person is exposed to the hazard and *how serious* any potential injuries.

Incident/accident

An incident is where a hazard results in injury, damage, or death. It may be totally random or an intentional act or action due to neglect or purposeful intention.

An accident is when something happens unexpectedly, without design, or by chance. An accident may cause injury, death, damage, or destruction.

To reduce incidents and accidents we need to first recognise the hazard and then rate the level of risk (**risk assessment**). The most common way of achieving this is by using a **risk matrix** as shown in Figure 1.7.

The following example will utilise the risk matrix from Figure 1.7 in a real-life context. On the job site,

Source: Small Business Development Corporation, Government of Western Australia

Consequence \ Likelihood		**Rare** The event may occur in exceptional circumstances.	**Unlikely** The event could occur at some time.	**Moderate** The event will probably occur at some time.	**Likely** The event will occur in most circumstances.	**Certain** The event is expected to occur in all circumstances.
		Less than once in 2 years	At least once per year	At least once in 6 months	At least once per month	At least once per week
	Level	1	2	3	4	5
Negligible No injuries, Low financial loss	0	0	0	0	0	0
Minor First-aid treatment Moderate financial loss	1	1	2	3	4	5
Serious Medical treatment required High financial loss, Moderate environmental implications, Moderate loss of reputation, Moderate business interruption	2	2	4	6	8	10
Major Excessive multiple long-term injuries, Major financial loss, High environmental implications, Major loss of reputation, Major business interruption	3	3	6	9	12	15
Fatality Single death	4	4	8	12	16	20
Multiple Multiple deaths and serious long-term injuries	5	5	10	15	20	25

Legend

Risk Rating	Risk Priority	Description
0	N	No Risk: The costs to treat the risk are disproportionately high compared to the negligible consequences.
1–3	L	Low Risk: May require consideration in any future changes to the work area or processes, or can be fixed immediately.
4–6	M	Moderate: May require corrective action through planning and budgeting process.
8–13	H	High: Requires immediate corrective action.
15–25	E	Extreme: Requires immediate prohibition of the work, process and immediate corrective action.

FIGURE 1.7 Risk matrix diagram

the foreperson sets up the compound mitre saw on the floor and asks you to cut timber to length; approximate time on the saw is 4 hours.

What are the hazards? Consider the following:

- hearing damage
- foreign objects being thrown into the eyes
- back strain
- breathing in timber dust
- amputating a hand or fingers.

Each of the above need to be considered using the matrix diagram; however, we will look at 'Amputating a hand or finger' only in our example.

So, ask the first question; *how serious* will an accident be if I put my hand through the saw? Multiple, Fatality, Major, Serious, Minor or Negligible? The answer is most likely 'Serious' or 'Major'.

Now the second question; *how likely is an accident to occur?* Well, with no training, impaired visibility because you are squatting over the top of the saw and the fact that you are more concerned about your back hurting all the time, then it all adds up to being 'Certain' that you will do damage.

So, using the matrix and plotting the result, the answer would be either Gold (level 10) 'High requiring immediate corrective action' or Red (level 15) 'Extreme – requires immediate prohibition of the work and immediate corrective action'.

So, what can be done to control the hazard and risk? Very simply, place the saw on a saw stand and be trained on how to use the saw correctly. This will reduce the likelihood of harm to 'Rare'. Now check the result on the matrix and it should be a level 2 or 3.

Workplace hazards

Some of the more common hazards that influence health and safety in the workplace are:

- lifting and handling materials
- falls – of objects and people
- machinery – power and hand tools
- chemicals and airborne dust
- noise
- vibration
- thermal discomfort (too hot or cold)
- illumination (visibility)
- fire and explosions.

Hazard categories

Hazards in the workplace can be divided into five distinct categories:

1 biological
2 physical
3 chemical
4 ergonomic
5 psychosocial.

By dividing hazards into these five major groupings the correct and most suitable control measures can be implemented to provide workers with the correct level of protection (Figure 1.8).

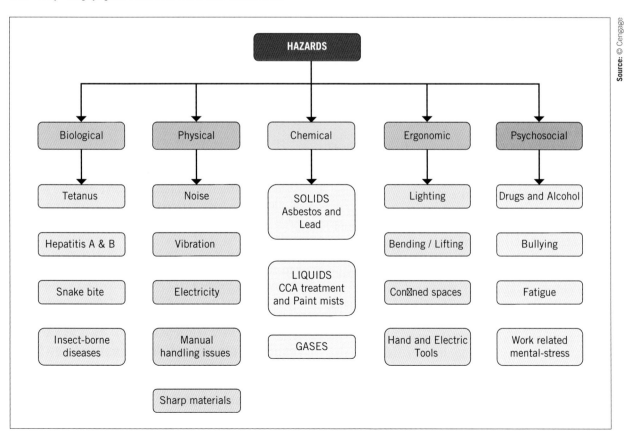

FIGURE 1.8 Hazard categories and subgroups

Source: © Cengage

Below is a list of the more common hazards found on building and construction sites.

Biological hazards

- Insect/animal – poor sanitation in dining areas or toilet areas and poor rubbish disposal practices increase the chances of the spread of disease.
- Microorganisms – when health and welfare facilities are not cleaned often, bacteria can grow and spread, causing disease. Some work materials, particularly soil, clay, plant materials and organic dusts may also harbour dangerous bacteria.

Physical hazards

- Noise – exposure to excessive noise can cause temporary or permanent hearing loss, stress and annoyance and tinnitus (ringing in the ears).
- Heat and cold – exposure to either can cause reduction in concentration and heat-related medical conditions such as heat exhaustion and heat stroke.
- Vibration – whole-body vibration injuries from working with heavy equipment; and hand and arm vibration injuries caused using vibrating tools.
- Manual handling – these injuries (lifting, carrying, pushing, pulling) are probably the most common injuries in the building industry.
- Electricity – underground and overhead supply cables, site wiring and power leads.
- Airborne materials (projectiles) – one of the most common causes of injury is falling or flying objects.
- Heavy equipment – heavy motorised equipment moving around construction sites is a major hazard.
- Sharp materials – materials with sharp edges can cause cuts and lacerations.

Chemical hazards

Many different chemicals that may be harmful to health are used in the construction industry in the forms of:

- Solids – dusts, fumes, and solid materials.
- Liquids – liquid materials and mists.
- Gases – gases and vapours. Some of these chemicals can cause acute or chronic injuries and severe medical conditions.
- Toxic chemicals – many chemicals are toxic or poisonous. The effects of chemicals on your health are difficult to identify at the source of exposure. Often the symptoms appear in parts of the body far removed from the point of absorption or perhaps do not appear until some years later.
Some examples of chemical hazards are asbestos, lead and lead-based paint.

Ergonomic hazards

Ergonomic hazards are where muscle, tendon and nerve damage may occur due to poor posture, repetitious actions, or poorly designed tools. Examples of these on the job site are:

- confined spaces may have a restricted means of entry and exit causing stress and strain on the muscles and/or joints
- hand tools may cause repetitive strain injuries to the hands, back and neck if they are uncomfortable to use or are used for extended periods of time
- electrical tools have similar issues to the hand tools; and they can also cause serious cuts, amputations, and electrocution if they are designed without the end-user in mind.

Psychosocial hazards

The following are some potential psychosocial hazards in the workplace.

- Drugs and alcohol in the workplace – in the construction industry there is a culture that is accepting of regular drug and alcohol use. Drugs and alcohol have a significant influence on the health and safety of workers.
- Bullying in the workplace – bullying is repeated, unreasonable behaviour directed towards a worker or group of workers; it may create a risk to health and safety.
- Work-related mental stress – stress is normally experienced as fatigue, anxiety, and depression.

 Be informed about materials, tools, and processes for your work so you can never say that you didn't have enough information to prepare yourself for a hazard. Get informed via the many websites that are available for you to get information on safety.

1.2 Report safety risks in the work area based on identified hazards, to designated personnel

Once you have identified and assessed hazards, it's time to report those hazards to your supervisor or whoever the designated personnel are at your workplace. A well thought through plan to control the hazard based on evidence needs to be recorded and communicated to all affected workers. In larger construction companies someone will have responsibility for recording and communicating safety issues; however, in smaller organisations it may just be a couple of workers. It is very important to have plans and actions in place so that any new workers onsite or visitors can be made aware of any hazards and the risks associated with those hazards.

1.3 Follow safe work practices, duty of care requirements and safe work instructions for controlling risks

Hazard control procedures

Once the hazard has been identified and its potential to do damage is measured, the hazard needs to be controlled. This means the hazard is eliminated or in some way lessened or reduced. This may be done in several different ways as seen in Figure 1.9. In many cases, more than one control method must be used to control a hazard. Various methods should be considered. The order of preference for controlling hazards is known as the 'Hierarchy of hazard controls'.

Depending on the hazard that has been identified, it may be necessary to stop work until the safety control has been implemented.

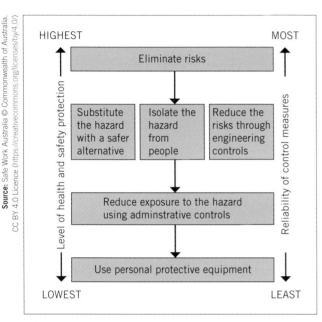

FIGURE 1.9 Hierarchy of hazard controls

Selection of controls

When selecting how to control a hazard/risk, the method/s used should be those that are least likely to be subject to human failure. **Personal protective equipment (PPE)** should be used as a last resort. But when used, PPE must be comfortable so as not to interfere with the wearer's activities.

Safety controls should be checked regularly to confirm that they are still effective.

The posting of safety signs, like the wearing of PPE, should be seen as a last resort, if it is not possible to eliminate the hazard at its source. Signs and PPE are not an alternative to the elimination or reduction of the hazard.

Only when it has been found to be impractical to eliminate the hazard at its source should PPE and safety signs be used as the main method of protection.

Elimination

One of the most common methods of risk elimination on a construction site is to reduce the amount of work carried out at height on ladders. By using an approved scaffold system workers can carry out tasks without risk of falling.

Substitution

Substitution occurs when a hazardous material is replaced by a safer, less hazardous, or non-hazardous

material. A common example in the construction industry is the substitution of asbestos cement sheeting and other asbestos containing materials (ACMs) by the less hazardous fibre cement sheeting.

Engineering controls

Engineering controls are devices or mechanisms built into the design of plant or equipment or are part of the safety process, such as a machine guard. They are an extremely dependable means to prevent workers or the public encountering serious hazards.

Administrative controls – sequence of work

This is a very simple and effective hazard control. All that is required is correct planning; for example, do *not* have the roof tiler working on the job at the same time as the bricklayer. If a tiler drops a tile onto the head of a bricklayer there is an accident – or even a death – on the site. Correct planning would see the bricklayer on the job before or after the tiler, not at the same time.

1.4 Contribute to WHS, hazard, accident or incident reports in accordance with workplace procedures, Australian government and state or territory WHS legislation, and relevant information

Current WHS Acts

Today the industrial workplace in Australia is governed by both federal and state legislation, followed by regulations and then codes of practice (see also Table 1.2). The main piece of legislation for all of Australia is the model *Work Health and Safety Act 2011* (WHS Act). Laws and regulations provide a set of minimum standards for the protection and the health and safety of workers. Codes of practice demonstrate practical methods for undertaking the work safely.

Safe Work Australia is the national body that works to coordinate and develop policy and assists in the implementation of the model Act. Safe Work Australia seeks to build cooperation between the three groups involved – governments, business, and unions – to bring them together to create strategies and decide on policy.

Current state and territory WHS acts and regulations can be accessed from the Safe Work Australia website via Law and Regulation links on the site https://www.safeworkaustralia.gov.au/law-and-regulation/whs-regulators-and-workers-compensation-authorities-contact-information.

Regulations, codes of practice and guidelines

Some workplace hazards have the potential to cause so much **injury** or disease that specific regulations or codes of practice are warranted. These regulations and codes,

TABLE 1.2 Current state and territory WHS Acts and Regulations

State/ territory	Current WHS act	Current WHS regulation	WHS regulating authority	Website and contact number
WA*	Occupational Safety and Health Act 1984	Occupational Safety and Health Regulations 1996	Worksafe WA	Website: https://www.commerce.wa.gov.au/ worksafe Contact: 1300 307 877
VIC	Occupational Health and Safety Act 2004	Occupational Health and Safety Regulations 2017	WorkSafe Victoria	Website: https://www.worksafe.vic.gov.au Contact: 1800 136 089
QLD	Work Health and Safety Act 2011	Work Health and Safety Regulations 2011	WorkCover Queensland	Website: https://www.worksafe.qld.gov.au Contact: 1300 362 128
NSW	Work Health and Safety Act 2011	Work Health and Safety Regulations 2017	SafeWork NSW	Website: http://www.safework.nsw.gov.au Contact: 13 10 50
SA	Work Health and Safety Act 2012	Work Health and Safety Regulations 2012	WorkCover SA	Website: https://www.safework.sa.gov.au Contact: 1300 365 255
TAS	Work Health and Safety Act 2012	Work Health and Safety Regulations 2022	Worksafe TAS	Website: https://www.worksafe.tas.gov.au Contact: 1300 366 322
ACT	Work Health and Safety Act 2011	Work Health and Safety Regulations 2011	WorkSafe ACT	Website: https://www.worksafe.act.gov.au Contact: 13 22 81
NT	Work Health and Safety (National Uniform Legislation) Act 2011	Work Health and Safety (National Uniform Legislation) Regulations 2011	NT WorkSafe	Website: https://worksafe.nt.gov.au Contact: 1800 019 115

*The current Act for Western Australia will be listed on the government website. At time of this publication it was *Work Health and Safety Act 2020*.

adopted under state and territory WHS Acts, explain the duties of groups of people in controlling the risks associated with specific hazards. Note that:

- regulations are legally enforceable
- codes of practice and guidelines provide advice on how to meet regulatory requirements. As such, codes are not legally enforceable, but they can be used in courts as evidence that legal requirements have or have not been met.

The basic purpose of these codes and guidelines is to provide workers in any industry with practical, common-sense, industry-acceptable ways by which to work safely.

They are now generally written and published by Safe Work Australia as 'Model Codes of Practice'. They are adopted by each state and territory's regulating authority and cover such areas as 'Managing the risk of falls at workplaces', 'How to safely remove asbestos', 'Labelling of workplace hazardous chemicals', 'Demolition work' and 'Excavation work' just to name a few. A typical example is shown in Figure 1.10.

Accident/incident reporting

All workers have a responsibility to report any illness, accident or near miss they are involved in or see at the workplace. They should report such incidents to their immediate supervisor, WHS **committee**, health and safety representative (HSR), a union delegate or first aid officer. If there is any plant or equipment involved, they must make the operator aware of the problem immediately.

FIGURE 1.10 Typical code of practice – 'How to safely remove asbestos' (front cover)

Once the relevant personnel become aware of an illness, accident or near miss, it must be determined whether it is a *reportable incident*. If so, the necessary forms must be completed and forwarded to the appropriate authorities (see Figure 1.11). This reporting procedure allows for steps to be taken on a state and national level to help reduce workplace accidents and illnesses in the future. It also means inspections of the workplace at which the accident or illness occurred will be made by the state's WHS regulating authority to ensure that steps are taken to prevent similar incidents from happening again.

An important part of any organisation's WHS procedures is that employees of all levels within the organisation have input into the process. For example, employees should be consulted and their input considered before a safe work procedure is implemented.

If an accident or incident occurs, all personnel who witnessed the event should contribute to the reporting process that follows. This ensures all employees have the opportunity to understand and contribute to the WHS procedures that are put into place for their own safety and the safety of others.

Reportable accidents/incidents

AS 1885.1 Measurement of occupational health and safety performance – Describing and reporting occupational injuries and disease (known as the National Standard for Workplace Injury and Disease Recording) provides a template for describing and reporting occupational injuries and disease. From this national standard, each state and territory developed an accident reporting register and reporting procedure to match its own WHS legislation.

Your state/territory regulating authority may require serious work-related illnesses, injuries or dangerous occurrences to be reported on an accident report form. Accidents that must be reported by law are known as *reportable accidents*.

Accident/incidents

An accident/incident is any event that causes human injury or property damage. Such events may occur because of unsafe acts, that may include the following:
- practical jokes gone wrong
- using tools and/or equipment in a manner for which they were not designed
- rushing and taking short cuts
- not using PPE or other safety devices
- throwing materials/rubbish from roofs or upper floors.
  Other causes of accidents and incidents include:
- little or no training in safety and proper use of tools and equipment
- poor housekeeping
- poor management of site safety issues
- poorly maintained tools and/or equipment
- damaged tools and/or equipment
- inadequate PPE or PPE not supplied or not used
- poor site conditions and congestion due to lack of preparation.

An example of an *illness* that must be reported is:
- an employee/worker has a medical certificate stating that he/she is suffering from a work-related illness that prevents the employee from carrying out his/her usual duties for a continuous period of at least seven days.
  An example of an *injury* that must be reported is:
- one resulting from a workplace accident where a person dies or cannot carry out his/her usual duties for a continuous period of at least seven days.
  Examples of *dangerous occurrences* that must be reported (even if no-one is injured) are:
- damage to any boiler, pressure vessel, plant, equipment, or other item that endangers or is likely to endanger the health and safety of any person at a workplace
- damage to any load-bearing member or control device of a crane, hoist, conveyor, lift, escalator, moving walkway, plant, scaffolding, gear, amusement device or public stand
- any uncontrolled explosion, or fire, or escape of gas or dangerous goods or steam, or where there is a risk that any of these events is likely
- a hazard exists that is likely to cause an accident at any moment and may cause death or serious injury to a person (e.g. an electric shock) or substantial damage to property.
  PCBUs/employers in control of a workplace are normally required to send an *accident report form* to their state or territory regulating authority even if the person injured or killed is not one of their employees (e.g. is a subcontractor or visitor to the site).

Near misses

These are accidents/incidents that didn't quite happen. They occur when the conditions are right for an accident, but people have not been hurt and equipment is not damaged.

Near misses usually indicate that a procedure or practice is not being carried out correctly or site conditions are unsafe. By reporting these near misses the problems may be looked at by management and rectified before someone is hurt or killed or equipment is seriously damaged.

Company accident report forms

As well as completing a state or territory accident/ incident notification form, a pro-forma company form, or Worker's Compensation Insurance Accident form *must* also be completed. This accident report normally requires the following information to be given:
  Information about the PCBU/employer:
- name of company
- office address
- address of site where accident happened
- main type of activity carried out at the workplace (e.g. building construction)

HAZARD/INCIDENT/ACCIDENT REPORT FORM

PART A – To be completed by employee

Name of employee:	Name Surname	**Date:**	01/01/05
Time of incident / accident:	Job Title		
Supervisor:	Host/Employer/Manager Name	**Work Area:**	

1. Describe the hazard / detail what happened – include area and task, equipment, tools and people involved.

2. Possible solutions / how to prevent recurrence – Do you have any suggestions for fixing the problem or preventing a repeat

PART B – To be completed by supervisor

3. Results of investigation – Determine whether the hazard is likely to cause an injury and explain what factors caused the event.

PART C – To be completed by supervisor

4. Action taken – Supervisor to identify actions to prevent injury or illness.

	ACTION	RESPONSIBILITY	COMPLETION DATE
4.1			
4.2			
4.3			
4.4			
4.5			

Feedback has been provided to person who reported the hazard / incident / accident.		(	
Employee representative (health and safety representative)	Name Surname	**Date:**	01/01/05
Business Manager	Name Surname	**Date:**	01/01/05

FIGURE 1.11 Sample accident report form

- major trades, services or products associated with this activity
- number of people employed at the workplace and whether there is a WHS committee/HSR at the workplace.
Information about the injured or ill person:
- name and home address
- date and country of birth
- whether the person is an employee of the company
- job title and main duties of the injured person.
Information about the injury/illness:
- date of medical certificate
- type of illness as shown on the medical certificate
- whether the injury resulted in death
- particulars of any chemicals, products, processes, or equipment involved in the accident.
Information about the injury or dangerous occurrence:
- time and date that it happened
- exact location of the event
- details of the injury
- type of hazard involved
- exactly how the injury or dangerous occurrence was caused
- how the injury affected the person's work duties
- details of any witnesses
- details of the action taken to prevent the accident from happening again.

Details of the person signing and the date of signing the accident report are also required. When the report is completed, copies are generally sent to the company's worker's compensation insurance organisation and a copy is kept by the PCBU/employer.

Accident report forms are a good way to stop the same injury from occurring more than one or two times. Regularly check what is causing injuries on your work site and design a safer process for doing that task or process.

Rights and responsibilities of PCBU/employers and employees/workers

In each state and territory there are specific rights and responsibilities for PCBU/employers and employees/workers under the WHS legislation.

WHS Acts aim to protect the health, safety, and welfare of people at work. They lay down general requirements that must be met at places of work. All states and territories have similar aims and regulations.

Requirements of state and territory regulations

Each state and territory have their own specific requirements, which may include any or all of the following.

PCBU/employers

PCBU/employers have a 'duty of care' and must provide for the health, safety, and welfare of

their employees/workers at work. To do this, PCBU/employers must:
- identify any foreseeable hazard that may arise from the work carried out by the company or partnerships
- provide and maintain equipment and systems of work that are safe and without risks to health
- plan to ensure the safe use, handling, storage and transport of equipment and substances
- provide the information, instruction, training, and supervision necessary to ensure the health and safety of employees at work
- maintain places of work under their control in a safe condition and provide and maintain safe entrances and exits
- make available adequate information about research and relevant tests of substances used at the place of work.

PCBU/employers must not require employees/ workers to pay for anything done or provided to meet specific requirements made under the Acts or associated legislation. They must also ensure the health and safety of people visiting their places of work who are not employees/workers.

Employees/workers

Employees/workers must take reasonable care of the health and safety of themselves and others. They must cooperate with their PCBU/employers and comply with WHS requirements.

Employees/workers must not:
- interfere with or misuse any item provided for the health, safety, or welfare of persons at work
- obstruct attempts to give aid or attempts to prevent a serious risk to the health and safety of a person at work
- refuse a reasonable request to assist in giving aid or preventing a risk to health and safety.

Subcontractors' responsibilities

The PCBU/principal contractor must ensure that each subcontractor/worker provides a written SWMS for the work to be carried out by the subcontractor/worker, which should include an assessment of the risks associated with that work (Figure 1.12).

If the subcontractor/worker does not comply with what has been outlined in the SWMS, the PCBU/ principal contractor has the right and responsibility to direct the subcontractor/worker to stop work and not resume work until the SWMS requirements are met. This could result in the subcontractor/worker losing money due to the implementation of safety requirements and/or equipment. If the subcontractor/ worker neglects to maintain or modify the SWMS or fails to inform the PCBU/principal contractor of the changes, then the subcontractor/worker may receive a fine for breaching the WHS legislation.

If the PCBU/principal contractor neglects to enforce the requirements of the legislation on the

APPENDIX A – SAMPLE SWMS TEMPLATE FOR HIGH RISK CONSTRUCTION WORK (HRCW)

DUTIES: 1) A SWMS **must** be prepared if proposed works involve any of the HRCW activities listed below and that work poses a risk to the health and safety of any person. **2)** Affected employees and their HSRs must be consulted in the preparation of the SWMS. **3)** Once a SWMS has been developed and implemented, the HRCW to which it relates must be performed in accordance with the SWMS. **4)** Duty holders (builder and sub-contractor) **must** stop the HRCW immediately or as soon as it is safe to do so if the SWMS is not being complied with; the HRCW **must** not resume until the SWMS is complied with or reviewed and revised as necessary. **5)** The SWMS **must** be reviewed and if necessary, revised whenever the HRCW changes, or after any incident that occurs during HRCW, or if there is any indication that risk control measures are not adequately controlling the risks. **6)** An employer must retain a copy of the SWMS for the duration of the HRCW.

Direct employer:	Principal contractor (PC) *(Name and contact details)*
Work supervisor: *(Name and contact details)*	**Date SWMS provided to PC:**
Work activity: *(Job description)*	**Workplace and works location:**

High risk construction work:		
☐ Where there is a risk of a person falling more than two metres.	☐ On or adjacent to roadways or railways used by road or rail traffic.	☐ In, over or adjacent to water or other liquids where there is a risk of drowning.
☐ At workplaces where there is any movement of powered mobile plant.	☐ Structural alterations that require temporary support to prevent collapse.	☐ In an area where there are artificial extremes of temperature.
☐ On or near energised electrical installations or services.	☐ Involving a trench or shaft if the excavated depth is more than 1·5 metres.	☐ On or near pressurised gas distribution mains or piping.
☐ Involving demolition.	☐ Involving a confined space.	☐ On or near chemical, fuel or refrigerant lines.
☐ Involving tilt-up or precast concrete.	☐ On telecommunications towers.	☐ Involving diving.
☐ Involving removal or likely disturbance of asbestos *(note: preparation of an asbestos control plan is taken to be preparation of a SWMS).*	☐ In an area that may have a contaminated or flammable atmosphere.	☐ Involving the use of explosives.
		☐ Involving a tunnel.

Person responsible for ensuring compliance with SWMS:	**Date SWMS received:**
What measures are in place to ensure compliance with the SWMS? *(eg direct supervision, regular spot checks)*	
Person responsible for reviewing SWMS control measures *(eg PC's representative):*	**Date SWMS received by reviewer:**
How will the SWMS control measures be reviewed?	
Review date:	**Reviewer's signature:**

>>

FIGURE 1.12 Typical SWMS format (This template should be used in conjunction with WorkSafe Victoria's publication 'Information about Safe Work Method Statements')

>>

Selecting risk controls:

*Any risk to health or safety must be **eliminated**, or if that is not reasonably practicable, **reduced** so far as is reasonably practicable by:*

- *implementing any mandated controls specified by law (eg the OHS Regulations 2007)*
- *substituting a new activity, procedure, plant, process or substance (eg scaffold in preference to ladders)*
- *isolating persons from the hazard (eg fence off areas for mobile plant operation)*
- *using engineering controls (eg guard rails, trench shields) - or a combination of the above.*

If any risk to health or safety remains, it must be reduced by using:

- *administration controls (eg activity specific safety training, work instructions, warning signs)*
- *PPE such as respiratory protection, hardhats, high visibility clothing - or a combination of the above.*

More Effective → **Less Effective**

What are the tasks involved?	What are the hazards and risks?	What are the risk control measures?
List the work tasks in a logical order.	*What aspects of the work could harm workers or the public?*	*Describe what will be done to make the activity as safe as possible?*

Name of Worker	Signature	Date	Name of Worker	Signature	Date	Name of Worker	Signature	Date

FIGURE 1.12 (Continued)

subcontractor/worker, then the PCBU/principal contractor may receive a fine for a breach of the WHS legislation.

It is therefore essential that all parties on a construction site comply with the requirements of the WHS legislation to create, implement and monitor SWMSs.

Offences and penalties

The national model WHS Act aims to harmonise the many conflicting aspects of the legislation of the various states and territories (refer to Table 1.1). For example, penalties for the same offence or infringement vary from state to state. The new model WHS Act will bring consistent rules and penalties across all states and territories.

The harshness of fines is influenced by the severity of the infringement. Also considered is whether the offender is an individual or a corporation and whether or not they are previous offenders. As well as imposing fines, courts may opt for a sentence of imprisonment. There is also the option to order offenders to undertake any or all the following:

- take steps to remedy or restore any problem caused by the offence
- pay the regulating/investigative authority (e.g. WorkCover) for the costs of the investigation
- publicise or notify other persons of the offence
- carry out a project for the general improvement of health and safety.

Note: It is recommended that you become familiar with your relevant state or territory infringement systems by either contacting your respective WHS regulatory authority, or by accessing their website, as detailed in Table 1.2.

 COMPLETE WORKSHEET 1

2. Identify hazardous materials and other hazards on work sites

2.1 Correctly identify and, if appropriate, handle and use hazardous materials on a work site in accordance with legislative requirements, and workplace policies and procedures

Assessing and controlling hazardous materials

The New South Wales government (Ministry of Health, n.d.) has revealed that high quantities of chemicals are used on construction sites and that these are causing major hazards for the workforce. The possible effects associated with these chemicals are a major threat to construction workers' health and safety. This situation would be similar in all states and territories.

Some of the effects of exposure to chemicals are well known, such as dermatitis from cement and epoxy resins

and nasal cancers from medium density fibreboard. The hazards of other materials are less well known, and the attitude in the industry is that most materials in use, such as adhesives, grouts, mastics, and powders, 'are not really chemicals'. This is wrong; and this incorrect assumption is causing most workers to take no or inadequate precautions when using hazardous materials.

Some chemical hazards include:

- compressed gases (e.g. nitrogen)
- flammable gases (e.g. acetylene)
- oxidising gases (e.g. oxygen)
- flammable liquids (e.g. petrol, solvents, thinners and paints)
- poisonous substances (e.g. two-pack products and isocyanate-containing paints; CCP-treated timbers)
- harmful substances (e.g. amine adduct adhesives; asbestos-containing materials (ACMs))
- irritant substances (e.g. methyl ethyl ketone peroxide (MEKP), a hardening agent used in fibreglass)
- corrosive substances (e.g. hydrochloric acid used for cleaning down brickwork)
- many other hazards (e.g. aerosols, insulation material including fibreglass batts and blankets).

It is very important for you to be able to identify chemical hazards as well as know exactly what PPE is needed while working with a particular chemical hazard. Therefore, a **safety data sheet (SDS)** is always required.

Dangerous goods

Many **dangerous goods**, mixtures or articles are also classed as hazardous substances because of their physical, chemical, or acute toxicity properties, which may cause serious hazard to people, property or even the environment.

Australia has adopted a system of classification and labelling for dangerous goods based on the United Nations' system used in other countries. This system determines whether a substance is classified as a dangerous good, and if so, it is recorded in the Transport of Dangerous Goods by Road and Rail Code (ADG Code). The ADG Code helps people recognise dangerous goods, their properties, and dangers quickly.

Dangerous goods can be identified by a diamond sign or label (Table 1.3). There are nine classes of dangerous goods under this system.

The diamond-shaped sign or label shows which of the nine classes that dangerous substance belongs to. These signs have distinctive symbols and colouring. Not all hazardous substances have dangerous goods labels because the dangerous goods diamond indicates only an immediate hazard, not necessarily a hazard that has only long-term health risks.

The details of the design and selection of signs and labels for the nine classes of dangerous goods are contained in AS 1216 Class labels for dangerous goods.

Correct labelling means dangerous goods don't have to be any more dangerous than they already are.

TABLE 1.3 Nine classes of dangerous goods

1	2	3	4	5
Explosives	Gases	Flammable liquids	Flammable solids	Oxidising substances
![EXPLOSIVES 1]	![NON-FLAMMABLE COMPRESSED GAS 2]	![FLAMMABLE LIQUID 3]	![FLAMMABLE SOLID]	![OXIDIZING AGENT 5.1]

6		7	8	9
Poisonous or infectious substances		Radioactive substances	Corrosives	Miscellaneous dangerous substances
![TOXIC 6]	![INFECTIOUS SUBSTANCE 6]	![RADIOACTIVE 7]	![CORROSIVE 8]	![MISCELLANEOUS DANGEROUS GOODS 9]

You should not only learn to recognise the various symbols but also learn about the actual properties of the substances you may be exposed to.

It is important to know which goods can produce toxic gases, which are highly flammable, which are dangerous when wet, which are dangerous on contact with air, and which are harmful when they come into contact with your skin. Taking an attitude of 'what you don't know cannot hurt you' is a fallacy and a dangerous way to work.

Not all dangerous goods have safe handling and storage instructions printed on them; they may have only warning diamond signs. The safe handling and storage instructions can be obtained from the SDS for the product, or in the relevant Australian Standard for the substance.

Beware: If you find a product or substance with a 'diamond' on the container, obtain and read the safety instructions for the material before storing, opening, or using it.

Storage

Chemicals brought onsite must be stored safely. The requirements for safe handling and storage can be read in the national Model Code of Practice – Managing Risks of Hazardous Chemicals in the Workplace available from Safe Work Australia.

Storage must be in properly constructed containers or storage areas. Correct signposting of the area should be carried out to warn of the hazards of the chemicals stored.

Stored chemicals can be dangerous to 'outsiders', such as rescue workers and fire fighters. In incidents where chemicals are spilt or involved in fires, toxic fumes and/or gases could be emitted. These situations can present a hazard to members of the public, as well as to the emergency personnel in attendance. Storage of specified quantities of substances that are classed as dangerous goods must be licensed in some states.

The purpose of licensing stored dangerous goods is to provide greater protection to people handling them, as well as to the public. The information obtained from the licensing is placed in a stored chemicals information database. This database can be accessed by emergency workers and government authorities in the case of an emergency, to find out what is being stored on the premises or transported in a vehicle.

Licensing of the storage of hazardous chemicals on building sites has not worked well due to the short nature of the storage and the mobility of the industry.

> **GREEN TIP**
>
> Chemicals are normally very concentrated and can cause a lot of damage to the environment. Make sure you work with chemicals according to the SDSs that are available from the chemical's manufacturer.

2.2 Apply measures for controlling risks and construction hazards effectively and immediately

Good work practices

In hazardous chemical areas, good hygiene is important. You should:
1. change clothes daily
2. shower after work
3. isolate or enclose dust-producing machines
4. check the workplace to ensure that
   i adequate PPE is provided and used
   ii there is medical monitoring of persons exposed for damaging effects, for example, air monitoring for asbestos dust, and/or silica dust.
   iii ventilation and other safety and health systems are effective.

2.3 Use appropriate signs and symbols to secure hazardous materials that have safety implications for self and other workers, immediately they are identified

Safety signs and tags

Safety signs are placed in the workplace to:

- warn of health and safety hazards
- give information on how to avoid hazards, thereby preventing incidents and accidents
- indicate the location of safety and fire protection equipment
- give guidance and instruction in emergency procedures.

Standards Australia has three Standards covering the use of safety signs in industry. These are:

1 AS 1216 Class labels for dangerous goods
2 AS 1318 Use of colour for the marking of physical hazards
3 AS 1319 Safety signs for the occupational environment.

AS 1319 identifies three main types of safety signs. These are:

- **Picture signs:** these use **symbols** (pictures) of the hazard, equipment or the work process being identified, as well as standardised colours and shapes, to communicate a message (see Figure 1.13).
- **Word-only signs:** these are written messages using standardised colours and shapes to communicate the required meaning (see Figure 1.14).
- **Combined picture and word signs:** these are clearer to understand with a picture and a short written message (see Figure 1.15).

To make sure the message reaches everyone at the workplace, including workers from non-English-speaking backgrounds and workers with low reading skills, picture signs should be used wherever possible. When this is not the case, it may be necessary to repeat the message in other languages.

FIGURE 1.13 Picture signs – smoking prohibited

FIGURE 1.14 Word-only messages

FIGURE 1.15 Combined picture and word signs

Colour and shape

There are seven categories of safety signs, and they are identified by colour and shape, as shown in Table 1.4.

Prohibition signs

Prohibition signs indicate that this is something you must not do: *red circle border with cross bar* through it, *white background,* and *black symbol* (see Table 1.4, Figure 1.16 and Figure 1.17).

FIGURE 1.16 Digging prohibited

TABLE 1.4 The seven categories of signs

Sign category	Picture	Sign category	Picture
1. Prohibition (must not do) signs	Fire, naked flame and smoking prohibited	5. Mandatory (must do) signs	Head protection must be worn
2. Restriction signs	40 km/hr restriction sign	6. Hazard warning signs	Fire hazard sign
3. Danger (hazard) signs	**DANGER** ASBESTOS REMOVAL IN PROGRESS — Danger warning sign – asbestos removal in progress	7. Emergency information signs	Emergency exit signs
4. Fire-fighting equipment signs	Fire safety – fire hose reel		

FIGURE 1.17 No pedestrian access

Mandatory (must do) signs

Mandatory signs tell you that you must wear some special safety equipment: *blue solid circle, white symbol*, no border required (see Table 1.4, Figure 1.18 and Figure 1.19).

Restriction signs

Restriction signs tell of the limitations placed on an activity or use of a facility: *red circular border, no cross bar, white background*.

Limitation or restriction signs normally have a number placed in them to indicate a limit of some type (e.g. a speed limit or weight limit, see Table 1.4).

FIGURE 1.18 Eye protection must be worn

FIGURE 1.19 Hearing protection must be worn

Hazard warning signs

Hazard warning signs warn you of a danger or risk to your health: *yellow triangle with black border, black symbol* (Table 1.4, Figure 1.20, Figure 1.21 and Figure 1.22).

(Danger) hazard signs

Danger signs warn of a particular hazard or hazardous condition that is likely to be life-threatening: *white rectangular background*, white/red DANGER, *black border*, and wording (Figure 1.23).

FIGURE 1.20 Fire hazard warning signs

FIGURE 1.21 Toxic hazard

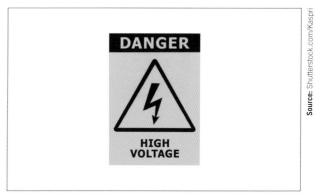

FIGURE 1.22 Electric shock hazard

FIGURE 1.23 Danger signs

Emergency information signs

Emergency information signs show you where emergency safety equipment is kept, *green solid square, white symbol* (Table 1.4, Figure 1.24 and Figure 1.25).

Fire signs

Fire signs tell you the location of fire alarms and fire-fighting facilities: *red solid square, white symbol* (Table 1.4 and Figure 1.26).

Sometimes a safety sign is required to be placed on a piece of equipment to indicate a problem or potential hazard if the machine is switched on. This is usually in the form of a tag.

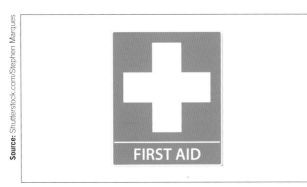

FIGURE 1.24 First aid

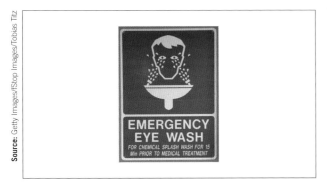

FIGURE 1.25 Emergency (safety) eye wash

FIGURE 1.26 Fire alarm call point

Safety and accident prevention tags for electrical equipment

Electrical wires and equipment that are being worked on or are out of service, or are live or may become live, must have Warning or DANGER safety tags fixed to them to help prevent accidents. Once the hazard has been removed, only the person who put the tag in place should remove it or authorise its removal.

Any standard safety sign may be made smaller and used as an accident protection tag (Figure 1.27). If words are to be used, this will generally be in the form of a danger sign. A tag should be at least 80 mm × 50 mm, plus any area required for tying or fixing the tag in place.

The background colour of the tag should be yellow for warning signs and white for danger signs.

FIGURE 1.27 Electrical safety signs and tags

For more picture safety signs and their meanings, see:
- Standards Australia: AS 1319 Safety signs for the occupational environment
- SafeWork South Australia: 'Work Safety Signs'
- safety sign manufacturers' catalogues: see Yellow Pages – Signs – Safety &/or Traffic.

2.4 Identify asbestos-containing materials on a work site and report to designated personnel

Asbestos – the silent killer

There is one product that is no longer used within the construction of buildings, but that is still considered a dangerous substance. It has no smell, no taste and in its raw form will not irritate the skin. This product is a naturally occurring rock that was mined from many locations around Australia, the most notable being a place called Wittenoom in Western Australia. There are an estimated 3000 products that were manufactured using this wonder product – asbestos. The difficulty with asbestos is that it was used in buildings from the 1950s up until the late 1980s, when it was finally banned from use. It will cause problems once it is disturbed or damaged in the removal and demolition process of a building.

The health risks involved with this material are not immediately obvious and related health problems take many years to develop. All are related to a person having breathed in the asbestos fibres. The three diseases associated with having breathed in asbestos fibres are:
- asbestosis
- lung cancer
- mesothelioma.

So, how to identify asbestos and asbestos-containing materials (ACMs)? To start with, consider the age of the structure being deconstructed or

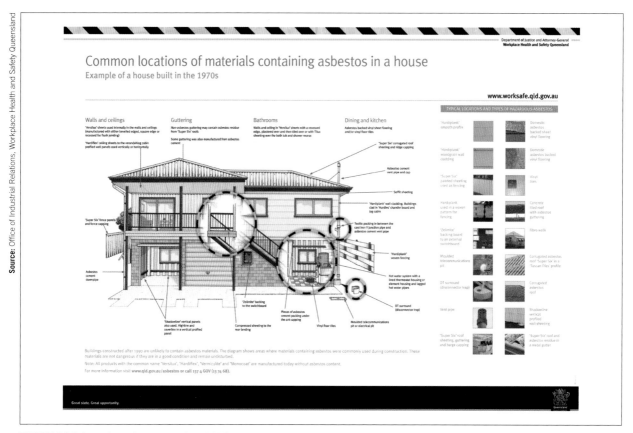

Source: Office of Industrial Relations, Workplace Health and Safety Queensland

FIGURE 1.28 Location of asbestos in a 1970s home

demolished. Was it built between 1950 and 1990? If the answer is yes, there is a high chance that it contains asbestos.

What are the products that may contain asbestos? Refer to **Figure 1.28** as this will give you a much better idea of what to look for.

It is one thing to identify asbestos material; it is another thing to remove it. Safe Work Australia and the model WHS Regulations have established two licences for asbestos removal – *Class A* and *Class B*.

Businesses with a *Class A* licence are permitted to remove all types of asbestos, including both friable and non-friable asbestos.

Businesses with a *Class B* licence can only remove non-friable asbestos.

It is important to consider the legislative requirements as set out by your state or territory, but in the national Model Code of Practice: How to Safely Remove Asbestos there is an allowance for an unlicensed person to safely remove up to 10 m² of non-friable asbestos or ACM. You must know and understand what your local requirements are for reporting of asbestos and how to document the safe removal of asbestos within your work region and workplace.

Silica dust is caused by cutting or grinding manufactured stone products. The latest regulation requires silica dust to either be collected with special extractors or use of a wet cutting technique that creates no dust. There are severe health risks and penalties for not controlling dust when working with silica products.

LEARNING TASK 1.1 GROUP ACTIVITY – LICENCES

Using the information in **Table 1.2**, visit the website of the regulating authority for your state or territory and search for 'Code of practice – asbestos'. Using this document, complete the following three (3) questions.

1 What type of licence is required to remove 'any amount of friable and non-friable asbestos or Asbestos-Containing Material (ACM)'?

....

2 What type of licence is required to remove 'any amount of non-friable asbestos ONLY or Asbestos-Containing Material (ACM)'?

....

3 If your state or territory allows: What is the maximum amount of non-friable asbestos or associated Asbestos-Containing Material (ACM) that may be removed *without* a licence?

.... m².

COMPLETE WORKSHEET 2

3. Plan and prepare for safe work practices

Guards and any safety device on plant and equipment needs to not only meet the manufacturer specifications in the original design of that equipment but also be safe for work site regulations and Australian Standards within the state and territory that you work in.

3.1 Identify, wear, correctly fit, use and store correct personal protective equipment and clothing for each area of construction work in accordance with workplace procedures

Personal protective equipment (PPE) and clothing

PPE is the last line of defence to protect your health and safety from workplace hazards. It is the PCBU/employer's responsibility to provide the PPE, clothing, and training to protect the worker. It is your responsibility to wear and look after the equipment provided.

PPE must be appropriate to each hazard. It is important to consider the following when assessing workplace PPE requirements:

1 *The workplace.* Could it be made safer so that you don't need to use PPE?

2 *PPE selection.* Is the PPE designed to provide adequate protection against the hazards at your workplace?

3 *PPE comfort and fit.* Is the PPE provided comfortable to wear? Even if the equipment theoretically gives protection, it won't do the job if it doesn't fit properly or is not worn because it is too uncomfortable. For example, close-fitting respirators give protection only if the person is clean-shaven. Those with a beard or 'a few days' growth' will need to use a hood, helmet, or visor-type respirator. The respirator must also be properly cleaned and maintained.

PPE can be grouped according to the part of the body it will protect:

- head – safety helmets, sun hats
- eyes/face – safety spectacles, goggles, face shields
- hearing – earmuffs, ear plugs
- airways/lungs – dust masks, respirators
- hands – gloves, barrier creams
- feet – safety boots and shoes, rubber boots
- body – clothing to protect from sun, cuts, abrasions, and burns; high visibility safety garments and fall protection harnesses.

Identify hazards

To decide what PPE and clothing is required, you must first be able to identify the hazards involved. Types of hazards commonly identified where PPE and clothing are a suitable means of protection are:

- physical hazards – noise, thermal, vibration, repetitive strain injury (RSI), manual and radiation hazards
- chemical hazards – dusts, fumes, solids, liquids, mists, gases, and vapours.

Once the hazards have been identified, suitable equipment and clothing must be selected to give the maximum protection.

PPE equipment

As mentioned previously, this PPE protection *must* only be used as the final defence against hazards if no other form of protection is available. Remember that the hazard is still present, and that diligence and care *must* be observed.

Safety helmets

Wearing safety helmets on construction sites may prevent or lessen a head injury from falling or swinging objects, or through striking a stationary object.

Safety helmets must be worn on construction sites when:

- it is possible that a person may be struck on the head by a falling object
- a person may strike his/her head against a fixed or protruding object
- accidental head contact may be made with electrical hazards
- carrying out demolition work
- instructed by the person in control of the workplace.

Safety helmets must comply with AS/NZS 1801 Occupational protective helmets and must carry the AS or AS/NZS label, and must be used in accordance with AS/NZS 1800 Occupational protective helmets – Selection, care and use (Figure 1.29).

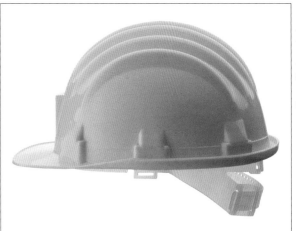

Source: iStock.com/malerapaso

FIGURE 1.29 Safety helmet

When wearing a helmet, the harness should be adjusted to allow for stretch on impact. No contact should be made between the skull and the shell of the helmet when subjected to impact.

Sunshade

The awareness of skin cancer for building workers is increasing. The neck, ears and face are particularly exposed. Workers should always wear sun protection when working outdoors (including in the winter).

Sunshades include wide-brimmed hats and foreign legion-style sun shields fixed to the inner liner of safety helmets, or safety helmet 'foreign legion sun brims' (**Figure 1.30**).

FIGURE 1.30 Fabric sun brim accessory for a safety cap and bucket hat

Eyes/face protection

The design of eye and face protection is specific to the application. It must conform to AS/NZS 1337.1 Eye protectors for industrial applications. The hazards to the eyes are of three categories:
1 physical – dust, flying particles or objects, molten metals
2 chemicals – liquid splashes, gases and vapours, dusts
3 radiation – sun, laser, welding flash.

The selection of the correct eye protection to protect against multiple hazards on the job is important. Most eyewear is available with a tint for protection against the sun's UV rays or may have radiation protection included (see **Figure 1.31** and **Figure 1.32**).

FIGURE 1.31 Clear wide-vision goggles

FIGURE 1.32 Clear-framed spectacles

Face shields

Face shields give full face protection, as well as eye protection. They are usually worn when carrying out grinding and chipping operations, when using power tools on timber. Shields are worn for full-face protection when welding (**Figure 1.33**). The shield may come complete with head harness (**Figure 1.34**) or be designed for fitting to a safety helmet.

FIGURE 1.33 Full-face welding mask

FIGURE 1.34 Face shield

Hearing protection

You should always wear ear protection in areas where loud or high-frequency noise operations are being carried out, or where there is continuous noise. Always wear protection when you see a 'Hearing protection must be worn' sign, and when you are using or are near noisy power tools.

The two main types of protection available for ears (Figure 1.35) are:

- ear plugs – semi- and fully disposable
- earmuffs – available to fit on hard hats where required.

FIGURE 1.35 Hearing protection

Choose the one that best suits you and conforms to AS/NZS 1270 Acoustic – Hearing protectors.

Disposable dust masks

Dust masks are available for different purposes, and it is important to select the correct type. If the work that you are undertaking is mowing or general sweeping, a nuisance-dust mask is appropriate and is designed to filter out nuisance dusts only. A nuisance-dust mask is easily recognised as it has only one strap to hold it onto the face (Figure 1.36). If, on the other hand, you are working with toxic dusts; for example, bonded asbestos or silica, you will require greater protection. A mask with two straps and labelled with either P1 class particle dust filter (minimum protection) or P2 class particle dust filter (mid-range protection) (Figure 1.37) will be required.

Respirators

Half-face and full-face respirators have cartridge-type filters that are designed to keep out dusts, smoke, metal fumes, mists, fogs, organic vapours, solvent vapours, gases, and acids, depending upon the combination of dust and gas filters fitted to the respirator. Cartridge type filters can be identified by the classification ratings from AS/NZS 1716 Respiratory protective devices.

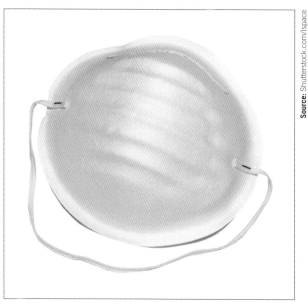

FIGURE 1.36 Mini dust mask warning: For nuisance dust only

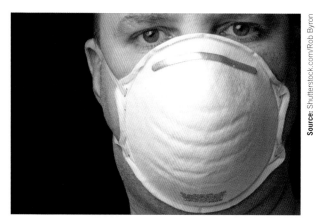

FIGURE 1.37 P1 and P2 disposable masks

Respirators fitted with P2 class particle dust filter (Figure 1.38) are suitable for use with the general low-toxic dusts and welding fumes that are commonly found on construction sites.

FIGURE 1.38 Half-face respirator with P2 class dust filters fitted

Some full-face or half-face respirators may also be connected to an air supply line or bottle that provides clean filtered air to the user. These are generally used for loose (friable) asbestos removal or in contaminated or low oxygen environments, where face and eye protection is also required.

Further information on respirators and dust masks should be obtained from the manufacturers. It is very important to be trained in the correct methods of selecting, fitting, wearing, and cleaning of the equipment in accordance with AS/NZS 1715 Selection, use and maintenance of respiratory protective equipment. When selecting, it is important that tight-fitting respirators and masks must have an effective seal on the face to ensure that all air entering your respiratory passages has been fully filtered.

Gloves

Gloves are used to protect hands and arms from both physical and chemical hazards. Stout leather gloves are required when handling sharp or hot materials. Rubberised chemical-resistant short or long gloves are used when handling hazardous chemical substances. Gloves should conform to AS/NZS 2161.1 Occupational protective gloves – Selection, use and maintenance (see Figure 1.39).

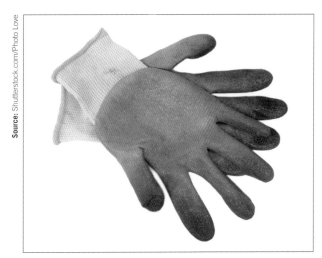

FIGURE 1.39 Gloves

Creams

Barrier creams may be used when gloves are too restrictive, to protect the hands from the effects of cement and similar low-toxic hazards (see Figure 1.40).

Foot protection

It is mandatory to always wear protective footwear at the workplace. *Thongs are not permitted at any time.* Footwear should conform to AS/NZS 2210.1 Safety, protective and occupational footwear guide to selection, care, and use.

FIGURE 1.40 Barrier cream

All safety footwear must have:
- stout oil-resistant, non-slip soles or steel midsoles to protect against sharp objects and protruding nails
- good uppers to protect against sharp tools and materials
- reinforced toecaps to protect against heavy falling objects.

Safety boots should be worn in preference to safety shoes on construction sites to give ankle support over the rough terrain. Safety joggers may be required when carrying out roof work or scaffold work; they must have reinforced toecaps. Rubber boots should be worn when working in wet conditions, in wet concrete, or when working with corrosive chemicals. They must have reinforced toecaps. See Figure 1.41 for examples of foot protection.

FIGURE 1.41 Foot protection

Clothing

Good quality, tough clothing is appropriate for construction work. It should be kept in good repair and cleaned regularly. If the clothing has been worn when working with hazardous substances it should not be taken home to launder but sent to a commercial cleaning company; this will prevent the hazards from contaminating the home and the environment.

A good fit is important, as loose-fitting clothing is easily caught in machine parts or on protruding objects. Work pants should not have cuffs or patch pockets, as hot materials can lodge in these when worn near welding or cutting operations.

Clothing should give protection from the sun's UV rays, cuts, abrasions, and burns.

Industrial clothing for use in hazardous situations should conform to AS/NZS 4501.2 Occupational protective clothing – General requirements.

Fall protection harnesses

In some instances when working at heights, it may be necessary to wear a harness. These are specialist items that need to be fitted to the individual. A harness *must be* correctly fitted, or serious injury may result if or when the person falls. All harnesses must comply with AS2626 Industrial safety belts and harnesses – Selection, use and maintenance.

Cleaning and maintenance

All PPE must be cleaned and maintained on a regular basis. This must be done by someone who has been trained in inspection and maintenance of such equipment. Remember, your life and wellbeing depends on this PPE, and if it is faulty or damaged or simply not functioning properly, you are at risk.

3.2 Select tools, equipment and materials, and organise tasks in conjunction with other personnel onsite and in accordance with workplace procedures

Selecting the correct tools, equipment and materials for a task is important so the task can be completed efficiently with good quality and safely. You need to coordinate with other personnel onsite considering your workplace policies and procedures or the standard safe ways of doing things for your workplace. Proper organisation and planning are important to have the tools at the right time and in the right place. Remember that all tools will have a safe operating procedure or user manual and you need to obtain these either from the original tool purchase records or from the manufacturer's website and use them in all your planning work.

3.3 Determine required barricades and signage, and erect at the appropriate site location

Prior to starting any work, the area needs to be isolated to stop other workers and/or the public from entering the work area. This is achieved using temporary fencing and physical barricades. The barricade and signage that you erect at a site location will be determined by the hazards that are likely to occur at that location and the risks of people entering and becoming injured. If your work site is in a remote location and there is no chance of people entering the site, then you may not need as strong barricades as you may if you're next to a primary school for example. The signage that is erected at a site needs to be determined using Australian standard signage at a level and location for people to have adequate warning before they encounter a hazard. Wherever possible, use symbols on signs so you overcome any language difficulties in being able to interpret the sign, and being warned before a hazard is encountered.

Your overall site safety management plan and your sketch of the site plan and where all the facilities and materials are going to be stored should also include information about the type and location of signage, the type and location of barricades, and any other safety related features such as emergency meeting points, fire services, and first aid supplies.

3.4 Apply safety data sheets (SDSs), job safety analyses (JSAs) and safe work method statements (SWMSs) relevant to the work to be performed

Safety data sheets (SDSs)

An SDS is a document that gives information on the properties of hazardous chemicals and how they affect health and safety in the workplace. SDSs usually include information on the:

- identity of the chemical
- health and physicochemical hazards
- safe handling and storage procedures
- emergency procedures
- disposal considerations.

The SDS should always be referred to when assessing risks in the workplace.

SDSs are prepared by the manufacturer of the products used at your workplace (Figure 1.42). They are available from the manufacturer or supplier.

Any hazardous material being delivered to a construction site should have an SDS for the product provided at the site before it is delivered. These sheets need to be kept onsite and should not be more than three years old. If you have not read the SDS for a substance you are going to use or will be exposed to when someone else uses it, make sure you obtain a copy and read it first. If any special training is required, you should complete the training before the material is handled.

Manufacturers and suppliers may be required by law to provide information 'about any conditions necessary to ensure that the substances will be safe and without risk to health when properly used'; this information is included on the SDS for that product.

SDS No. CASDS23 | Issue Date: 1 October 2017

SAFETY DATA SHEET

Concrete Mix

Section 1: Identification of the Material and Supplier

Company Details

Cement Australia Pty Limited

ABN 75 104 053 474

18 Station Avenue
Darra, Queensland 4076

Tel: 1300 CEMENT (1300 236 368)
Fax: 1800 CEMENT (1800 236 368)
Website: www.cementaustralia.com.au

Emergency Contact Number:

Contact Person: Technical Manager
Telephone: 1300 CEMENT (1300 236 368 - Business Hours) or
Poisons Information Centre 13 11 26

Manufacturing Plants

Geelong: 292 Thompson Road, Geelong North VIC 3215
Brisbane: 77 Pamela St, Pinkenba QLD 4008
Auburn: Highgate Street, Auburn NSW 2144
Townsville: Benwell Road, Townsville QLD 4810

Product

Name: Concrete Mix

Other Names: Tradies Own Concrete
Extra Strength Concrete Mix
Extra Strength PRO-50 Concrete Mix

Use: Concrete Mix is used to produce concrete.

Section 2: Hazards Identification

Hazardous Substance. Non-dangerous Goods
A low proportion of the fine dust in the supplied dry product will be respirable crystalline silica. Once wetted, in the wet or final set form, risk of any airborne respirable dust will be low, but dry residues, or dust from cutting, grinding, abrading or finishing the set product may contain respirable crystalline silica.

Skin Corrosion/Irritation
(Category 1c) – When Water is added

Sensitisation – Respiratory
(Category 1)

DANGER
CAN CAUSE SKIN BURNS (When water is added) & EYE DAMAGE: Avoid contact with the eyes and skin from both wet and dry cement. Wet cement can be corrosive to the eyes and skin and may cause skin sensitisation (dermatitis). Safety: Wear suitable protective clothing, gloves (AS2161), and eye/face protection (AS/NZS1337.1).

FIGURE 1.42 Page 1 of a sample SDS

IF ON SKIN: Wash thoroughly after handling. Wash clothes before re-use and separately from other clothing.

IF IN EYES: Rinse cautiously with water for several minutes. Remove contact lenses, if present and easy to do so. Continue rinsing.

IF SWALLOWED: Rinse mouth. Do NOT induce vomiting.

For more information
call **1300 CEMENT**
(1300 236 368) or visit
**www.cementaustrali
a.com.au**

CEMENT
AUSTRALIA®

mix it with the best.

FIGURE 1.42 (Continued)

Employers and PCBUs may be required by law to provide such information, or such instruction, training, and supervision as is necessary to ensure the health and safety of employees/workers.

SDSs, user manuals and standard operating procedures can also be used to help you to:

- be aware of any health hazards of a product
- check that the site emergency equipment and procedures are adequate
- store the chemicals properly
- check that a chemical is being used in the right way for the right job
- decide whether any improvements or changes should be made to machinery or work practices
- decide whether any environmental monitoring should be done.

GREEN TIP

If environmental monitoring needs to be done on a site, call in the experts to set up their equipment in the right place to obtain correct readings.

When you first view an SDS, ensure that the following headings (at least) are complete and you understand the information it provides:

1 Identification
2 Health hazards
3 Precautions for use
4 Safe handling.

Health hazards should include at least two subheadings:

1 Acute health effects
2 Chronic health effects.

Acute health effects

These relate to short-term exposure to the chemical. For example, if you swallow a poison, you are either dead or very ill within 48 hours. Similarly, if you have acid splashed onto your skin, you might suffer burns immediately or within the next 48 hours.

Chronic health effects

Chronic health effects relate to the long-term effects of exposure to a chemical. In general use, these effects may take years to become apparent. The chronic health effects may be just as serious as the acute health effects in the long term; for example, exposure to cancer-causing agents may take 20 years to become apparent but may still end up killing you.

Disposal of chemicals

It is important to dispose of chemicals safely, as prescribed in the SDS for the product. Always wash yourself carefully with soap after handling any chemical. If you spill any chemical on your clothing, remove the clothing and wash the body part affected. If you experience skin problems or difficulty in breathing, *always* seek medical advice or contact the Poison Information Centre (ph.: 13 11 26) immediately.

Safe work practices

A job safety analysis or JSA is a good tool to use to identify and evaluate potential hazards for a job or a task that you are going to do. This is a process where you look at all potential hazards and the risks that may occur from these hazards and build a trail of evidence of research from all the available documentation for the particular task or tools or materials you're going to use. By doing this, you can address any potential hazards before they cause harm to workers, material, equipment or others.

Having identified the hazards and then assessed the risks, it is now time to plan how the work can be carried out safely. One method that has great merit is to write a document outlining the steps to be carried out, the hazards associated with each step, followed by the controls that are to be used to eliminate or reduce the hazards. This document is called a **safe work method statement (SWMS)**.

Safe work method statements

Definition:

A Safe Work Method Statement (SWMS) is a type of risk assessment that:

- lists the types of high-risk construction work being done
- states the health and safety hazards and risks arising from that work
- describes how the risks will be controlled, and
- describes how the risk control measures will be put in place.

Source: WorkSafe Victoria https://www.worksafe.vic.gov.au/resources/safe-work-method-statements-swms

Since the introduction of the national model WHS regulations, only high-risk construction work legally requires a SWMS. High-risk construction work is as follows:

a   involves a risk of a person falling more than 2 metres
b   work carried out on a telecommunications tower
c   demolition of an element of a structure that is load bearing
d   removal of asbestos
e   structural alterations or repairs that require temporary support to prevent collapse
f   confined spaces work
g   is carried out in or near a:
   i   shaft or trench with an excavated depth greater than 1.5 m
   ii   tunnel
h   use of explosives
i   carried out on or near pressurised gas distribution mains or piping
j   carried out on or near chemical, fuel, or refrigerant lines
k   carried out on or near energised electrical installations or services
l   carried out in an area that may have a contaminated or flammable atmosphere
m   tilt-up or precast concrete
n   carried out on, in or adjacent to a road, railway, shipping lane or other traffic corridor that is in use by traffic other than pedestrians
o   carried out in an area at a workplace in which there is any movement of powered mobile plant
p   carried out in an area in which there are artificial extremes of temperature
q   carried out in or near water or other liquid that involves a risk of drowning
r   involves diving work.

Source: Adapted from Work Health and Safety Management Systems and Auditing Guidelines (edition 5) May 2014 © NSW Government.

Even though a SWMS is only required for 'high-risk construction work' it is still advisable to prepare a SWMS for all work (see Figure 1.12); thus controlling all potential

hazards. Conducting a risk assessment for all tasks that contain potential risk on a construction site is an effective method of ensuring tasks are organised in conjunction with others onsite and all risks are mitigated.

Another type of risk assessment that is commonly used in the construction industry for tasks that are not considered 'high-risk construction' is a job safety analysis (JSA).

JSAs are completed in a similar fashion as to a SWMS by first establishing the steps of the tasks, identifying the potential hazards involved, implementing controls to reduce the potential risk of the hazard and then reviewing the hazard once the controls have been implemented.

LEARNING TASK 1.2

GROUP ACTIVITY – DISCUSS THE HEALTH ISSUES INVOLVED WITH USING GENERAL PURPOSE CEMENT AND THE PPE REQUIRED

Obtain a full copy of the SDS for Blue Circle – general purpose cement (see Figure 1.42). You can obtain a PDF copy online by typing into Google 'Blue circle – general purpose cement SDS'.

Read the SDS and discuss the safety issues with your group.

COMPLETE WORKSHEET 3

4. Apply safe work practices

4.1 Carry out tasks in a manner that is safe for operators, other personnel and the general community, in accordance with legislative requirements, and workplace policies and procedures

In order to carry out tasks in a way that is safe for operators, other personnel and the general community, while keeping in mind legislative requirements, workplace policies and procedures, it's important to adopt a mindset of 'safety first'. Always have the attitude of wanting and being able to finish work each day in the same safe and healthy state you were in when you started work.

Some key steps to achieve this can involve some of the following processes:

■   Familiarise yourself with any legislative requirements or workplace policies and procedures that may be relevant to help you work safely. Ensure you are up to date with codes of practice that may be applicable to the tasks on your construction site and stay updated with any changes or amendments to ensure compliance. Subscribe to your state or territory's safety email subscription system to receive alerts and updates about any changes or any incidents that have occurred on construction sites that you may learn from.

- Ensure that everyone on the construction site has been trained in hazard identification, using personal protective equipment, emergency procedures, and proper use of all tools and materials. Ensure that training has been recorded in a site diary for verification if needed. Record all toolbox talks and regularly participate in safety talks on the construction site.
- Participate in risk assessments and hazard identification before you begin a task thereby ensuring you have the most up-to-date current and applicable documentation to advise you as you're seeking to avoid any potential hazards. Develop control measures for any hazards that you have identified and document your work.
- Ensure an adequate supply of personal protective equipment and that all safety equipment is operating correctly.
- Work to your safe work procedures and regularly review your procedures to make sure that all workers are following your plans consistently.
- Spend time making sure you have a clean and organised site with materials and equipment stored in safe locations and all paths of travel are clean and free from hazards to avoid slips, trips and falls.
- Regularly check signage and warnings and any safety equipment and perform emergency practice simulations and ensure any new visitors or workers are trained correctly and this training is recorded.
- Implement a policy of regularly inspecting equipment and making sure that your company has a policy for faulty and damaged equipment to be serviced immediately and not used until it is safe to do so.
- The cost of an accident can put a business into bankruptcy or close it down so make sure it doesn't happen to you by ensuring you adopt a safety culture within your organisation, and everybody knows that safety comes first. Implement ongoing monitoring and improvement in all your processes.

Every construction site is different, so you need to adapt your methods for each different situation and seek technical professional advice where necessary, so your business operates safely.

4.2 Use plant and equipment guards in accordance with manufacturers' specifications, work site regulations and Australian Standards

Guards on tools and equipment

A guard on a power tool, static machine or any equipment with a moving blade is a form of protective equipment. Its purpose is to prevent material and/or

material waste from being projected towards the operator, as well as preventing fingers or hands from being drawn into moving parts or blades. Guards are also used to prevent pieces of shattered blade or abrasive disc from striking the operator when they are faulty or disintegrate if they jam.

A guard is fitted as the last line of protection for an operator, and therefore should *never* be removed or tied back while the tool or machine is in use. In fact, the only time the guard should be allowed to move from its safety position is when the tool is in use, and it retracts as it is fed into the material or when the power has been disconnected from the tool to allow the blade or disc to be removed.

Many portable hand-held power tools and static machines have guards fitted, for example:

- circular power saw (Figure 1.43)
- drop or drop/slide saw
- angle grinder
- bench or table saw.

FIGURE 1.43 Circular saw fitted with a guard

Many pieces of large machinery and plant will also have mechanisms in place to ensure the safety of the operator and others in the area. For example, some functions of an excavator may be restricted if the operator is not sitting in the seat of the machine.

Safety features like this can become disruptive for the operator and people around the plant, however these are in place for everyone's safety. It is important that safety features implemented by the manufacturer are not disengaged or 'worked around' by people working with the plant.

If an accident does occur and it is found a worker purposely 'worked around' a safety feature there will be significant ramifications, including fines or even jail time.

4.3 Follow procedures and report hazards, incidents and injuries to relevant authorities

Well before an incident or injury occurs make sure you have identified the procedures and the reporting methods for your workplace and for your local relevant authorities so that you can follow those procedures if any incident or injury happens. The safe work regulating authority in your state or territory will have requirements for notification of near miss incidents or accidents or fatalities and these phone numbers, email addresses or websites should be noted in your company policies and procedures.

The purpose of reporting hazards, incidents and injuries is to track those that are occurring frequently and to put plans into place to stop the same accident occurring. By monitoring reports, you can improve your business and minimise the chances of accidents happening again.

4.4 Recognise and do not use prohibited tools and equipment in areas containing identified asbestos

There are important considerations when removing asbestos:

- Make sure you report any suspected or confirmed asbestos locations to your supervisor and any required authorities according to the law and company policies and procedures
- Wearing the correct PPE (see **Figure 1.44**)

FIGURE 1.44 PPE that must be worn when removing non-friable asbestos sheeting

- Have the area barricaded and use the appropriate **signage** (see **Figure 1.45**)
- *Never* use power tools that cut, scrub, sand or grind down the surface of the ACM sheet as this action releases asbestos fibres into the air (see **Figure 1.46**)
- Contain the asbestos fibres by wetting the material and removing it as a full sheet (*do not break it up*) and dispose of it correctly.

Note: All asbestos removal and disposal must be done in accordance with the national Model Code of Practice: How to Safely Remove Asbestos or your state or territory requirements.

FIGURE 1.45 Asbestos removal signage

FIGURE 1.46 Asbestos warning sign and tools not to use on ACM

4.5 Identify and follow requirements of work site safety signs and symbols

Placement of safety signs

Signs should be located where they are clearly visible to all concerned, where they can easily be read and so that they will attract attention. If lighting is not adequate, illuminated signs can be used.

Signs should not be located where materials and equipment are likely to be stacked in front of them, or where other obstructions could cover them (e.g. doors opening over them). They should not be placed on movable objects such as doors, windows or racks so that when the object is moved, they are out of sight or the intention of the sign is changed.

The best height for signs is approximately 1500 mm above floor level. This is at the normal line of sight for a standing adult. The positioning of the sign should not cause the sign itself to become a hazard to pedestrians or machine operators.

Regulation and hazard-type signs should be positioned in relation to the hazard to allow a person plenty of time to view the sign and take notice of the warning. This distance will vary; for example, signs warning against touching of electrical equipment should be placed close to the equipment, whereas signs on construction work may need to be placed far enough away to permit the warning to be understood before the hazard is reached.

Care should be taken where several signs are intended to be displayed close together. The result could be that so much information is given in one place that little or no notice is taken of it, or that it creates confusion.

LEARNING TASK 1.3

GROUP ACTIVITY – DRUGS AND ALCOHOL

1 Do you know someone on the job site who may have a problem with drugs or alcohol?
2 If a fellow worker asked you if there is 'help' available, what would you say?
3 Visit https://www.safework.nsw.gov.au/hazards-a-z/alcohol-and-other-drugs. Read as much as you can on this website and discuss the issues it raises with your group. Remember someone under the influence of drugs is also a hazard on the job site.

4.6 Clear and maintain work site area to prevent and protect self and others from incidents and accidents, and to meet environmental requirements

A clear and tidy site with open walkways and access areas will help prevent accidents and will protect yourself and others from incidents. Before you start a job, it's important to make sure you have checked your walking and working area to make sure it is clear. When you put tools or materials down, make sure you don't put them in a walking area. Always keep hazardous materials locked away from people who should not use them. Also check that your company's policies and procedures for each site includes the procedures to meet the environmental requirements for your area.

On a construction site there is always waste material that accumulates as the work gets completed. If this material is not disposed of correctly, it can very quickly create a slipping and tripping hazard.

To eliminate this hazard it is important that the site is maintained and cleaned frequently. Waste material should be put into waste or recycling bins instead of left on the ground. This removes the hazard and the double handling of the material.

Any harmful or toxic chemicals must be stored or disposed of according to the relevant SDS to reduce the potential hazard to people onsite and the environment if a spill does occur.

COMPLETE WORKSHEET 4

5. Follow emergency procedures

5.1 Identify designated personnel in the event of an emergency for communication purposes

Roles

Many companies will have selected and trained competent individuals to take charge during an emergency; as a worker it is important to recognise the various roles that these responsible personnel perform.

Emergency coordinator

- Determines the nature of the emergency and the course of action to be followed.
- Sets off any alarm or siren to warn persons of an emergency.
- Contacts appropriate emergency services such as police, fire, or ambulance.
- Initiates the emergency procedure and briefs the emergency services when they arrive.

Warden or controller

- Assumes control of the occupants/workers until the emergency services arrive.
- Notifies all persons regarding the nature of the emergency.
- Gives clear instructions and makes a record of what was carried out.
- Reports all details to the emergency coordinator as soon as possible.

Casualty control

- Attends to casualties and coordinates first aid.
- Coordinates the casualty services when they arrive.
- Arranges for further medical or hospital treatment.

Injury management

To keep workers and visitors safe on a site it's important to identify and assess any emergency risks that may occur. Some steps to do this are:

- Identify hazards by having a competent person or persons conduct a risk assessment looking particularly at unstable structures, excavation materials, power tools or likely environmental events.
- Assess the likelihood and how severe the event could be that you have identified and consider any consequences and how severe they may be.
- Determine the risk level and use the matrix shown earlier in Figure 1.7 so you can rate the action required.
- Develop an emergency plan for each hazard that you identified including procedures, evacuation routes, communication methods and emergency contacts.
- Communicate the plan to all workers and make sure any visitors to the site are made aware of how to learn the plan so they can be kept safe.
- Regularly review and update the plan especially when you have new workers on site, after any incidents, or after the plan is put into practice in an emergency.

The loss or disruption a company can experience because of a hazardous incident may be multiplied tenfold when that incident leads to a worker being injured.

A comprehensive risk management system should include a well thought out plan to maximise the opportunity for injured workers to remain at work. This allows the worker to be productive in some capacity and assists with the recovery and rehabilitation process.

Therefore, the risk management system should cover the following points:

- early notification of the injury
- early contact with the worker, his or her doctor and the builder's insurance company
- provision of suitable light duties as soon as possible to assist with an early return to work
- a written plan to upgrade these duties in line with medical advice.

Payment of compensation

Worker's compensation insurance is a system that provides payment and other assistance for workers injured through work-related accidents or illnesses. It may also provide their families with benefits where the injury is very serious, or the worker dies.

PCBU/employers must take out worker's compensation insurance to cover all workers considered by law to be their employees/workers.

Eligibility for compensation

To claim worker's compensation a worker must have suffered an injury or disease. The injury or disease must be work-related: that is, it must have happened while working, during an allowed meal break, or on a work-related journey.

The injury or disease must result in at least one of the following:

- death of the worker
- the worker being totally or partially unable to perform work
- the need for medical, hospital or rehabilitation treatment
- the worker permanently losing the use of some part of the body.

A claim for compensation is made by:

- informing the PCBU/employer and lodging a claim as soon as possible
- seeing a doctor and obtaining a medical certificate.

If any problems arise with the compensation claim the worker should contact their state or territory compensation board or authority, or their own union.

A record of *all* injuries that occur at a workplace must be entered in an injuries register kept at the job site.

5.2 Follow safe workplace procedures for dealing with accidents, fire and other emergencies, including identification and use, if appropriate, of fire equipment within scope of own responsibilities

Evacuation process

It is vital that PCBU/employers prepare an emergency plan for the workplace, and each worker needs to be aware of what to do in the event of an emergency evacuation.

An emergency plan is a written set of directions that outlines what workers and others at the workplace must do in an emergency. An emergency plan must provide the following information:

- emergency procedures, including:
  - an effective response to an emergency
  - evacuation procedures
  - notifying emergency service organisations at the earliest opportunity
  - contact details for medical treatment and assistance
  - effective communication between the person authorised to coordinate the emergency response and all people at the workplace
- testing of the emergency procedures – including the frequency of testing
- information, training, and instruction to relevant workers on implementing the emergency procedures.

Another important part of the evacuation process is the evacuation diagram. This diagram shows important information, such as the location of evacuation assembly points, fire hose reels, fire extinguishers and where the Fire Indicator Board (FIB) is. These diagrams should be placed at strategic locations in the structure allowing all workers easy access to the information shown on them (Figure 1.47).

Fire

Identifying hazardous materials on a site begins by obtaining the work site's hazardous materials register and becoming familiar with where those materials are stored and how to access them if you need to. If you are not sure if the material is hazardous or not, then obtain the safety data sheets or the instructions that came with that material when it was delivered to the site.

There are many other hazards on a work site and it is important for you to obtain the site safety management plan and study it to identify hazards and to learn from previous notes in this chapter about how to identify and control hazards.

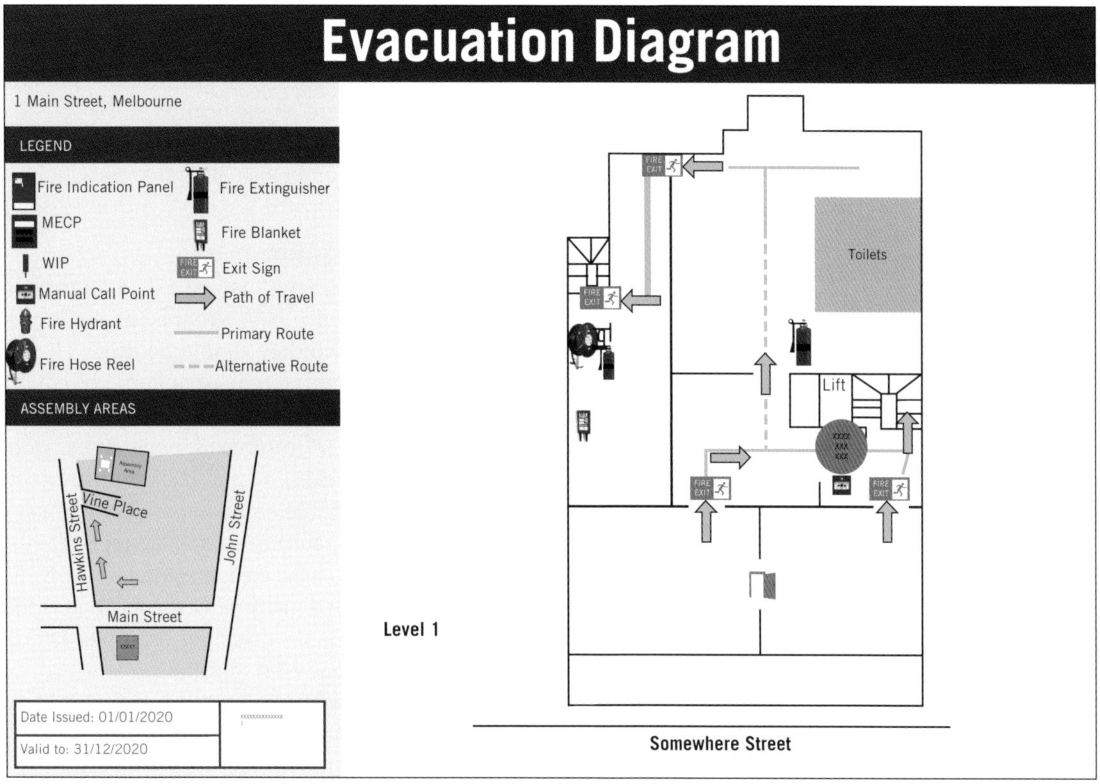

Evacuation Diagram

1 Main Street, Melbourne

LEGEND

- Fire Indication Panel
- MECP
- WIP
- Manual Call Point
- Fire Hydrant
- Fire Hose Reel

- Fire Extinguisher
- Fire Blanket
- Exit Sign
- Path of Travel
- —— Primary Route
- --- Alternative Route

ASSEMBLY AREAS

Assembly Area

Hawkins Street · Vine Place · John Street · Main Street

Date Issued: 01/01/2020

Valid to: 31/12/2020

Toilets

Lift

FIRE EXIT

Level 1

Somewhere Street

FIGURE 1.47 Evacuation diagram

One of the biggest hazards on any construction site is fire – it produces heat that can burn, and smoke that can asphyxiate people. Fire can have such a high degree of heat that it can bring down steel or brick structures.

Being able to prevent fires and fight fires is part of every worker's responsibility. It is important for the safety of every worker on a job to understand the procedures to follow in the event of a fire.

Large construction sites and buildings should have fire-fighting teams responsible for each floor or the whole building. The fire-fighting team must be specially trained staff members who can direct the evacuation and fire-fighting operation, until the fire brigade arrives.

It is therefore important to have a fire plan set up on the construction site. It may be quite simple; a person discovering a fire should:

- rescue anyone in immediate danger if it is safe to do so
- alert other people in the immediate area
- if equipment is available, take action to extinguish the fire before it takes hold
- confine the fire by closing any doors
- dial 000 for the fire brigade to attend
- contact the emergency coordinator, or warden, as soon as possible.

Understanding fires and what causes them

To understand how a fire can start and how to fight a fire, you must first understand the three basic elements necessary for combustion to begin:

- **Fuel** can be any combustible material; that is, any solid, liquid or gas, that can burn. Flammable materials are any substances that can be easily ignited and will burn rapidly.
- **Heat** that may start a fire can come from many sources; for example, flames, welding operations, grinding sparks, heat-causing friction, electrical equipment, or hot exhausts.
- **Oxygen** comes mainly from the air. It may also be generated by chemical reactions.

If any one of the three elements is taken away, the fire will be extinguished (see **Figure 1.48**).

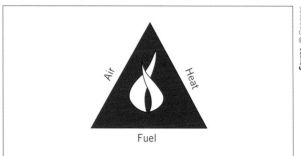

FIGURE 1.48 The elements necessary for a fire

Some of the more common examples of fires on building sites are:

- *burning off rubbish* – site rubbish should be cleared away regularly and no burning off should take place, in accordance with EPA requirements
- *electrical fires* – caused by overloading equipment, faulty equipment, faulty wiring, etc. All equipment should be carefully checked, maintained, and used correctly.
- *contractors using naked flames* – such as plumbers, structural steel workers, etc. These contractors must ensure that they do not carry out naked flame operations within the vicinity of stored rubbish, paints, sawdust, or any other highly flammable material.
- *smokers* – carelessly disposing of cigarettes, matches, etc. Butane lighters may also be a source of ignition and should not be exposed to naked flames or other situations where ignition could occur.

Preventing and fighting fires

The first rule and probably the best rule is *don't give fires the chance to start*.

Remember, fires need fuel, heat, and oxygen; remove any one of these and the fire will go out. So look for and act on the following items, as prevention is better than cure:

- remove unwanted fuel from the workplace (e.g. rubbish and waste materials)
- store fuels and combustible materials carefully; use safe carrying and pouring cans
- use only approved electrical fittings, and keep them in good order
- do not overload electrical circuits
- do not smoke at the workplace
- take special care if working with flammable liquids or gases
- be careful of oily rags, which can ignite from spontaneous combustion (e.g. turps- or linseed oil-soaked rags)
- avoid dust hazards. Many types of dust are so highly flammable that they can explode when mixed with air or when they are exposed to flame or sparks.

In the event of a fire

- Don't panic – keep calm and think.
- Warn other people in the building/work site.
- Workers who are not needed should leave the building at once and assemble at the designated emergency evacuation point.
- Arrange for someone to phone the fire brigade.
- Have the power and gas supplies turned off if it is appropriate (some lighting may still be required). Close doors where possible to contain the fire.
- Stay between a doorway and the fire.
- Be aware of containers of explosive or flammable substances. Remove them from the area only if it is safe to do so.

- If it is safe to fight the fire, select the correct type of extinguisher, and have others back you up with additional equipment.
- Know how to use the extinguisher; be confident and attack the fire energetically.
- If the fire is too large for you to extinguish, get out of the building and close all doors. Assemble at the designated area.

Classes of fires and extinguishers

In the event of a fire, you may be called on to fight it. Fire extinguishers have been grouped according to the class of fire on which they should be used. The class of fire is determined by the type of material or equipment involved in the fire. The five main classes of fires are:

- Class A – ordinary combustible materials, wood, paper, plastics, clothing and packing materials
- Class B – flammable and combustible liquids; for example, petrol, spirits, paints, lacquers, thinners, varnishes, waxes, oils, greases, petrol or diesel-driven motor vehicles, and many other chemicals in liquid form
- Class C – flammable gases; for example, LPG, acetylene, gas-powered forklifts
- Class E – electrically energised equipment; for example, electric motors, power switchboards, computer equipment
- Class F – cooking oils and fats.
- Each class of fire must be fought using different extinguishers; each type of extinguisher has its own colour code, as seen in Figure 1.49.

Extinguishers to use on a Class A fire are:
- water-type extinguishers are the best type to use. There are two types of water extinguisher: gas pressure and stored pressure. Water-type extinguisher details are:
  - range of operation: up to 10 metres
  - methods of activating the extinguishers are different for each type; it is therefore important to read the instructions on the container before attempting to use it
  - once activated, the jet of water should be directed at the seat of the fire
- any other type of extinguisher may be used except B(E) dry chemical powder.

Extinguishers to use on a Class B fire are:
- foam-type extinguishers are the best type to use. There are two types of foam extinguisher: gas pressure and stored pressure. Foam-type extinguisher details are:
  - range of operation: up to 4 metres
  - methods of activating the extinguishers are different for each type; it is therefore important to read the instructions on the container before attempting to use it
  - once activated, the jet of foam is directed to form a blanket of foam over the fire. This stops oxygen from getting to the fire long enough to

Portable Fire Extinguisher Guide

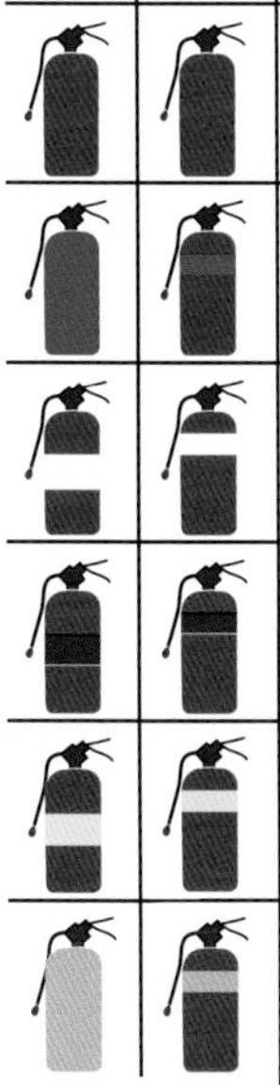

Two colour schemes exist for the extinguishers		Extinguishant	Class A	Class B	Class C	Class E	Class F	Red text indicates the class or classes in which the agent is most effective.
Pre 1999	Post 1999		Wood, Paper, Plastic	Flammable and combustible liquids	Flammable gases	Electrically energised equipment	Cooking oils and fats	
		WATER	YES	NO	NO	NO	NO	Dangerous if used on flammable liquid, energised electrical equipment and cooking oils and fat.
		FOAM	YES	YES	NO	NO	LIMITED	Dangerous if used on energised electrical equipment.
		POWDER	YES (ABE) / NO (BE)	YES (ABE) / YES (BE)	YES (ABE) / YES (BE)	YES (ABE) / YES (BE)	NO (ABE) / LIMITED (BE)	Look carefully at the extinguisher to determine if it is a BE or ABE unit as the capability is different.
		CARBON DIOXIDE	LIMITED	LIMITED	LIMITED	YES	LIMITED	Limited outdoor use.
		VAPORISING LIQUID	YES	LIMITED	LIMITED	YES	LIMITED	Check the characteristics of the specific extinguishing agent.
		WET CHEMICAL	YES	NO	NO	NO	YES	Dangerous if used on energised electrical (power on) equipment.

LIMITED indicates that the extinguishant is not the agent of choice for the class of fire, but may have a limited extinguishing capability

FIGURE 1.49 Portable fire extinguisher guide

allow the flammable substance to cool below its re-ignition point

- dry chemical and carbon dioxide (CO_2) may also be used but are not as effective as foam.

Extinguishers to use on a Class C fire are:

- ABE dry chemical powder extinguishers are the best type to use. Dry chemical powder extinguisher details are:
  - small sizes have a range of 3 metres, with larger types effective up to 6 metres
  - powder is discharged through a fan-shaped nozzle. It should be directed at the base of the fire with a side-to-side sweeping action
- carbon dioxide (CO_2) extinguishers are not as effective on Class C fires so their use should be limited.

Extinguishers to use on a Class E fire are:

- carbon dioxide (CO_2) extinguishers are the best type to use. Carbon dioxide extinguisher details are:
  - small sizes have a range of 1 metre, with larger types effective to 2.5 metres
  - useful for penetrating fires that are difficult to access
  - use as close to the fire as possible. Aim the discharge first to the rear edge of the fire, moving the discharge horn from side to side, progressing forward until flames are out.

Warning:

- never use carbon dioxide in a confined area as this could cause suffocation. Move clear of the

area immediately after use and ventilate the area once the fire is out

- never direct the nozzle at a person as carbon dioxide gas under pressure will freeze their skin causing frost bite-like burns
- never use water or foam extinguishers on Class E fires. AB and ABE dry chemical powder extinguishers will have limited effectiveness.

Note: Class F fires usually occur in kitchens and have not been covered here.

Operating and using fire extinguishers

As mentioned earlier, safe work practices involve being familiar with how to do a task, knowing which documentation or codes of practice can help you, and creating a SWMS so you have considered all the possible hazards for each step and the controls that you're going to put in place to work safely.

You should be familiar with your site's emergency evacuation procedures and your emergency contacts and what to do with even a simple first aid requirement.

One work practice that you need to be familiar with before an incident occurs is using fire extinguishers or implementing your fire plan.

In general, all fire extinguishers operate on the same principle, and you can use the acronym PASS to help you remember how to use them (refer to Figure 1.50).

FIGURE 1.50 Using a fire extinguisher – PASS

1 Pull (pin)
- Pull pin at the top of the extinguisher, breaking the seal. When in place, the pin keeps the handle from being pressed and accidentally operating the extinguisher. Immediately test the extinguisher (aiming away from the operator). This is to ensure the extinguisher works and shows the operator how far the stream travels.
2 Aim
- Approach the fire, stopping at a safe distance. Aim the nozzle or outlet towards the base of the fire.
3 Squeeze
- Squeeze the handles together to discharge the extinguishing agent inside. To stop the discharge, release the handles.
4 Sweep
- Sweep the nozzle from side to side as you approach the fire, directing the extinguishing agent at the base of the flames. After a Class A fire is extinguished, probe for smouldering hot spots that could reignite the fuel.

Other fire-fighting equipment

Fire blankets can be used on most types of fires (Figures 1.51 and 1.52). The function of the blankets is to smother the fire. Blankets are very useful on burning oils and electrical fires. Keep the blanket in place until the fire is out and for long enough

to allow the flammable substance to cool below its re-ignition point.

Hose reels are the most effective means of fighting Class A fires (Figure 1.53). Extreme care is required to ensure that no live electrical equipment is in the area. Hose reels should not be used on liquid fires, as the water may spread the burning liquid.

FIGURE 1.51 A typical fire blanket packet

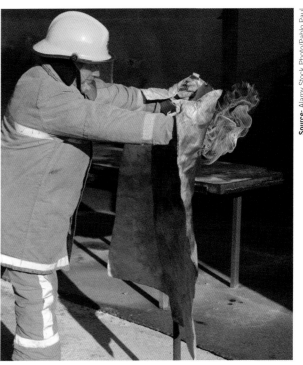

FIGURE 1.52 A fire blanket in use

Source: iStock.com/christianpound

FIGURE 1.53 Hose reel

LEARNING TASK 1.4

VIDEO ACTIVITY – FIRE SAFETY

View the YouTube video 'What is Fire?' uploaded by RespectTheFlame at https://www.youtube.com/watch?v=DIsIIeE2irE.

After viewing this video, discuss with your group what elements are necessary to produce fire.

LEARNING TASK 1.5   VIDEO ACTIVITY – FIRE AWARENESS

View the video 'Fire Awareness' related to fires and their containment, produced by Safetycare. It is available from Safetycare Australia Pty Ltd, ph: (03) 9569 5599 or talk to your institution's library to see if these videos are available at your school.

After viewing the video, complete the following questions.

1 List the three (3) essential fire safety management elements.

....

....

....

2 What is the name given to the chemical reaction that results in fire?

....

3 What are the two (2) forms of energy created by fire?

....

....

4 What are the four (4) main elements of fire?

....

....

....

....

5 List four (4) sources of heat for fires:

....

....

....

....

6 List the five (5) main principles of fire:

....

....

....

....

....

7 Why should you close windows and doors behind a fire?

....

8 What is the most common method of reducing the heat of a fire?

....

9 What are the three (3) main causes of death from fire?

....

....

....

LEARNING TASK 1.6 VIDEO ACTIVITY – FIRE FIGHTING AND ACCIDENT REPORTING

Option 1
View the following YouTube videos:
- Title: 'Fire fighting – Using a water extinguisher' uploaded by OTEN Maritime Studies (https://www.youtube.com/watch?v=BmNRdgn9YF0)
- Title: 'Fire fighting – Dry powder & foam on oil spill' uploaded by OTEN Maritime Studies (https://www.youtube.com/watch?v=iLa6JZ8y2W4)
- Title: 'Fire fighting – CO$_2$ Extinguisher' uploaded by OTEN Maritime Studies (https://www.youtube.com/watch?v=W_jgJzb8IxM)

Task: After viewing these videos, discuss with your group what processes were used while putting the fires out. Was the PASS process used?

Option 2
View the video 'Accident Investigation' related to accidents and reporting produced by Safetycare Australia.

Available from Safetycare Australia Pty Ltd, ph: (03) 9569 5599 or talk to your institution's library to see if these videos are available at your school.

Task: Engage in a class discussion, directed by your teacher/instructor, to identify important issues.

5.3 Describe, practise and effectively carry out emergency response and evacuation procedures when required

Emergency procedures

An emergency may develop due to any number of reasons, such as a fire, gas or toxic fumes leak, improper use of flammable materials, partial collapse of a building, bomb threat, crane overturning, unstable ground, materials improperly stored or a trench collapse. Therefore, every organisation must have an emergency procedure in place and personnel appointed to control the safe exit of persons from the workplace as well as to assist anyone injured or seriously hurt.

It is the responsibility of the PCBU/employer to ensure that in the event of an emergency the following arrangements have been made:
- safe and rapid evacuation of persons from the place of work to a designated assembly area on the site
- emergency communications, such as a landline phone or mobile with emergency phone numbers are always clearly visible and accessible
- provision of appropriate medical treatment of injured persons by ambulance, medical officer, or access to a suitable first aid kit.

Note: If a person is seriously injured the relevant state or territory regulating authority must be notified. If a person is killed, the relevant state or territory regulating authority and the police must be notified.

Responsible personnel

On a large building site, the responsible personnel may include the:
- head contractor/PCBU
- safety officer
- head foreperson or site supervisor.
  Small building site personnel may include the:
- builder
- foreperson
- leading hand or a nominated tradesperson.

These people are responsible for following set procedures to get all other persons onsite out of the danger area to a predetermined emergency evacuation point, so that during an incident all persons may easily be accounted for. The nominated responsible personnel are indemnified against liability resulting from practice evacuations or emergency evacuations from a building, where the persons act in good faith and in the course of their duties.

On large sites, these persons are identified by a coloured helmet they wear, which would be determined as part of the organisation's emergency plan.

5.4 Carry out emergency first aid treatment of minor injuries and, as soon as possible, accurately report treatment details to designated personnel

First aid

A PCBU/employer must ensure that employees have access to first aid facilities that are adequate for the immediate treatment of common medical emergencies; if more than 25 persons are employed at the workplace there should be trained first aid personnel. Trained first aid personnel may include a person with a current approved first aid certificate, a registered nurse, or a medical practitioner.

The following first aid kit recommendation should be referred to as a guideline only. It is highly recommended that you refer to your own state or territory WHS legislation in relation to this. Alternatively, visit the St John Ambulance Australia website at http://www.stjohn.org.au.

The following, therefore, is to be referred to as an example only; requirements may change from jurisdiction to jurisdiction:

- *First aid kit A* – Construction sites at which 25 or more persons work or other places of work at which 100 or more persons work
- *First aid kit B* – Construction sites at which fewer than 25 persons work or other places of work at which fewer than 100 and more than 10 persons work
- *First aid kit C* – Places of work (*other than construction sites*) at which 10 or fewer persons work.

Most residential building sites would fall into the 'kit B' category, as it would be rare to have 25 or more persons on the site at any one time.

Note: 'Kit C' is not suitable for construction sites of any kind, as it lacks many of the items required for first aid treatment of common building site injuries.

Contents of a first aid kit B

The contents of a type B kit are listed in Table 1.5 and the first aid kit and stores are shown in Figure 1.54 and Figure 1.55.

TABLE 1.5 Contents of a first aid kit B

Item	Quantity
Adhesive plastic dressing strips, sterile, packets of 50	1
Adhesive dressing tape, 25 mm × 50 mm	1
Bags, plastic, for amputated parts:	
Small	1
Medium	1
Large	1
Dressings, non-adherent, sterile, 75 mm × 75 mm	2
Eye pads, sterile	2
Gauze bandages:	
50 mm	1
100 mm	1
Gloves, disposable, single	4
Rescue blanket, silver space	1
Safety pins, packets	1
Scissors, blunt/short-nosed, minimum length 125 mm	1
Sterile eyewash solution, 10 mL single-use ampoules or sachets	6
Swabs, prepacked, antiseptic, packs of 10	1
Triangular bandages, minimum 900 mm	4
Wound dressings, sterile, non-medicated, large	3
First aid pamphlet (as issued by the St John Ambulance or the Australian Red Cross Society or as approved by WorkCover)	1

Note: Remember that the above list is a recommendation only and it is very important that you refer to your own relevant state or territory obligatory requirements.

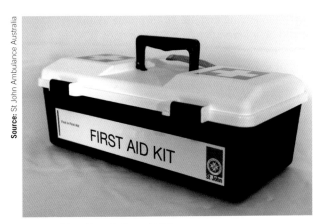

FIGURE 1.54 Type B first aid kit

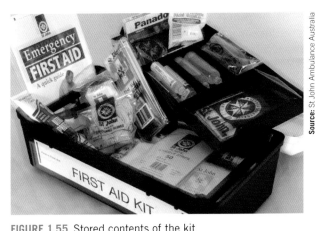

FIGURE 1.55 Stored contents of the kit

COMPLETE WORKSHEET 5

SUMMARY

In Chapter 1 you have learnt how to carry out work health and safety (WHS) requirements through safe work practices in all onsite or offsite construction workplaces.

You have learnt how to perform work in a safe manner through awareness of risks and work requirements, and the planning and performance of safe work practices with concern for personal safety and the safety of others.

You have also learnt about fundamental WHS requirements necessary to undertake work tasks within any sector in the construction industry. You also learned about the identification of hazardous materials, including asbestos, and compliance with legislated work safety practices. We did not cover removal of asbestos, which is a licensed activity. You have also learnt to:

- identify and assess risks
- identify hazardous materials and other hazards on work sites
- plan and prepare for safe work practices
- apply safe work practices
- follow emergency procedures.

REFERENCES AND FURTHER READING

Acknowledgement

Reproduction of the following Resource List references from DET, TAFE NSW C&T Division (Karl Dunkel, Program Manager, Housing and Furniture) and the Product Advisory Committee is acknowledged and appreciated.

Texts

Comcare government website, search for General construction induction training card information,

Graff, D.M. & Molloy, C.J.S. (1986), Tapping group power: a practical guide to working with groups in commerce and industry, Synergy Systems, Dromana, Victoria

Ministry of Health (n.d.), Building and industry hazards, NSW Government, Sydney, retrieved from **http://www.health.nsw. gov.au/environment/hazard**

NSW Health, *building hazards*, NSW Government, Sydney, retrieved from **http://www.health.nsw.gov.au/**

National Centre for Vocational Education Research (2001), *Skill trends in the building and construction trades*, National Centre for Vocational Education Research, Melbourne, available from the ncver.edu.au website

Safe Work Australia (2012), *Emergency plans fact sheet*, Safe Work Australia, Canberra

Safe Work Australia (2014), Safe work method statement for high risk construction work: Information sheet, Safe Work Australia, Canberra

Safe Work Australia (2022), *Work-related traumatic injury fatalities, Australia 2021*, Safe Work Australia, Canberra

TAFE Commission/DET (1999 / 2000), *Certificate 3 in General Construction (Carpentry) Housing*, course notes (CARP series)

Safework NSW, search for general construction induction

Safework NSW search for 'pocket guide to construction

Safework NSW (2020), *Consumer building guide*, Safework NSW, Sydney

WorkSafe Victoria (2023), search for safe work method statements

Web-based resources

Regulations/Codes/Laws

http://www.austlii.edu.au search for Australian laws

http://www.safeworkaustralia.gov.au, search for Model WHS Regulation 2011

https://www.safeworkaustralia.gov.au/doc/model-code-practice-construction-work, Safe Work Australia – Model Code of Practice – Construction Work

https://www.worksafe.vic.gov.au

www.safeworkaustralia.gov.au has the new model WHS Regulations amendment that deal with Engineered Stone hazards. Silica comes under that category and anyone who works with manufactured benchtops, or similar needs to be familiar with the regulations.

Resource tools and VET links

http://training.gov.au, training package information

https://www.safework.sa.gov.au, search for WHS toolbox

Industry organisations sites

https://www.worksafe.qld.gov.au, search for asbestos

https://www.safeworkaustralia.gov.au, search for Emergency plans fact sheet

https://www.boral.com.au, search for General Purpose Cement safety data sheet

Audiovisual resources

Title: 'Accident Investigation'

Running time: approx. 10 minutes

Author: https://www.safetyhub.com/au/ Safetycare Australia Pty Ltd

Available from: https://www.safetyhub.com/au/ Safetycare Australia Pty Ltd, Ph: (03) 9569 5599

Title: 'Fire Awareness'

Running time: approx. 16 minutes

Available from: https://www.safetyhub.com/au/ Safetycare Australia Pty Ltd, Ph: (03) 9569 5599

AS/NZS 1801 Occupational protective helmets
AS 1800 Occupational protective helmets – Selection, care and use
AS/NZS 1337.1 Eye protectors for industrial applications
AS/NZS 1270 Acoustic – Hearing protectors
AS/NZS 1716 Respiratory protective devices
AS/NZS 1715 Selection, use and maintenance of respiratory protective equipment
AS/NZS 2161.1 Occupational protective gloves
AS/NZS 1269 Occupational noise management
AS/NZS 2210.1 Safety, protective and occupational footwear guide to selection, care and use
AS 2626 Industrial safety belts and harnesses – Selection, use and maintenance
AS 1216 Class labels for dangerous goods
AS 1318 Use of colour for the marking of physical hazards
AS 1319 Safety signs for the occupational environment
AS 1216 Class labels for dangerous goods
AS 1885.1 Measurement of occupational health and safety performance

GET IT RIGHT

In the photo below, the person is using the drop saw unsafely.

Identify the unsafe practices and provide reasoning for your answer.

Source: Richard Moran

_____

_____

_____

_____

_____

_____

WORKSHEET 1

Student name: _____

Enrolment year: _____

Class code: _____

Competency name/Number: _____

Task: Read through sections beginning at *Identify and assess risks* then complete the following questions.

1   What is a hazard?

_____

2   What is a chronic hazard?

_____

3   What is risk?

_____

4   WHS self-regulation wasn't working. Approximately how many people were dying per year throughout
Australia before WHS legislation came into being? (Circle the correct answer below.)

200   250   300   4400   500   600

5   Name the organisation that oversees WHS at the national level:

_____

6   Name the organisation that controls WHS in your state or territory.

_____

7   What is required by the WHS Regulations 2011 (Commonwealth) prior to a person carrying out
construction work?

_____

WORKSHEET 2

To be completed by teachers

Student competent ☐

Student not yet competent ☐

Student name: _____

Enrolment year: _____

Class code: _____

Competency name/Number: _____

Task: Read through the sections beginning at *Identify hazardous materials and other hazards on work sites* then complete the following questions.

1   List four (4) workplace hazards.

    1 _____

    2 _____

    3 _____

    4 _____

2   List the five (5) hazard categories.

    1 _____

    2 _____

    3 _____

    4 _____

    5 _____

3   What must be done once the hazard is identified and its potential to do damage is measured?

    _____

    _____

4   List the five (5) levels under the hierarchy of hazard control:

    1 _____

    2 _____

    3 _____

    4 _____

    5 _____

WORKSHEET 3

Student name: _____

Enrolment year: _____

Class code: _____

Competency name/Number: _____

Task: Read through the sections beginning at *Plan and prepare for safe work practices* then complete the following questions.

1 Which hierarchy of hazard control do guards on power tools belong to?

_____

2 Which hierarchy of hazard control does using scaffolding to work at heights belong to?

_____

3 When should PPE and safety signs be used as the main method of protection against a hazard?

_____

_____

4 List the body parts that may be protected by using PPE.

_____

_____

5 List four (4) *physical hazards* that PPE may be used to protect the wearer from.

1 _____

2 _____

3 _____

4 _____

6 List four (4) *chemical hazards* that PPE may be used to protect the wearer from.

1 _____

2 _____

3 _____

4 _____

7   List the three (3) hazard categories that eye protection is designed for:

1 _____

2 _____

3 _____

8   List the two (2) main PPE methods used to protect your hearing:

1 _____

2 _____

9   How can you recognise a disposable nuisance-dust mask?

_____

_____

10   Write down a suitable use for P2 class filters:

_____

_____

11   What hazards do rubberised chemical-resistant long gloves protect the wearer from?

_____

_____

12   List the main requirements that safety footwear must have to provide maximum protection:

_____

_____

_____

13   What do the letters SDS mean?

S _____

D _____

S _____

14   List the six (6) reasons for using an SDS (what can an SDS help you with)?

1 _____

2 _____

3 _____

4 _____

5 _____

6 _____

15 What are the two (2) subheadings on an SDS under the heading 'Health hazards'?

1 _____

2 _____

16 Using the *full* copy of the 'Blue circle – general purpose cement' SDS, name the 16 major headings:

1 _____

2 _____

3 _____

4 _____

5 _____

6 _____

7 _____

8 _____

9 _____

10 _____

11 _____

12 _____

13 _____

14 _____

15 _____

16 _____

17 Using the *full* copy of the 'Blue circle – general purpose cement' SDS, name the four (4) items of PPE that must be used when handling this product:

1 _____

2 _____

3 _____

4 _____

18 What does the acronym SWMS mean?

S _____

W _____

M _____

S _____

19 A SWMS is a document that:

1 _____

2 _____

3 _____

4 _____

20 Since the introduction of the national model of SWMS, what type of work legally requires a SWMS?

_____

21 List three (3) examples of work where a SWMS is required.

1 _____

2 _____

3 _____

22 When filling in a SWMS document, what are the headings or names of the three main columns in the first section?

1 _____

2 _____

3 _____

23 What needs to be listed in the column 'Personal qualifications, experience'?

_____

_____

_____

 **WORKSHEET 4**

Student name: _____

Enrolment year: _____

Class code: _____

Competency name/Number: _____

To be completed by teachers
Student competent ☐
Student not yet competent ☐

Task: Read through the sections beginning at *Apply safe work practices* then complete the following questions.

1   List two (2) reasons why a safety helmet must be worn on a construction site:

1   _____

2   _____

2   When wearing a safety helmet, is it acceptable to wear a helmet without the helmet harness fitted? Circle the correct answer.

YES   NO

3   How is the back of the neck protected from sunburn when wearing a safety helmet?

_____

_____

4   Briefly describe the three (3) main categories of common safety signage used in the building industry:

1   Picture signs

_____

_____

2   Word-only signs

_____

_____

3   Combined picture and word signs

_____

_____

5   Briefly describe the shape, colours and the detail/symbol found on a sign used to indicate a toxic hazard warning.

Shape: _____

Colours: _____

Detail/symbol: _____

6   Briefly describe the background, detail/symbol and border found on a sign used to indicate an emergency exit sign.

Background: _____

Symbol: _____

Border: _____

7   What is the purpose of fitting a Warning or DANGER safety tag/accident prevention tag to an electric tool machine?

_____

_____

_____

8   Signs should be placed in a position that allows them to be clearly seen. State the preferred position for safety signs:

_____

_____

9   Identify the following signs, stating what they represent:

Source: iStock.com/alessandro0770

Source: Shutterstock.com/kevin brine

Source: Shutterstock.com/voylodyon

Source: Shutterstock.com/northallertonman

TOXIC HAZARD

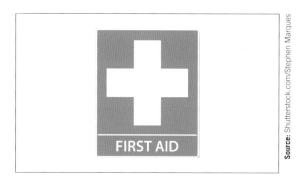

FIRST AID

10 How many classes of dangerous goods are there? (Circle the correct answer below.)

7   8   9   10   11   13

11 Link the dangerous good to its correct class number by drawing a line from one to the other.

1   flammable liquids

2   flammable solids

3   gases

4   explosives

12 In hazardous chemical areas it is important to have good hygiene. Name three (3) of the four (4) precautions that you should adopt.

1 _____

2 _____

3 _____

13 In hazardous chemical areas it is important to have good hygiene. Describe what should happen in regards to asbestos dust.

_____

_____

_____

14 Name the three (3) diseases that may be caused by breathing in asbestos dust.

1 _____

2 _____

3 _____

15 You are undertaking an extension on a house built in 2005. Is it likely that you will find asbestos-containing material in the section of the building you have to deconstruct? (Circle the correct answer below.)

YES   NO

16 List four (4) products, and where they are located on or in a house, that may contain asbestos.

	Product	Location
1		
2		
3		
4		

17 When removing asbestos cement sheeting, what must *never* be used on the sheets?

_____

_____

_____

 **WORKSHEET 5**

Student name: _____

Enrolment year: _____

Class code: _____

Competency name/Number: _____

Task: Read through the sections beginning at *Follow emergency procedures* then complete the following questions.

1   If a person is seriously injured, who must be notified immediately?

_____

2   In the event of an emergency on a small building site, who in the building company must be notified? (List 2 possible people.)

1   _____

2   _____

3   In the event of an emergency (e.g. a fire) where should all the workers on that site go? (Circle the correct answer/s.)

A   Go to the local shops

B   Get in the car/truck and go home

C   Assemble in the middle of the street

D   Assemble at evacuation assembly point

4   What are four (4) items that are shown on an evacuation diagram?

1   _____

2   _____

3   _____

4   _____

5   What type of first aid kit is required for construction sites at which fewer than 25 persons work? (Circle the correct answer/s.)

A   First Aid Kit A

B   First Aid Kit B

C   First Aid Kit C

6   What are the three (3) basic elements necessary for combustion to begin?

1   _____

2   _____

3   _____

7   List three (3) common causes of fires on a site:

1   _____

2   _____

3   _____

8   List three (3) ordinary combustible materials that may be used as fuel for a Class A fire:

1   _____

2   _____

3   _____

9   State the source or fuel and most suitable types of fire extinguishers for use on the following classes of fires:

Class A

Source/Fuel _____

Most suitable extinguishers _____

_____

Class B

Source/Fuel _____

Most suitable extinguishers _____

_____

Class E

Source/Fuel _____

Most suitable extinguishers _____

_____

10   What type of extinguishers should never be used for Class E fires?

_____

11   Fire extinguishers come in a variety of easily identifiable colours. State the common colour used to identify a foam-type extinguisher:

_____

12 Relating to fire extinguishers, what does the acronym PASS mean?

P _____

A _____

S _____

S _____

13 Name two (2) other items that can be used to fight a fire:

1 _____

2 _____

PERFORMANCE OF SKILLS EVIDENCE

Student name: _____

Enrolment year: _____

Class code: _____

Competency name/Number: _____

Task: Working with your teacher, prepare to demonstrate your skills by completing the following task.

Materials needed:

- Simulation site – virtual or physical – with emergency plans
- Sample of personal protective equipment
- Access to signs and symbols representing hazards
- Access to real or simulated hazardous materials and the appropriate safety data sheets
- Access to emergency response equipment such as fire extinguisher and/or first aid kit
- Access to communication devices as appropriate for the industry workplace.

To demonstrate competency, a candidate must apply WHS requirements, policies and procedures on three separate and different occasions in the construction industry. While doing so, the candidate must meet the performance criteria for this unit by:

1   Identify and assess risks.

   a   Participants walk through the simulation site and identify various hazards and risks and report the hazards to a designated person. Demonstrate how to control the risks and then complete accident reports in accordance with workplace procedures or other requirements.

2   Identify hazardous materials and other hazards on work sites.

   a   Correctly identify hazardous materials and discuss how to apply measures to control the risks of those materials and what type of signs may be required to notify others of those hazards. Identify asbestos containing materials or discuss what asbestos may look like.

3   Plan and prepare for safe work practices.

   a   Identify various types of PPE and fit them correctly.

   b   Discuss how to select tools equipment and materials in the workplace and how to organise the tasks.

   c   Discuss types of barricades and signage and where they should be located.

   d   Locate and explain various safety data sheets, job safety analysis forms and safe work method statements.

4   Apply safe work practices.

   a   Carry out a sample of simple tasks as outlined by your teacher in a safe manner using plant and equipment guards in accordance with manufacturers' specifications.

   b   Describe the procedures to report hazards and recognise prohibited tools and equipment when working with asbestos containing materials.

   c   Describe how to clear and maintain a work site to protect people from incidents.

5   Follow emergency procedures.

   a   Identify the designated personnel onsite to communicate emergency information and demonstrate how to follow safe workplace procedures involving accidents, fire and other emergencies, including demonstrating the use of fire equipment.

   b   Describe, practise and effectively carry out a simulated emergency evacuation.

   c   Describe how to carry out emergency first aid treatment of minor injuries and then report the details to designated personnel.

2

WORK EFFECTIVELY AND SUSTAINABLY IN THE CONSTRUCTION INDUSTRY

This chapter covers the following topics from the competency CPCCOM1012 – Work effectively and sustainably in the construction industry:
- work effectively in a team
- investigate construction industry employment pathways
- identify and follow environmental and resource efficiency requirements

This unit of competency specifies the skills and knowledge required to work effectively and sustainably in the construction industry.

The unit is suitable for those with basic skills and knowledge undertaking routine work tasks under the direction of more experienced workers.

1. Work effectively in a team

1.1 Participate in planning work tasks with team members

Participating in planning work tasks with team members involves all of you working together to figure out what needs to be done, how you might do it, and when you might do it. Participating means you will be discussing ideas, sharing information and resources, and allocating particular roles or tasks to each person. To do this successfully you need to make sure that everyone has a chance to have some input, everyone needs to be adaptable, the overall goals for the project need to be kept in mind, particularly in relation to immovable items like Australian Standard qualities, and a timeline needs to be organised and agreed upon so that progress to achieve the work task can be planned for, monitored and reviewed.

When many people are working together in an organisation there is a need to divide the work among them and be sure that they understand what they must do and to whom they are responsible. An organisation chart is drawn up to give a pictorial representation of the business structure. Senior positions are shown at the top, descending to the lower positions; they can extend horizontally with multiple positions of equal status (Figure 2.1).

There are four main levels of authority, with each person being responsible to the person or persons above them on the next level.

This arrangement of site personnel can be referred to by several descriptions, such as:
- site hierarchy
- site lines of authority
- site organisation diagram.

A large building firm would be involved with multi-storey construction, civil construction and so on and is structured in the same way as are individual sites; that is, with the person having the most authority at the top of the firm and the people with the least authority at the bottom. It is also possible to have several people with the same authority or responsibility at the one level.

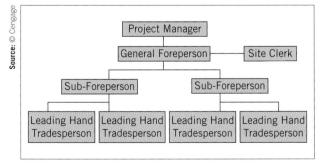

FIGURE 2.1 Hierarchy chart showing onsite organisation

Workplace committees and meetings

There are three main ways of dealing with a situation or settling a problem on a job site:
1 imposing a solution either from above or below
2 accepting a compromise reached through bargaining
3 agreeing to the best answer determined by joint discussion.

The first method implies the use of force, which emphasises a permanent division, as it represents either a victory to be maintained or a defeat to be avenged at the next opportunity. Equally, the second method is an uneasy armistice decided by relative powers, rather than a true assessment of the facts.

By contrast, the use of discussion ensures a genuine solution based on conviction, with the maximum chance of acceptable and harmonious action. Thus, committees or informal group meetings can play an important part in the organisational structure. Committees can be used for different purposes, with appropriately varied membership as follows:
- as an advisory body, such as a site WHS committee/health and safety representatives (HSR) elected from each section or workplace group to deliberate on particular problems and to assist management by combining the total group's knowledge and experience
- as a means of *consultation*; for example, a planning meeting, where members are appointed by reason of their individual functions and contribute their different viewpoints, ensuring that all aspects are properly considered
- as a channel for information and a method of communication, usually between a supervisor and his or her managers, where reports are tabled and resources are pooled to find a suitable solution to a problem
- as a process of coordination, usually between a specialist and several production units, such as a transport manager allocating his or her daily vehicle tasks to meet the requirements of the respective general foreperson.

All these elements go into forming an effective working committee to resolve problems and make judgements.

Consultative committees

These committees consist of employees/workers and middle management, and consider problems and make suggestions for policies in areas such as safety, health, social activities and amenities.

Works committees (employee/worker negotiation teams)

It may be appropriate to have a team of employee/worker negotiators representing different areas of the workplace. The team might consist of a full-time union official as team leader supported by employees/workers from different work areas. If a union is not involved, the team leader and other team members should be elected from among the employees/workers to be represented by the team.

Work health and safety committees/HSRs

All Australian states and territories now legislate to provide for workplace consultation via health and safety representatives (HSRs) or health and safety committees. Legislation requires a systematic approach in which management and worker representatives regularly discuss work health and safety issues. The objective of these meetings is to prevent and resolve WHS problems in a self-regulatory manner.

At workplaces that consist of 20 or more employees, permanent safety committees are normally established with representatives from management and the workforce. These safety committees are responsible for carrying out workplace safety inspections; however, with the introduction of the model *Work Health and Safety Act 2011*, HSRs have also been introduced. HSRs are elected to represent particular work groups, which they must be a member of, at their respective workplaces.

Workplace meetings

There are two main methods used to raise issues or discuss problems on a building site: through forming committees (as seen above) or holding site meetings. Committees have a limited membership that is elected or appointed to perform specified functions, such as a site WHS committee and HSRs. A site meeting on the other hand, consists potentially of all members of an association or their representatives. Special meetings may be held onsite involving the PCBU/builder, client, architect and subcontractors in order to consider important matters relating to the project. Where deals involving variations are made, all affected parties may sign a statement on the spot, confirming changes that have been agreed to. Copies of the meeting report should be made available to all parties who attended.

Formal minutes should be kept of all site meetings arranged and conducted by the project manager or general foreperson. Copies of these minutes, which represent a report of proceedings, can be distributed to all persons present at the meeting and any other people who need to be informed.

Trade unions also have meetings, such as annual, delegate, branch and committee meetings. The branch meeting may take place once a month, but the executive committee may meet once a week.

Toolbox meetings or *toolbox talks* are more informal group discussions that focus on workplace activity and safety issues arising from daily activity. These talks may be daily, weekly or fortnightly depending on the complexity of the work site. Even though toolbox meetings are informal, a recording sheet of what was discussed and who was present should always be kept (see Figure 2.2).

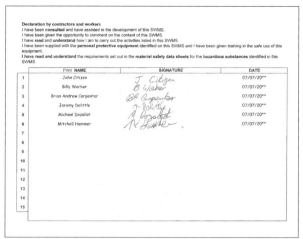

FIGURE 2.2 Toolbox talk record sheet

Source: Ed Hawkins

Representation and participation

Representation and participation are seen as a means of finding out what people want and of communicating information and decisions to them. This process aids understanding as the elected representatives discuss matters with other team members, promoting decision making and allowing actions to flow more easily and quickly. This can be a matter of persuasion as much as sharing in making decisions.

It is important to note that representation and participation are different. Voting for a representative from your group to do what he or she thinks would be best for you is different from being there yourself to have your own say on the matter. Sending a delegate who is instructed to vote in a specific way is different again.

Rights

In a democratic society we have the right to be represented in our national government and to be consulted in making decisions, such as voting. These are rights people have argued and fought for, which we continue to assert, which are accepted by our society and which are continually subject to readjustment as our society changes. Therefore, each person or representative of a committee or part of a meeting has the right to voice an opinion on the subjects raised. You should not sit and say nothing if you disagree strongly, as you may have to live with the decision. Your say may sway the argument in your favour by convincing others who were undecided.

Efficacy

People are more likely to implement a decision effectively if they have shared in making it, possibly because they feel that their contribution makes it a better decision, they are responsible for it, trust it and understand it.

Group dynamics

To gain the greatest input of people at a formal meeting, it is important to have a room layout where everybody can see and be seen easily. This will promote a more inclusive environment than having individuals pushed into a corner, which may make them feel insignificant. Two of the most important requirements for effective meetings are that:

- communication takes place in an orderly manner
- content is relevant.

Too often it is assumed that both of these requirements are the responsibility of the chairperson, who achieves them by controlling the meeting. There are obviously instances where such control is necessary, but responsibility for orderly communication and relevance lies with every individual attendee. The chairperson does not need to intervene if every person attending adopts the self-discipline needed to achieve an effective meeting. It is easy to get off the track when many people have input, so it is necessary to outline a plan at the beginning of the meeting that everybody agrees to. Everyone should have the opportunity to contribute to the discussion, which may mean that some people will have to be restrained while others may need to be encouraged.

Effective participation

If the meeting or committee is to function effectively, the following areas must be addressed.

- The aims and objectives of the meeting should be clearly formulated and understood. They must also be accepted by all members.
- Discussion should be relevant, with virtually everybody contributing.
- Everyone should be prepared to listen to each other and consider the points made. Members should not be afraid to put forward views and ideas.
- The atmosphere should be informed and relaxed, with all members of the group being involved and interested.
- Only constructive criticism should be offered and accepted.
- Disagreements should be dealt with openly and examined to find a compromise.
- Members should feel free to express their own feelings and attitudes towards a problem.
- Decisions should be arrived at by consensus; individuals should not be afraid to disagree and should be given fair consideration.
- Decisions and follow-up action should be determined, with all people kept fully aware at all times. Jobs should be allocated clearly and appropriately.
- The chairperson should not dominate; the members should not be subservient to the chairperson.
- The group should be aware of and be prepared to discuss its own deficiencies.

Limiting factors

Things that may limit a person's performance at a meeting or within a committee are:

- experience
- knowledge
- fear

- behaviour
- dominance.

People's lack of experience and knowledge of how meetings and committees operate may inhibit their performance. Also, fear of conflict, hidden hostilities and underlying personal factors are difficult to overcome.

Another inhibiting factor in the success and effectiveness of a meeting is the misapprehension that the effectiveness of a group rests solely with the leader. Research indicates that skilful membership behaviour is the operative factor. The greatest danger of all is that the person with the loudest voice or strongest personality will become the chairperson and totally dominate the meeting, so that those who have relevant knowledge and opinions are not heard.

Purpose of meetings

It is important to be quite clear about the purpose of the meeting or committee discussion so that each member is fully aware of the reason for his/her participation. Common reasons for calling a meeting are to:

- pass on information
- seek ideas to solve a problem
- coordinate information gathered over a period of time
- negotiate changes to contracts
- resolve disputes
- form new policies or revise old ones to improve the running of the site
- plan for future work
- decide on proposals or outstanding issues
- act on decisions made as quickly as possible to ensure that enthusiasm and credibility are ongoing.

LEARNING TASK 2.1

IDENTIFY RESPONSIBILITIES WHILE WORKING IN A TEAM

After reading through the section 1.1 'Participate in planning work tasks with team members', complete the following project under the direction of your teacher/instructor.

1 As a team, identify all of the team members required to complete a construction job.
2 Allocate a role to each person in the group, with one person nominated as the team leader.
3 Identify the responsibilities for each team member. Suggested activities might include:
    - moving and stacking materials ready for transport by forklift truck
    - cleaning up a designated work area.
4 Set ground rules for the team to follow.
5 Rotate roles at least twice.
    These team functions will be used throughout your training and in the workplace.

1.2 Work with team members to review team purpose, roles, responsibilities, goals, plans and objectives

Working with team members to review team purpose, roles, responsibilities, goals, plans and objectives involves a group coming together to discuss and understand the reason why the team exists and what each person's role and responsibilities are. The team needs to decide on what task they are aiming to achieve and how they going to do it in a way that makes use of everyone's individual strengths and skills and allocating specific tasks related to those individual strengths and skills. The team must keep in mind the goals and objectives that they want to accomplish. Depending on the size of the task there may be a need to meet regularly to review progress, making sure that everyone is on the same page and focused towards achieving the work task. Table 2.1 shows some of the roles within the team.

Work team

Teams of workers, which may consist of two, three, four or more, are formed when building site tasks require *rapid completion*, such as constructing and covering a new roof on an existing cottage; *completion within a given timeframe*, such as completion of floor and wall framing so the bricklayers may commence the external brick veneer; a *number of related parts or elements*, such as a series of individual trusses that require placement in a set sequence to provide support for others.

Once the team members understand the role they play in the overall construction of the building, they can contribute more effectively to meet the 'site goals' and to work more **efficiently** with other teams or individuals (Figure 2.3).

Source: Shutterstock.com/Andreas G. Karelias

FIGURE 2.3 Two working as one

TABLE 2.1 Roles and responsibilities

Role	Responsibilities
Client	Finances the project. This may be the developer, or persons involved in a joint venture who require a building for a specific purpose.
Architect	Engaged by the client to design and control the building process on the client's behalf. Other responsibilities include lodging the development application with designs or models and supervising the drafting of plans and the specification, preparing tender documents to engage a builder, engaging consultants for the structural, mechanical and other specialist areas.
Quantity surveyor	Responsible for preparation of a Bill of Quantities for the architect, to enable the building to be priced by an estimator. Also required to prepare quantities for variations to the contract and assessing progress claims.
Land surveyor	Responsible for site setting out and control of the vertical alignment of a building.
Draftsperson	Engaged by the architect to prepare plans and details before and during construction.
Builder	The main contractor selected by a tender process to carry out construction as per plans and specifications. The builder will also engage subcontract labour to deal with specific parts of the construction.
Construction manager	Responsible for all building contracts the company has won or is tendering for. Usually working from the main office, the construction manager forms tender documents and controls the building activities for the company.
Project manager	Usually employed on very large sites to control the running of the site and to make sure the project or projects run to schedule and budget.
Site manager	Runs the day-to-day activities and liaises directly with the foreperson and the project manager.
General foreperson	Responsible for delegation of duties to workers onsite and also responsible for site safety. Other responsibilities are coordination of subcontractors and progress of work.
Quality assurance officer	Engaged to control the quality of construction onsite and ensure that the work meets the standards laid down in the specification and tender documents.
Building inspector	For domestic construction, the inspector visits the site on behalf of the council to check structural work and pass completed stages of a job.
Principal certifying authority (PCA)	For domestic construction the PCA carries out similar duties to the local government building inspector, but is engaged directly by the client to conduct the inspections.

To allow smooth operation and cohesion within a team, each person should be allocated a job or function that will complement the roles of other team members and not duplicate their effort. Therefore, a simple meeting is required to establish the role of each member within the team. Ideally, one person in the team will be nominated and appointed as the 'team leader' to ensure consistency of effort and communication.

The old adage of 'many hands make light work' applies to a team effort; however, 'too many cooks spoil the broth' also applies if there is more than one team leader.

Regular rotation of roles within the team is also important, so that there will not be just one or two members doing all the work. This rotation of roles and responsibilities also reduces the risk of individual egos getting in the way of a productive team effort.

Individual strengths and weaknesses

Rotation of roles and responsibilities is important to effective function, but there also needs to be acceptance for specialisation within the team. If an individual has special skills or talents they should be able to maintain a particular role or position in the team. This may occur where only one member of the team holds a qualification or has a licence to do specific work; for example, an EWP (elevated work platform) licence.

On the other hand, where a member of the team lacks skill or knowledge to carry out a task on their own, then this person may 'buddy up' with another member who can supervise, support and mentor them. This may occur where one or more members of the team are labourers and the others are tradespersons. In this way individuals in the team may enhance or increase their function or role within the team and be given more responsibility. By encouraging and focusing on their strengths these individuals can become an even greater asset to the team, to the building company and ultimately to the construction industry. They may even enrol in a trade course or post-trade course, or attend a skills recognition assessment program to gain formal recognition of the skills they have learned within the team.

Team rules

Each team should discuss and set ground rules by which all members must abide. This is important as it maintains harmony and allows the team to function effectively. Any changes to these rules, and/or improvements made, must be discussed and agreed on by all the members. All members should have equal rights and responsibilities and be able to suggest changes and improvements without fear of being ridiculed or vilified by others in the group.

1.3 Work with team members following guidelines, directions and instructions to complete work tasks

For a task to be done correctly a team must come together and decide on what the guidelines, directions and instructions need to be to achieve the task to the standard that is required. This will involve cooperating with your teammates and having a thorough understanding of the guideline's directions and instructions for the job. Constant communication with supervisors and team leaders and any specialist advisers is important. Make sure you communicate with each other and ask for clarification if something is unclear because the aim is to avoid mistakes and achieve a satisfactory completed task as efficiently as possible. Different team members will have different roles and responsibilities in relation to achieving the task and may only work on part of the instructions to get a job done. For example, one person's task may be to safely erect scaffold for a working platform while another person's task may be to work on that platform fitting eaves to a house. Both people are important to get the job done and need to work together.

The guidelines, directions and instructions that may be used to complete work tasks may come in the form of user manuals, technical installation instructions, work process steps from codes of practice, or a competent person's SWMS outlining the steps to achieve a task. Each member of the work team needs to be aware of the overall guidelines for completing a task and where they fit within those guidelines.

1.4 Work with team members to resolve problems that impede the team's performance

It is inevitable that a team will experience problems sooner or later as they work on jobs. It's important that all team members remember to work together to try and find solutions once it's recognised that problems have been encountered. Identifying and addressing issues and obstacles involves communication and honesty, with every person on the team making sure everyone feels confident to share concerns and ideas and look for the best possible solution. The important thing to remember here is that the focus needs to be on the problem not on people. Quite often a problem will be caused by one or more people but there's usually more involved in what caused the problem than just those people. It takes a level of maturity and focus on the job to listen to everyone's perspectives and work collaboratively together to find the best course of action. A good team can problem solve and get back on task, recognising that there will always be problems to solve. Of course, we shouldn't be having the same problem too many times because a solution to a particular problem in one situation will probably be very applicable to the same problem in another situation. Always try and learn from mistakes.

Disharmony

Where **conflict** within a team arises, it should be dealt with immediately and not allowed to fester and grow into a full-blown confrontation. Ideally, all conflicts should be resolved within the team so that other unwanted persons are not involved, as this may lead to individuals left feeling resentful. If a resolution cannot be found, another person from outside the team should be brought in to mediate the situation. This person should be someone whom all the team members know and respect; otherwise an agreeable resolution may not be found and resentment may result.

The key to successful **teamwork** lies in the selection of members with suitable skills, who are willing to do their bit and abide by the team rules. Putting a team structure in place that encourages participation and appreciation of each member, allowing each person to have an equal say in the running and organising of the team and to hold skills that are valued and complement the other team members will ensure successful teamwork.

COMPLETE WORKSHEET 1

2. Investigate construction industry employment pathways

Data from the Australian Bureau of Statistics indicates that the total value of building work done in the December quarter 2022 was nearly $30.1 billion. New dwelling commencements were approximately 41 000 dwellings in the December quarter 2022. Over 80 per cent of land parcels zoned residential were approximately 600 to 800 square metres in size. Nearly 60 per cent of new dwellings approved for residential land were approximately 200 to 400 square metres in size.

Source: Adapted from Australian Bureau of Statistics, Land and Housing Supply Indicators. https://www.abs.gov.au/statistics/industry/building-and-construction

The construction industry plays a significant role in the Australian economy and society, providing homes, workplaces, and infrastructure such as schools, recreation facilities, hospitals, roads, electricity supply and telecommunications. The demand for, and supply of, construction is influenced by a variety of factors including interest rates, tax changes and changes in population.

The construction industry, and its activities, are strongly linked to other parts of the Australian economy such as manufacturing, wholesale trade, retail trade, and finance and insurance. In addition, industries such as property operators, real estate, architecture and engineering are closely allied with the construction industry.

Source: Australian Bureau of Statistics (1301.0), © Commonwealth of Australia CC BY 4.0 International (https://creativecommons.org/licenses/by/4.0/)

Another way that we can measure the construction industry's contribution to Australia is to consider how many people are directly employed within this industry using data from the Australian Bureau of Statistics. In 2008–2009, the construction industry employed a total of one million people, which was approximately 9 per cent of all people employed in Australia, and in 2010–2011 the industry employed 1.037 million people (9.1 per cent). A construction industry survey from August 2013 estimated that 1.062 million people were employed in the industry at that time – again, around 9.1 per cent of the total workforce (PricewaterhouseCoopers, 2013). This means that in 2013 the construction industry was the third largest employment sector in the country.

The construction industry in 2010–2011 contributed $101 868 million (approximately 7.7 per cent) to the Australian GDP (gross domestic product); by August 2013 the construction industry accounted for more than 10 per cent of GDP (PricewaterhouseCoopers, 2013).

This information shows that the construction industry has and will continue to have a very strong influence on the Australian economy.

GREEN TIP

A marketing advantage for your construction company could be that you are a green company. You could provide links between an original dwelling and a newly renovated dwelling by making a feature out of some of the original parts of the dwelling. This can provide you with a unique selling proposition for your business.

2.1 Describe the process for becoming a tradesperson or skilled operator in the construction industry

The building industry can be divided into three main industry sectors, which incorporate a wide range of skills carried out in an equally wide range of environments. These three sectors are:

1 residential construction
2 non-residential construction
3 engineering construction.

Residential construction

This sector is involved with the construction of domestic buildings (homes) (see **Figure 2.4**). These include dwellings of one to three storeys, flats, duplex dwellings, villas, townhouses, home units, boarding houses, guest houses and hostels. Generally, such work involves the use of concrete, light-weight framing in timber and steel, cladding in brick, timber or composite sheeting, and fitting out with linings, cupboards, ceramic tiles and basic furnishings. Jobs may involve specialisation, subcontracting or multi-skilling.

FIGURE 2.4 Home under construction

Non-residential construction

The non-residential sector can be divided into two main areas:

■ The *commercial sector* is involved with construction of shopping centres, hotels, schools, office blocks, hospitals and multi-storey and high-rise construction. Work may involve reinforced concrete-framed construction, pre-cast concrete panels, pre- and post-tensioned concrete, steel-framed construction, etc. Construction involves mainly large buildings and structures that involve a wide variety of skills areas such as special hydraulics, lifts, telecommunications, electronics and a vast variety of designs. Jobs range from simple and repetitive to difficult and complex.

■ The *industrial sector* is mainly involved with construction of factories, warehouses, storage areas and small industrial complexes (see Figure 2.5). Work may involve pre-cast tilt-up concrete panels, portal frame steel construction and combinations of these. This also requires many skills, which may range from labour-based tasks to highly mechanised tasks. Formwork and steel-tying skills are highly regarded in both industrial and commercial work.

FIGURE 2.5 Factory unit under construction

Engineering construction

The *engineering* or *civil* sector is involved with large projects such as bridge building, road building, tunnelling, airport construction and dam building. Again, a variety of jobs and skills are required, from operating heavy earth-moving equipment to detailed construction of formwork (Figure 2.6).

FIGURE 2.6 Roadway under construction

Another way that the construction industry can be divided is using public and private sectors. The *public sector* includes government projects which involve mainly engineering construction – that is infrastructure (road and rail systems) and non-residential construction associated with services (hospitals, schools and other government buildings). The *private sector* is involved across all three sectors: residential, non-residential and engineering.

An alternative view in looking at the construction industry is to examine the relationship between employment within the industry and training. We will look at this later in the chapter.

Identifying your role in the construction industry

Being employed in the construction industry means that you need to be aware of the legal requirements and obligations associated with your employment, including the activities you are permitted to undertake, and those that you may not, and which governmental or non-governmental organisation to approach to obtain specific information relating to your employment conditions.

In this section we will examine the following items:

■ construction job roles
■ industrial relations
■ legal and ethical obligations of businesses
■ work health and safety requirements.

Construction workers generally fit into one of the trade categories shown in Table 2.2. These trade categories include the workers who are 'on the tools' and do not include *management* roles, which will be discussed later in this chapter. This list is not exhaustive and there are roles other than these. Another aspect that needs to be considered when discussing construction roles is the division between 'direct' and 'indirect' employment – is the worker paid weekly or fortnightly by his employer (direct labour or employee) or paid as a contractor or subcontractor at the completion of the task or work period (indirect labour or contractor)?

TABLE 2.2 Construction job roles

machine/plant operator	demolition worker
traffic controller	road worker
construction cleaner	insulation installer
construction carpenter/ formworker	steel fixer
concreter	carpenter and joiner
stonemason	bricklayer/blocklayer
painter and decorator	plasterer (render and plasterboard)
roof tiler	wall and floor tiler
metal worker	plumber/drainer/gas fitter
electrician	dogger/dogman
rigger	scaffolder
waterproofer	glazier
carpet layer	trade assistant/labourer

There are many different roles within the construction industry and choosing the right one may seem difficult. It is important to research the roles available and get good advice from someone who has knowledge of the industry. The Australian government has developed websites to help people make good choices. Refer to https://www.australianapprenticeships.gov.au for more information.

Industrial relations

Another area of change has been **industrial relations**. Industrial relations is 'the relationship between management and workers in industry' (*Macquarie Dictionary*).

Industrial relations is about people and organisations working together within the social and political systems of our society. Employment makes up a large part of our lives and determines our living standards. The industrial relations process determines the employment conditions of the environment in which employers and employees work.

There are two industrial relations levels or systems in Australia: the federal system and the state system. Up until 2009 each state/territory government maintained its own industrial relations system with the federal Australian Industrial Relations Commission (AIRC) as a backup to overview major disputes between employers and employees. This meant that each state had its own laws and ran its own arbitration system to deal with industrial issues; in effect, each state was replicating the others. In 2009 the Australian government introduced the *Fair Work Act 2009*, which was introduced and accepted by each state and territory, allowing all private sector employers and employees to be covered by the national workplace relations system. The Fair Work Ombudsman is the national office that oversees this Act.

Changes included under the Fair Work system are:

■ national employment standards
■ national minimum wage
■ unfair dismissal laws
■ modernisation of existing awards or modern awards.

National employment standards

There are 10 minimum standards of employment which apply to all employees under the Fair Work system.

National minimum wage

A minimum wage is an employee's base rate of pay for ordinary hours worked and is generally dependent on the industrial instrument that applies to their employment whether it is a modern award, enterprise agreement, transitional pay scale or national minimum wage order. Each year on 1 July minimum wages are reviewed by the Fair Work Commission and then adjusted accordingly.

Employers and employees can not agree to a rate of pay which is less than the applicable minimum wage.

For employees who are not covered by an award or agreement, a national minimum wage order is made by the Minimum Wages Panel of the Fair Work Commission. In this way *all* employees are covered.

Unfair dismissal laws

According to Australian law, 'unfair dismissal is when an employee is dismissed from their job in a harsh, unjust or unreasonable manner.' (Fair Work Ombudsman, © Commonwealth of Australia 2015 / CC BY 3.0 AU)

As defined by the Fair Work Commission, dismissal is classified as unfair when the:

■ dismissal was harsh, unjust or unreasonable
■ dismissal was not a case of genuine redundancy
■ dismissal was not consistent with the Small Business Fair Dismissal Code, in cases where the employee was employed by a small business.

Source: Fair Work Commission © Commonwealth of Australia (https://www.fwc.gov.au/termination-of-employment/unfair-dismissal)

If someone believes that they have been unfairly dismissed they may notify the Fair Work Commission within 21 days after their dismissal.

The basis for the commission to evaluate if there is a case or not is set by the employment period. So, workers for large companies must have been employed for at least six months before they can apply for unfair dismissal. Employees of small businesses must have been employed for a minimum of 12 months before they can apply.

Source: FairWork Ombudsman, © Commonwealth of Australia 2015 / CC BY 3.0 AU (https://creativecommons.org/licenses/by/3.0/au/legalcode)

Modern awards

An award is the law that establishes the minimum wages and conditions of employment in defined industries or occupations. An award provides for minimum wages and conditions (e.g. overtime, sick leave, annual leave loading and work health and safety requirements). It is the minimum set standards of entitlement due to a worker for their labour. As of 1 July 2009, under the *Fair Work Act 2009* most private sector awards started to be 'modernised'. The process of modernising awards was started by the AIRC in March 2008 and then continued by Fair Work Australia from 2009. This process involves reviewing and condensing of more than 1500 awards to create 122 modern industry and occupation awards. These awards came into effect on 1 January 2010. However, a five-year transitional period occurred and the majority of the modern awards were fully implemented by 1 July 2014.

Enterprise agreement

An enterprise agreement is the contract that has been negotiated and determined between the employer and an employee of a particular workplace; this *enterprise bargaining* is based on determining wages and conditions of work in the business. Enterprise agreements, or workplace agreements, fall into one of the following categories:

- a collective agreement involving the employer and a group of employees
- an individual agreement or contract between an employer and an employee covering specific employment matters on an individual basis
- an independent contractor agreement, which is a special form of individual contract covering independent contractors who are sometimes engaged by a business to do work that otherwise would be done by employees/workers.

For an enterprise agreement to be ratified it must be lodged with the Fair Work Ombudsman.

National minimum wage order

The national minimum wage order is set by the Minimum Wages Panel of the Fair Work Commission. This order is set for employees not covered by an award or enterprise agreement. It must be noted that an employee cannot be paid less than the national minimum wage order.

Workplace organisations

There are support organisations available to help employees and employers. The employee organisations are referred to as 'unions' and the employer organisations are referred to as 'employer associations'.

Unions

A trade union is an association formed by workers to act for them inside and outside the workplace. Historically, trade unions arose in Australia as organisations for the defence and improvement of the conditions of various sections of the workforce. Unions also exert influence in environmental issues and the provision of public facilities. Unionism is a useful tool for negotiations between workers and employers.

Unions usually employ organisers whose job covers promoting membership, contact with shop stewards, assisting shop stewards in difficult negotiations at the workplace, representing workers in negotiations, court appearances on award matters and liaising with the employer and employer associations. The union involved in the construction industry is the CFMMEU (Construction, Forestry, Maritime, Mining and Energy Union).

Employer associations

Employer associations are organisations formed by management to act for them outside the workplace and to provide information, support and advice. The Australian Chamber of Commerce and Industry (ACCI) is the largest single organisation representing industry and commerce in Australia. The three main bodies for the building industry are:

- Housing Industry Association (HIA)
- Master Builders Australia (MBA)
- Master Plumbers Association (MPA).

Ethical standards

Ethics or morals are how people define what is right and wrong or what is good or bad behaviour. Society's leaders use laws and codes of practice in an attempt to give direction and guidance to the society under their care or control. There are, however, individuals who do not comply with social norms of what is considered good or right, who take advantage of others for personal gain and who believe that morality is not their concern. Such individuals may think that the statement 'do to others as you would have them do to you' is not relevant or does not apply to them. Within the construction industry, ethical and moral practices vary. Many workers in the construction industry try to do what is morally right and to provide good quality and service, while others do not. It is for this reason that legislation, codes of practice, codes of ethics and codes of conduct have been produced by federal, state and territory governing bodies as well as industry bodies like the Housing Industry Association (HIA) and Master Builders Australia (MBA).

Access and equity

Access involves making sure that government and corporate services, resources and programs are available to all, including disadvantaged or minority communities that may not have the same level of access as the mainstream population.

Equity involves ensuring that these services, resources and programs deliver satisfactory outcomes for all people and their diverse communities.

These services, resources and programs include training programs, pre-employment programs and employment support services, and are in place to support people in accessing resources that will help them; for example, to secure a position in the workforce.

Employee assistance programs (EAP) are just one example of an access scheme that corporations and businesses can use to provide employees with counselling and psychological support to workers. This may include grief or trauma counselling or behaviour management.

Anti-discrimination and equal employment opportunity (EEO)

Under the *Fair Work Act 2009*, it is unlawful for an employer to discriminate or take adverse action against a person who is an employee or prospective employee because of their personal attributes, such as:

- race
- colour
- sex
- sexual preference
- age
- physical or mental disability
- marital status

- family or carer responsibilities
- pregnancy
- religion
- political opinion
- nationality
- social origin.

Equal employment opportunity (EEO) means that all people regardless of personal attributes have the right to be given fair consideration for a job or position. There is to be no bias.

Bullying and harassment

Workplace bullying is a risk to health and safety because it may affect the mental and physical health of workers.

Workplace bullying is defined as repeated and unreasonable behaviour directed towards a worker or a group of workers that creates a risk to health and safety.

Not all behaviour that makes a person feel upset or undervalued at work is classified as workplace bullying. Examples of behaviour, whether intentional or unintentional, that may be considered to be workplace bullying if they are repeated, unreasonable and create a risk to health and safety include, but are not limited to:

- abusive, insulting or offensive language or comments
- unjustified criticism or complaints
- deliberately excluding someone from workplace activities
- withholding information that is vital for effective work performance
- setting unreasonable timelines or constantly changing deadlines.

A single incident of unreasonable behaviour is not considered to be workplace bullying. However, it may have the potential to escalate and should not be ignored.

Harassment is any of form of unwanted physical or verbal behaviour that offends or humiliates a person. Generally, harassment takes place over a period of time but there can be serious one-time incidents too.

Harassment may include, but is not limited to:

- behaviour that threatens or intimidates a person
- unwelcome remarks or jokes about a person's race, religion, sex, age or disability
- unwelcome physical contact with a person, such as touching, patting, pinching or punching, which can also be considered assault.

Your employer has a legal responsibility to provide a safe work environment free of violence, bullying and harassment.

Implications of workplace bullying:

■ failure to manage the risk of workplace bullying is a serious breach of WHS laws

■ education, prevention and a quick response (proactive approach) should a case arise are the best ways to handle workplace bullying

■ relationships within the workplace, and health and safety factors will be greatly impacted if ignored or the response is too slow

■ degradation of the victim's mental health (stress, depression, anxiety, etc.) will impact their work (increase in sick days and decrease in work performance) and in turn, affect productivity overall. Bullying has a flow-on effect.

Work health and safety requirements

In each state and territory there are specific rights and responsibilities for PCBU and employers, employees/ workers and contractors under federal, state and territory WHS legislation.

WHS legislation aims to protect the health, safety and welfare of people at work. This legislation lays down health and safety requirements that must be met at places of work. Health and safety in the workplace is a very important employment condition that needs to be considered to ensure the rights and responsibilities of employers, employees and contractors are exercised in accordance with federal, state and territory WHS laws.

PCBU/employer responsibilities under the Act

PCBU/employers must provide for the health, safety and welfare of their employees/workers. To do this, PCBU/employers must:

■ provide and maintain equipment to ensure safety and avoid risks to health

■ ensure work systems comply with safety legislation

■ make arrangements to ensure the safe use, handling, storage and transport of equipment and substances

■ maintain places of work under their control in a safe condition and provide and maintain safe entrances and exits

■ provide adequate information about substances used at the place of work, including research and relevant tests.

PCBU/employers must not request that employees/workers pay for anything in order to meet specific requirements under WHS legislation. They must also ensure the health and safety of people visiting their places of work who are not employees or workers.

Workers/employees' legal responsibilities

Workers and employees must take reasonable care for the health and safety of others and cooperate with PCBU/employers in their efforts to comply with work health and safety requirements.

Employees/workers must not:

■ interfere with or misuse any item provided for the health, safety or welfare of persons at work

■ obstruct attempts to give aid or attempts to prevent a serious risk to the health and safety of a person at work

■ refuse a reasonable request to assist in giving aid or preventing a risk to health and safety.

Offences and penalties

The national model for WHS legislation has harmonised many conflicting aspects of past state and territory OHS/WHS legislation. One of these aspects is the varying penalty and infringement systems in place pertaining to similar offences committed across the various national jurisdictions. The new model for work health and safety ensures that on a national platform penalties applied to comparable infringements are regularised.

The harshness of fines is influenced by the severity of the infringement, and takes into account whether an individual or corporation is guilty of the offence and whether or not they are previous offenders. It should also be noted that, as well imposing fines, courts may opt for a sentence of imprisonment. They also have the option to order offenders to:

■ take steps to remedy or restore any damage or injury caused by the offence

■ pay the regulating/investigative authority for the costs of the investigation

■ publicise or notify other persons of the offence

■ carry out a project for the general improvement of health and safety.

Note: It is recommended that readers become familiar with their relevant state or territory infringement systems by either contacting their WHS regulatory authority, or accessing their website (details for all state and territory regulatory bodies can be found in Table 1.2 in Chapter 1).

Site amenities

Under national WHS regulations it is the duty of a PCBU to maintain and provide adequate facilities for workers (good working order, clean and safe), as far as reasonably practicable, for any workplace within their control. Facilities include toilets, drinking water, washing and eating facilities (see Figure 2.7).

Source: Ed Hawkins

FIGURE 2.7 Work site toilet

Emergency procedures

Another duty or responsibility of the PCBU is the implementation and maintenance of emergency plans and procedures. These include:

- evacuation planning, process and practice (see Figure 2.8)
- fire control planning, training and practice
- first aid provision – being adequate and available to workers on the job site.

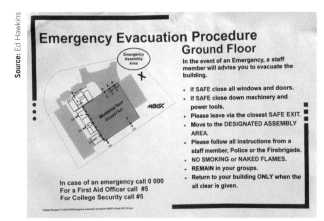

Source: Ed Hawkins

FIGURE 2.8 Sample evacuation plan

Licensing

In order to carry out residential building work, most states and territories require that you must either be the holder of a contractor licence, or employ someone who is the holder of a qualified supervisor's certificate. Issuing a licence ensures that all qualified builders and contractors are registered with their relevant state or territory's licensing regulator. This gives the client the ability to check on the credibility of the builder or contractor. It also provides the client with some protection against faulty production and poor workmanship.

In February 2007, new legislation was passed that allowed licences granted in a particular state or territory to be mutually recognised nationally. This means that a person licensed to carry on a particular occupation (such as a carpenter) in a certain state or territory can also practise the equivalent occupation in another state or territory, provided they notify the local regulating authority. It is the licence holder's responsibility to undertake work *only* for the purpose for which the licence is held and they must also show their respective licence numbers on all advertising, stationery and signage. Failure to follow these conditions could be deemed a direct infringement of the legislation and result in fines being issued to the respective licence holders.

Note: It is recommended that readers become familiar with the relevant state or territory licensing requirements by either contacting the appropriate regulating authority directly or visiting the regulating authority's website as detailed in Table 2.3.

Personal responsibility

It is the worker or contractor's responsibility to pay attention to the details and quality outlined by the employer's directions and instructions when completing a job. Remember that the employer has the knowledge to understand the requirements laid out within the specifications and work documents. It is your responsibility to complete the work as he/she has asked. Remember, having the right attitude will affect the world around you. Problems can be prevented by maintaining standards and building quality into procedures, products and design at the start of a job. Rather than fixing mistakes, one should take the time to analyse failures and succeed the next time.

Take the initiative when it comes to accepting new challenges and stick with them. Set goals to do your job better and aim to do it quickly and accurately. Share any ideas you might have to improve procedures, save money or increase efficiency or productivity with your employer. All of these personal responsibilities will lead to improved quality and results for everyone.

TABLE 2.3 Regulators for carpenters, joiners, builders and bricklayers (as at September 2019)

State / Territory	Regulator	Location	Contact Details	Website
WA	Building and Energy Division of the Department of Mines, Industry Regulation and Safety	Mason Bird Building 303 Sevenoaks St CANNINGTON WA 6107	1300 489 099	https://www.commerce.wa.gov.au/building-and-energy
VIC	Victoria Building Authority	Goods Shed North 733 Bourke Street DOCKLANDS VIC 3008	1300 815 127	www.vba.vic.gov.au/
QLD	Queensland Building and Construction Commission	299 Montague Road WEST END, QLD 4101	13 93 33	www.qbcc.qld.gov.au/
NSW	NSW Fair Trading	27–31 Argyle Street PARRAMATTA NSW 2150	13 32 20	www.fairtrading.nsw.gov.au
SA	Consumer and Business Services	Ground Floor, 91 Grenfell Street ADELAIDE SA 5000	131 882	https://www.cbs.sa.gov.au/
TAS	WorkSafe Tasmania	30 Gordons Hill Road ROSNY PARK TAS 7018	1300 366 322	www.worksafe.tas.gov.au/
ACT	Environment, Planning and Sustainable Development Directorate	8 Darling Street MITCHELL ACT 2911	(02) 6207 1923	www.actpla.act.gov.au
NT	Building Practitioners Board	Level 3, NAB House 71 Smith Street DARWIN NT 0800	(08) 8936 4082	https://bpb.nt.gov.au/

LEARNING TASK 2.3

LICENSING FOR CARPENTERS

Go to the website of the regulatory authority in your state or territory and view the information relating to how to apply for a carpentry licence.

Answer the following questions.

1 How much will it cost to obtain an individual licence for carpentry?

   $ ....

2 What is required before you can obtain a carpentry licence?

   ....

3 List two reasons why someone may be refused a licence.

   a ....

   b ....

2.2 Identify own existing skills and the additional skills required for a tradesperson or skilled operator role in the construction industry

Identifying responsibilities of individuals for their own workload

Managers have the responsibility to plan the work activities and also the timeframe in which to complete each task or stage. Time is, therefore, a very important commodity, simply because it relates directly to money.

As an individual involved in this process, it is important for you to know how long it should take you to complete a set task and the steps involved in completing that task. You have an important role as an individual as well as part of a team to work as closely as possible to the manager's time estimates.

Planning work activities

Correct planning, involving sound decisions made by the building manager, leads to an orderly progression of the tasks or steps to be completed: from the decision to tender to the final completion of a project. Taking the time to plan correctly is extremely important and must begin before site operations commence so that the proper methods and equipment are decided on and the correct materials are ordered. Planning has a direct effect on the whole job.

The output of subcontractors and tradespeople will determine the schedule of a job. Stages of construction and sequencing of trades should commence as soon as possible, without necessarily waiting for the completion of the preceding work, and should continue without interruption. The labour and time allocated should be proportional to the extent of work being performed. For example, brickwork is a labour-intensive trade, as is the erection of formwork for reinforced concrete, so buildings that require these trades will require detailed planning for the job to be

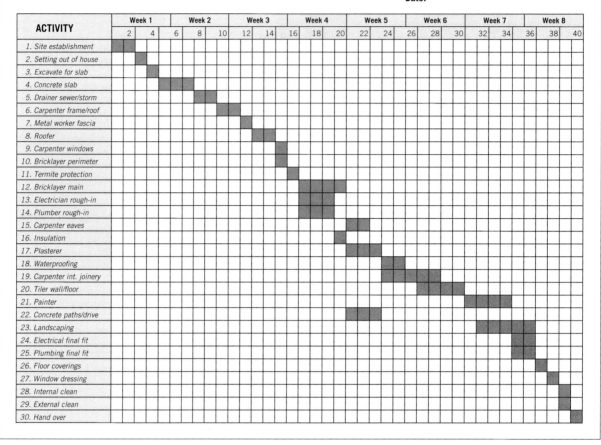

Construction Schedule for a simple brick veneer cottage on a concrete slab

Project
Location:
Schedule by:
Date:

FIGURE 2.9 Typical Gantt or bar chart for a brick veneer cottage on a concrete slab

done on schedule. To this end, many managers will use a *Gantt chart* (see **Figure 2.9**) to schedule when subcontractors will be needed, when orders are to be placed and when supplies are to be delivered. This also shows the **sequence** of operations and provides a guide for progress and costing. These charts can be used to show time, progress and financial position. They can be weekly or monthly charts, or be for the duration of the total contract. Gantt charts can be adjusted to take into account any changes in the contract or any other disruptions to the onsite work. They are used to assess planning prior to the commencement of the contract and for monitoring the delivery of materials and commencement of different stages of the job.

Table 2.4 shows a typical sequence of events for a construction schedule with the time given in days. It also shows the relationship of activities to each other.

Figure 2.9 shows the same information graphically in the form of a Gantt chart; it highlights the overlaps in activities.

As a worker it is vital that you understand that the tasks you carry out are done within a team structure, and that you are not alone. You are part of a team which includes your managers, supervisors, work colleagues, union representatives (if you are an employee) and employer associations (if you are a contractor).

Identifying individual career path and development needs

In recent years the type of work performed on building sites has changed, reflecting the changing needs of business and the growth of the global economy. Many industries are meeting these changes by varying employment patterns, such as changing from the traditional full-time, 38-hour week to offering part-time, casual or contract positions, and outsourcing or using labour hire as alternatives.

Career pathways for workers within the construction industry have also changed and mapping out a career with potential for advancement

TABLE 2.4 Schedule for a brick veneer cottage on a concrete slab

Construction of brick veneer cottage on a concrete slab			
Ref	Activity	Time (Days)	Preceding Activity
1	Site establishment	2	–
2	Setting out of house	1	1
3	Excavate for slab	1	2
4	Concrete slab	3	3
5	Drainer sewer/storm water	2	4
6	Carpenter: frame/roof	2	5
7	Metal worker fascia/gutter	1	6
8	Roofer	2	6, 7
9	Carpenter windows	1	8
10	Bricklayer perimeter course	1	8
11	Termite protection	1	10
12	Bricklayer main	4	11
13	Electrician rough-in	3	6
14	Plumber rough-in	3	6
15	Carpenter eaves	2	12
16	Insulation	1	8, 10, 13, 14
17	Plasterer	3	16
18	Waterproofing	2	17
19	Carpenter internal joinery	5	17
20	Tiler walls and floor	4	18
21	Painter	4	17, 19, 20
22	Concreter paths and driveway	3	5
23	Landscaping/fencing	5	22
24	Electrical final fit	2	21
25	Plumbing final fit	2	21
26	Floor coverings	1	24, 25
27	Window dressings	1	26
28	Internal clean	1	26, 27
29	External clean	1	23
30	Hand over	1	28, 29

has become increasingly difficult with many of the traditional pathways no longer existing. Some employers expect their employees/workers to contribute new knowledge and innovation to their organisation to assist in the growth, effectiveness and competitiveness of the business. Therefore, individuals may initially need to seek advice from a range of personnel to determine what they need to do to move in a new or different direction. Good advice is best gained from experienced personnel with a good understanding of the trade or industry. Suitable people may include:

- the foreperson on the job
- architects and/or engineers
- progressive builders
- teachers/lecturers from registered training organisations

- environmental experts
- designers
- building material manufacturers.

Once a person has determined a pathway or direction, there are two steps they need to take:

- practical industry involvement
- training or educational pathway.

Practical industry involvement

Experiential learning is the process of learning through experience, and is more specifically defined as 'learning through reflection on doing'. For many within the construction industry this method of learning is the central way that they learn; that is, by doing. Many people learn how to be productive within the industry by being guided by someone who demonstrates the practical skills they require to complete tasks.

A classic example of this is the tradesperson–apprentice relationship where the tradesperson takes on an apprentice and teaches them how to become a tradesperson.

This method of learning has existed for many years; today, however, without formal recognition this alone is not considered sufficient training.

Training or educational pathway

The first item that needs to be addressed for anyone starting in the building industry is general induction training.

General construction induction training

Employers and principal contractors must ensure that persons carrying out any form of work on the construction site have undertaken the relevant general construction **induction training**. This is an existing obligation outlined within the relevant WHS legislation. It applies to all persons carrying out work in the residential, commercial and high-rise construction sectors. It is also a requirement under the relevant Acts and Regulations that employers and principal contractors must not direct or allow a person to carry out construction work unless that person has completed the relevant WHS induction training as follows:

- WHS general construction induction training (meeting the criteria for the card system relevant to the state or territory requirements), consisting of a broad range of safety awareness instruction
- work activity WHS induction
- site-specific WHS induction.

The regulation also requires that self-employed individuals must not carry out construction work until they have personally completed WHS general construction induction training. This induction training must be provided for all workers at no charge.

The person responsible for the general construction induction training must also provide a written statement for each person being inducted, stating that the person concerned has satisfactorily completed the training; listing the activities covered in the training; specifying the dates on which the training occurred; and specifying the name and qualifications of the person who conducted the training. This statement must be signed by the person who conducted the training. Each person who has successfully completed the general construction induction training should keep the written statement until they receive their plastic induction card (see **Figures 1.2–1.5** in Chapter 1) to be produced if a safety officer asks for it.

Site-specific induction must be provided for every construction site, as there may be site-specific safety issues or hazards that need to be identified. This means that each worker must undertake a site-specific safety induction prior to entering that site. (For further details and information, it is recommended that readers either contact their state or territory WHS regulating authority or access their website, as detailed in **Table 1.2** in Chapter 1).

Training packages

A training package is a set of nationally endorsed standards, guidelines and qualifications that are used for recognising and assessing people's skills. They describe the skills and knowledge needed to perform effectively in the workplace.

Training packages set a national industry standard for skills and are used as the basis for most of the programs delivered in the VET (vocational education and training) system, including Australian Apprenticeships and VET-in-schools programs. These programs are delivered by registered training organisations (RTOs).

Training qualifications

The training package for the building industry is CPC08 Construction, Plumbing and Services Training Package. It specifies the combination of units of competency required to achieve a particular qualification (see **Table 2.5**). Learners who complete some, but not all, units for a qualification are awarded a statement of attainment. When they are assessed as competent in the remaining units, they attain the qualification.

In broad terms, AQF (Australian Qualifications Framework) level 1 equates to Certificate I, AQF level 2 to Certificate II, and so on. The AQF provides the characteristics, distinguishing features and criteria for each qualification level to ensure consistency across sectors and from one qualification to the next. It also establishes the principles for the issuing of qualifications by RTOs. Under the AQF, a learner can start at the level that suits them and then build up their qualification as their needs and interests develop and change over time. The framework assists learners to plan their career progression regardless of their life stage or location, by supporting movement across the sectors and encouraging lifelong learning.

Examples of some of the training package qualifications are:

- Certificate I in General Construction
- Certificate II in General Construction
- Certificate III in General Construction (Bricklaying/Blocklaying)
- Certificate III in General Construction (Carpentry)
- Certificate III in General Construction (Carpentry – Formwork/Falsework).

TABLE 2.5 Construction, plumbing and integrated services framework job pathways

| Potential Australian apprenticeships jobs outcomes guide examples | | | | Further job & study options | |
Stream	Certificate II	Certificate III	Certificate IV	Diploma	Advanced Diploma
Bricklaying/Blocklaying	A range of trades assistant jobs – such as: Trades Assistant Builder's Labourer Site/Field Labourer Carpenter's Assistant Plasterer's Assistant Tiler's Assistant Joiner's Assistant	Bricklayer – Heritage Bricklayer Blocklayer	Trade Contractor Building Estimator Building Contractor Builder/Construction Manager Building Site Manager Sales Consultant (Building & Construction) Building Contract Administrator	Builder Construction Manager Project Manager Estimating Manager Sales Manager	Construction Manager
Wall & floor tiling		Tiler			
Wall & ceiling lining		Wall & Ceiling Liner Wall & Ceiling Liner – Commercial			
Solid plastering		Solid Plasterer			
Painting and decorating		Painter & Decorator			
Roof tiling		Roof Tiler			
Carpentry		Carpenter & Joiner Carpenter			
Shop fitting & stairs		Joiner Shopfitter Carpenter/Joiner Carpenter/Joiner (Stairs) Stairbuilder			
Specialist trades			Crane Operator Building Manager & Supervisor Swimming Pool & Spa Builder		
Stone work	Stoneworker	Stonemason Stonemason (Monumental)			
Signage		Sign Writer			
Concreting	Concrete Trades Assistant	Concreter Concrete Pump Operator			
Low rise structural framing		Framer			
Formwork/falsework		Formworker			
Demolition		Demolisher			
Rigging		Rigger			
Scaffolding		Scaffolder			
Dogging		Dogger			
Waterproofing	Waterproofing Assistant	Water proofer			
Steel fixing	Steel Fixing Assistant	Steel Fixer			
Building surveying				Building Surveyor	Senior Building Surveyor

Source: The Australian Apprenticeships & Traineeships Information Service (www.aatinfo.com.au/jpc) funded by the Australian Government Department of Education and Training

Units of competency

Units of competency are developed by industry to meet identified skill needs. Each unit of competency identifies a separate workplace requirement and includes the knowledge and skills that underpin competency as well as language, literacy and numeracy and work health and safety requirements.

Each unit of competency describes:

- a specific work outcome
- the conditions under which it is conducted
- the knowledge and skills required to achieve the work outcome to the standard
- the evidence that may be gathered in order to determine whether the activity is being performed in a competent manner.

Assessment process

Skills are very important in the construction industry. There are skills you already have and skills you need to acquire, which can be assessed onsite or at an RTO. Competency-based training involves making sure people have the right skills to do a particular job competently. It does not matter if they spend six months or three years in training, as long as they are competent at the job. This is where assessment is important as it ensures that an individual is competent to perform a job.

Assessment may be undertaken by practical demonstration, third-party reporting or recognition of prior learning (RPL). RPL skills assessment may be undertaken onsite (while working on the job) by an accredited workplace assessor. This assessment option is being performed on many sites. Skills can also be assessed off-site. There are a number of centres, or RTOs, across Australia that offer recognition of current skills through an 'assessment pathway only'. For example, in New South Wales, numerous institutes offer a skills assessment recognition program, commonly referred to as 'Skills Express', which allows adult learners with no formal training to obtain a recognised qualification in a number of trade areas. Other similar programs are in place to perform skills assessment for licensing purposes.

Continuing professional development

In 2004, various states across Australia introduced a new system that affected building contractor licence renewals. The system is known as continuing professional development (CPD) and requires building contractors to undergo an upgrading process by participating in an assortment of designated learning activities. Each of these activities equates to a certain number of credit points. The minimum number of credit points a building contractor must accumulate is 12 points per annum (or 36 points when licences are renewed for a three-year period). The types of activities that generate credit points include, but are not limited to:

- employing an apprentice
- attending industry forums
- gaining additional licences
- undertaking academic training related to building activities
- completion of certificate training
- subscribing to industry magazines.

CPD must be relevant to the building contractor's area of practice and ultimately enable the building contractor to:

- extend or update their knowledge, skill or judgement
- become more productive
- understand and apply advances in technology
- face changes in the industry
- improve their individual career paths and opportunities for advancement
- better serve the community.

While the CPD program is currently administered in Victoria, New South Wales and Tasmania (with other states currently considering its introduction), there are significant differences with how each state has decided to regulate their particular version; for example, in Victoria the program is voluntary, whereas in New South Wales and Tasmania it is compulsory.

Note: It is recommended that readers become familiar with their relevant state or territory CPD system, by either contacting their appropriate regulating authority, or accessing their website as detailed in Table 2.3.

LEARNING TASK 2.4

TRAINING COMPONENTS

1 Go to the training.gov.au website (National Register of VET).
2 Once on the website, quick search 'CPCCOM1012' under 'Title or code' in the 'Nationally recognised training' section.
3 Export a PDF or Word summary version of the document and read through it.
4 Discuss the information you have found with your group and with your teacher or facilitator.

 COMPLETE WORKSHEET 2

---

3. Identify and follow environmental and resource efficiency requirements

3.1 Identify environmental and resource efficiency requirements that apply to entry level roles in the construction industry

The building industry today is in a state of constant change, particularly regarding the use of technology and improved work processes. Technological changes include:

- improved tools, plant and equipment
- new materials and composite materials
- new building methods
- improved computer software.

All of these technological advancements have allowed buildings to be constructed more quickly and efficiently than ever before, but have also created a need for tradespeople to be trained both formally and informally on the work site.

Tools, plant and equipment

To gain an understanding of why these changes are so readily accepted we must consider a brief period of building history.

Machine technology began to impact on building during the Industrial Revolution from the late eighteenth century. While the technology from that time appears simple and even crude by today's standards, this was a huge improvement on the hand tools that had previously been used. This trend towards mechanisation has continued through to today where building operations have become largely mechanised through the use of machines such as excavators (see Figures 2.10 and 2.11), tower cranes, concrete mixers and power tools.

Source: From the collection of the National Archives of Australia

FIGURE 2.10 Early road-making in Canberra — Keystone steam shovel with horse and dray

Source: Ed Hawkins

FIGURE 2.11 Modern-day excavator

The acceptance of new technology in the Australian construction industry is partially due to 250 years of tool and machine development as well as a willingness to embrace new technology. We will cover the specifics of tools, plant and equipment in Chapter 8.

Materials and composite materials

Another area where technological improvement has occurred is demonstrated in the wider range of materials and fabrics used in the construction industry. These include a variety of patent-type floor systems, such as suspended steel framing and reinforced concrete slab-on-ground (see Figure 2.12 for an example of a waffle pod slab); pre-cast tilt-up slab walling systems; autoclaved aerated concrete blocks and panels; light-weight timber trusses; and a variety of improved roof coverings such as concrete tiles, terracotta tiles and Colorbond® metal sheeting. We will cover the specifics of materials and composite materials in Chapter 8.

Source: Ed Hawkins

FIGURE 2.12 Waffle pod slab construction

Building methods

Over the past 80 years building methods have radically changed. The first factor to cause dramatic change was the Great Depression (1929–1932). During this time there was little to no construction being undertaken in Australia, except for construction of the Sydney Harbour Bridge. This major project is still an engineering marvel as the world's tallest steel arch bridge (see Figure 2.13).

FIGURE 2.13 Sydney Harbour Bridge under construction

This period was quickly followed by the Second World War, during which construction work that had previously only been done by men was being done by women due to a shortage of available men. During this period there was a great need for light-weight prefabricated, transportable buildings that could be used to house the thousands of defence force personnel or to store machines and supplies. There was little or no domestic construction during this time.

From 1945, following the end of the Second World War, new homes, facilities and infrastructure were needed for the returning soldiers and their families and to accommodate the influx of European immigrants wishing to start a new life in Australia. However, with the loss of many skilled tradespeople and the shortage of traditional building materials, alternative solutions were needed. The shortage of skilled labour was rectified by the 'populate or perish' immigration policy that saw many English, German, Maltese, Greek, Yugoslavian and Italian workers arrive in Australia. Materials shortages led to the design of more simplified cottages. Timber-framed structures clad with asbestos cement sheets (the wonder product of the day) for both walls and roofs were built until the 1960s. Expensive double-brick homes were replaced with cheaper brick veneer constructions, starting in Melbourne and then spreading to the other major Australian cities. It was also during this time that the state and federal governments started infrastructure projects such as the Snowy Mountains Hydro-Electric Scheme, which required many thousands of workers as well as the use and development of new tunnelling techniques and reinforced concrete.

During the 1960s, concrete roof tiles were increasingly used to cover roofs rather than terracotta tiles, asbestos roof sheets or corrugated iron. Walls and ceilings were lined with plasterboard instead of fibrous plaster and towards the end of the 1960s brick veneer construction was the preferred cladding on new homes and is still the most commonly used method of construction throughout all major centres of Australia.

Builders also started building homes on concrete slabs rather than bearers and joists at this time.

Another aspect of building that has changed over time is the use of specialist tradespeople. Prior to the 1970s, a carpenter was involved with all aspects of timber work including the manufacture and erection of the wall and roof frames, manufacture and installation of stairs, fixing out the building and even installing the kitchens. From the 1970s onwards the carpentry trade has been broken up into specialist areas. Today we see teams of carpenters who specialise in the erection of wall frames and trusses; other teams that specialise in fitting out the building; specialist stair manufacturers and kitchen manufacturers and installers. The reasoning behind this change is that a specialist in one area can complete the task faster and, therefore, cheaper than a carpenter undertaking a wide range of carpentry tasks.

Other twentieth-century advances in building methods and technology are exemplified by the construction of multi-storey skyscrapers. These constructions could not have been built without the invention of the elevator as well as the advent of steel-reinforced concrete and improvements in glass manufacture (see Figure 2.14).

FIGURE 2.14 Melbourne's Eureka Tower – the tallest building in Australia to roof (2006)

Computer software

Another, more recent advance is the use of computers to design, calculate and solve engineering problems, such as those encountered during construction of Jørn Utzon's Sydney Opera House (see Figure 2.15), the Sydney and Melbourne football stadiums and the new Parliament House in Canberra. Computer-controlled or programmed machines make light work of tedious jobs such as pre-cutting timber frames and trusses, polishing or grinding pre-cast concrete facings to buildings, and programming for laser-cutting equipment. CAD (computer-aided design) as well as the more current BIM (building information modelling), have allowed the designer the luxury of seeing the building or components of the building in three-dimensional form before the designs are finalised and construction has even started.

Source: iStock.com/Simon Bradfield

FIGURE 2.15 Sydney Opera House – designed by Danish architect Jørn Utzon and built from 1959 to 1973

Legal and ethical obligations of businesses

PCBU and employers have legal and ethical or moral obligations to:

- business partners
- shareholders (in the event that the business is a public company)
- clients
- suppliers
- contractors and subcontractors
- employees
- the environment.

Some of these legal and ethical obligations are outlined below.

Quality

Any product that is manufactured or processed has to maintain a certain **quality** or tolerance that is consistent over time. Inspection and testing is carried out to ensure that this happens.

There are two terms associated with quality: *quality control* and *quality assurance*.

Quality control is the system that is used to test a product by sampling and then checking it against set specifications. An example is when concrete is poured

and a small amount is set aside in test cylinders, which are then crushed after a prescribed time to test the compressive strength of the concrete. This ensures that the strength of the concrete can be checked to see if it complies with the engineer's specifications (see Figure 2.16).

Source: Ed Hawkins

FIGURE 2.16 Concrete being collected for testing

Quality assurance is preserving or keeping a set level of quality in the overall finished product. This is achieved by considering every stage of the construction. Quality assurance requires planning and documentation from the very beginning of the job so that the structure complies with *all* relevant specifications and standards as set out in the building contract to ensure that the client actually gets what they pay for. This means that the contractor and subcontractor must take more responsibility for the end product and in doing so they help to ensure better quality control for the building industry generally.

It is important, therefore, that as an employee you know what the specifications and standards are so that you can complete your set tasks to the agreed quality required. In this way you are 'doing the job right the first time' and 'using the right tool for the job'.

Quality can be expressed as the level of excellence that goes into a product or service. It is the skill and commitment you and everyone you work with bring to the job each time, all the time. A commitment to quality will help you to perform error-free work on every job you do.

Quality is of benefit to everyone: from consumers, who expect and demand quality in the goods and services they receive; to management, who know it will increase profits; to the employee, who benefits from job security and better pay; and to the national industry, which will be more competitive in the world market.

Quality assurance and quality management is addressed in AS/NZS ISO 9001 Quality management systems – Requirements.

The eight key principles of quality control

1 All systems exhibit variation.
2 High quality does not cost anything – it pays for itself (in the long run).
3 People work within systems.
4 Everyone serves a customer.
5 Improvements should be plan-driven, not event-driven.
6 Improvement should be a way of life.
7 Management should be controlled by facts and data.
8 Control the process, not the output.

The six key terms you need to know

1 *Quality* – fitness for purpose.
2 *Reliability* – will continue to work for its guaranteed life.
3 *Maintainability* – parts and services are readily repaired if necessary.
4 *Availability* – the system/goods work instantly when required.
5 *Supplier* – a person giving customers goods and materials with which to work.
6 *Customer*
    a internal – the person who receives your work next in the work process, e.g. frames by carpenter received by brickworker
    b external – the person who buys goods and services, e.g. the client or owner of the building.

Standards

Standards ensure that goods, services and products are accurate and safe to use. They are put together by a panel of experts in the specific area that the standard applies to. This panel may include representatives of safety committees, manufacturing industries, suppliers, customers and government departments. Standards Australia was founded in 1922 as the Australian Commonwealth Engineering Standards Association and was renamed the Standards Association of Australia in 1929; an independent non-profit body incorporated by Royal Charter. Standards Australia operates in the national interest and its principal functions are to prepare and publish Australian Standards and to promote their adoption.

Examples of Australian Standards relevant to the construction industry are:

- AS 2870 Residential slabs and footings – Construction
- AS 1684 Residential timber-framed construction
- AS 3600 Concrete structures
- AS 3660.1 Termite management – new building work.

Standards within the building industry are extremely important as many building products that are manufactured overseas are now being used *illegally* within domestic and multi-storey structures.

An example of the illegal use of materials that did not comply with Australian Standards became apparent in 2014 when a fire started on an eighth-floor balcony of a Docklands apartment block in Melbourne. The fire spread up the side of the building to the twenty-first floor in less than 11 minutes, fuelled by the aluminium composite panel cladding product Alucobest, which was being used on some high-rise apartment buildings. This product was imported from China and does not comply with Australian safety and fire standards. So, while the cost of this product was significantly cheaper than the non-combustible Australian-made Alucobond®, the long-term costs and safety risks were far greater (Towers, 2015).

Identifying and complying with environmentally sustainable work practices and techniques

Sustainability and sustainable housing is about designing and building homes that are comfortable and practical to live in, economical to maintain and that have the least impact on the environment.

This balanced approach takes into account the economic, environmental and social aspects of housing development, which are intrinsically linked to each other as well as being co-dependent, as indicated in Figure 2.17.

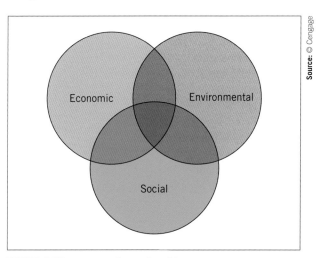

FIGURE 2.17 Aspects of sustainability

Economically sustainable homes are cost-efficient over the lifespan of the building. They are well designed with consideration to orientation, insulation and ventilation to provide an energy-efficient home, as well as using low maintenance materials to reduce the running costs of the home.

In effect, the design of the building balances construction costs against ongoing running and maintenance costs.

Environmentally sustainable homes are **resource efficient** in terms of materials, **waste**, water and energy. They are designed for water efficiency in the house and garden, and **waste minimisation** during construction, occupancy and demolition. Environmentally sustainable homes are energy efficient in terms of energy consumption with household running costs reduced by up to 60 per cent, saving over 3 tonnes of greenhouse gases and more than 100 000 litres of water a year. Environmentally sustainable elements include water-efficient and energy-efficient appliances, solar hot water systems, insulation and efficient lighting.

Socially sustainable homes are designed with all people in mind. Features such as flexibility, comfort, access, safety and security are addressed so that homes accommodate the changing needs of the owners and occupants.

Does a sustainable home cost much extra?

Current estimates suggest that for a new home of average size an initial outlay of approximately $6000 to $8000 will allow a new home builder to install:

- a solar or electric reverse cycle hot water system instead of electric
- low-flow taps and fittings
- water tank for the garden
- dual-flush toilet
- bulk ceiling and wall insulation.

Besides the eventual financial savings the householder will receive, they can also enjoy greater living comfort with the knowledge they are helping the environment. At the same time they are adding to the value of their home and improving its future saleability.

Sustainable alternatives

Brisbane City Council in Queensland (https://www.brisbane.qld.gov.au), like many other councils, recommends you consider the following design features, product decisions and site management practices when designing or renovating a home.

General

- An open plan and northerly orientation will maximise breezes and avoid the western sun.
- Bathroom, kitchen and laundry should be located close to the hot water system.
- Ensure living areas are positioned to capture winter sun and summer breezes.
- Plan window size, style and location to optimise protection against summer sun and access to winter sun.
- Minimise windows on the western side to avoid the afternoon sun.
- Use materials with low long-term maintenance costs.

- Install awnings and eaves to reduce heat.
- Install insulation in roof, ceiling and walls.
- Consider an insulated skylight to let in natural light and not heat.
- Install compact fluorescent lighting including down-lights with efficient 12-volt task lighting.
- Incandescent lighting can be used for shorter duration lighting in selected areas.
- Paint the exterior of your house and roof in a light colour to help cooling.

Kitchen/Bathroom/Laundry

- Install double sinks so you can rinse in a second sink and not under a running tap.
- Install AAA-rated water-efficient taps.
- Provide task lighting over sink, stove and work surfaces.
- Choose water-efficient and energy-efficient white goods: oven, dishwasher, refrigerator and freezer. Look for the AAA rating on water products and the highest star rating on energy-efficient appliances (see Figure 2.18).
- Place your fridge in a cool spot away from the stove and direct sunlight.
- Stove range-hoods should be vented to the outside.
- Use rain water for toilet flushing, hot water, washing and showering.
- Grey water from laundry and bathrooms may be used for the garden irrigation system (pending state and territory legislative requirements).
- Use AAA-rated taps and shower roses for water efficiency.
- Install mixer taps in showers to reduce hot water loss while you adjust the temperature.
- 6-litre/3-litre dual flush toilets will reduce water use.
- Choose an AAA water conservation rating and high star energy rating front-loading washing machine.
- If you must install a clothes dryer choose an energy-efficient one, but it's best to use the outdoor clothes line.

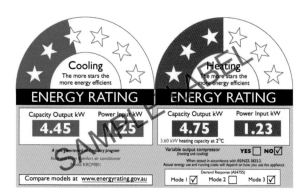

FIGURE 2.18 Energy rating label

Source: www.energyrating.gov.au

Sorting your waste as you clean up the site can save you a lot of money when you dispose of that waste. Local councils charge different fees for different types of waste.

Finishes

- Use non-toxic paints, renders and floor finishes with either no or low volatile organic compound (VOC) emissions to give superior air quality compared to a standard house.
- Consider using floor tiles in rooms reached by winter sun.
- Bamboo flooring is an efficient renewable resource with low VOC emissions.
- Ensure your carpet underlay is fully recyclable and the carpet has some natural fibre.

Hot water systems and energy supply

- Install a gas, solar or heat-pump hot water system for the greatest energy efficiency.
- A solar photovoltaic (PV) electricity system converts sunlight into electricity. This will eliminate electricity bills for the life of the system, and you can sell any excess electricity. The Australian government currently offers a range of rebates.

Garden and outdoor areas

- Position trees to maximise shade on your property.
- Local, native plants in well-mulched gardens will minimise the need for external watering.
- An automatic underground irrigation system will also minimise water use.
- Where practical, create porous surfaces outside the house to allow storm water to soak into the soil.
- Use recycled timber for outside decking.
- Install an external clothes line.
- Compost bins and worm farms encourage recycling of all food wastes.
- Use pervious materials such as rocks and pebbles for driveways and paths to slow water run-off into gutters and stormwater drains.

Rainwater tanks

- Install a rainwater tank to supply water for purposes such as toilet flushing, hot water, washing and garden irrigation (see **Figure 2.19**).

During construction

- Use renewable resources and materials with low VOCs.
- Work around established trees rather than cut them down.
- A site management plan will help control storm water and waste, minimise soil loss, and ensure materials are handled efficiently and that the site is clean and safe.
- **Recycle** construction waste where possible.
- Direct storm water to a stormwater drain, not to the sewerage system.

Source: Ed Hawkins

FIGURE 2.19 Rainwater tanks

3.2 Follow requirements to identify and report environmental hazards

To identify and report environmental hazards it's important to understand what the requirements are and who the appropriate authorities are in your state or territory regarding those hazards. Be observant while you work, stay focused and consider what might happen if something went wrong, and you should be able to identify hazards such as pollution, hazardous substances or materials or activities that may harm the environment. Reporting the hazards will involve a process that the relevant authority has made available on their website or in their documentation. This enables the authorities to check, monitor and remediate any hazard.

Environmental aspects

Estimates suggest that when renovating, demolishing or constructing a house from new, up to 200 tonnes of waste could be generated. Furthermore, approximately 80 per cent of this waste material could be reused by means of adopting thoughtful *reduce*, *reuse* and *recycle* processes and procedures (the three Rs of waste minimisation). In doing so, vast quantities of energy, water, resources and money could be saved.

Averaged nationally, each Australian produces, per year, in excess of one and a half tonnes of so-called 'waste', with approximately 40 per cent of this waste resulting directly from renovation, demolition and/or new construction activities. What's more, it would be safe to suggest that a further 50 per cent of this specifically discarded construction waste, if prudently managed, could be segregated for future recycling and reuse.

By recycling we are saving the Earth's resources, i.e. the raw materials of all the products we buy – the minerals, oil, petroleum, plants, soil and water. We also reduce our consumption of energy, limit pollution and lessen global warming by cutting down on the harvesting, construction, transportation and distribution of new products.

For example, let's look at aluminium. Recycling 1 tonne of aluminium saves 4 tonnes of bauxite and 700 kilograms of petroleum. It also prevents the associated emissions (which would include 35 kilograms of the toxic air pollutant aluminium fluoride) from entering our air. As an added bonus, it would reduce the load on the number of landfill sites.

Remember, one person's junk is another person's treasure, so skip the skip bin and start recycling and reusing your building materials.

The first step in a construction waste reduction strategy is good planning. Design should be based on standard sizes and materials should be ordered accurately. This approach can reduce the amount of waste material and increase profitability and economy for the builder and customer.

The next step in reducing the waste that goes into the skip bin and then into landfill is to follow the very simple, but very effective waste minimisation hierarchy process, as demonstrated in Figure 2.20.

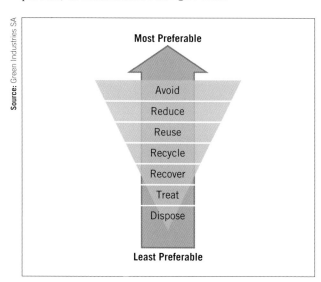

Source: Green Industries SA

FIGURE 2.20 Waste minimisation hierarchy diagram

Avoid

There are two ways that the builder can avoid waste:
- by correctly calculating the amount of material required for the job plus a small allowance for waste. This avoids over-ordering of material that is either thrown out at the end of the job due to damage or is moved to the next job or builder's yard
- by covering, securing and correctly storing material, which prevents building and construction materials, such as bricklayer's sand, being blown around the job site by the wind or washed into the stormwater system by a sudden downpour of rain. (This stormwater runoff in turn chokes the drainage pipes and pollutes the waterways.)

Correctly securing building waste such as insulation, sarking, general rubbish and other light-

weight products also prevents this waste from being blown or washed off the site. This helps to:
- minimise the unnecessary contamination of our water system
- reduce the need to replace building materials already purchased.

Reduce

The general trend today is to build from boundary to boundary. Over time the typical house in Australia has evolved from having three bedrooms, one bathroom and separate living areas, into a more open plan, including extra bedrooms, ensuites, rumpus, walk-in wardrobes and pantries, studies, specific theatre rooms and the like. All this has had obvious consequences, notably the overall increase in size of the modern home. In fact, according to the Australian Bureau of Statistics residential houses in Australia have grown in size by approximately 40 per cent (from 162.2 m² to 227.6 m²) since the mid-1980s. This simply equates to approximately 40 per cent more energy, resources and materials being required to construct a house today.

You can considerably reduce waste by planning carefully and sensibly. Determine exactly what it is that you require within your own residential home. Don't over indulge or fall into the trap of 'keeping up with the Joneses'; design to accommodate standard sizes of materials and utilise prefabricated frames and trusses as this is a proven method of reducing waste. Obviously, reducing the size of your construction will see a reduction in the use of energy, resources, materials and generated waste.

Reuse

Undertake a commitment to use, where possible, materials that have previously been used and reclaimed. The following items are the most common reclaimed materials and are therefore generally easy to locate:
- steel
- second-hand or **plantation timber**
- recycled or crushed concrete and masonry
- second-hand bricks
- soil and fill.

Remember there are many fittings and fixtures (such as doors, windows and kitchen cupboards) that are also available second hand. A list of recycled building products and/or suppliers can be sourced by an internet search (e.g. Yellow Pages – www.yellowpages.com.au/, under 'Building Materials – Secondhand').

Buying and incorporating recycled or reclaimed products into buildings increases the market for them, making it more viable for businesses to supply them as well as reducing the amount of building materials and products being turned into landfill.

Recycle

Some materials can be recycled or reclaimed directly into the same product for reuse. Others can be reconstituted into another usable product.

Unfortunately, recycling that requires reprocessing the material is not usually economically feasible, unless a facility using recycled resources is located near the material source.

Many construction waste materials that are still usable can be donated to non-profit organisations. This keeps the material out of the landfill and supports a good cause.

The most significant method of recycling construction waste is onsite separation. Initially, this will take some extra effort and training of onsite personnel; but once separation habits are established, onsite separation can be done at little or no additional cost.

Table 2.6 details the most common building products that can be recycled and suggests particular pathways for which they could be reused.

Also consider the food and drink containers used by onsite workers – one recycled aluminium can saves half a can of petroleum and 20 litres of water.

You can also take the materials yourself to your local recycling centres or transfer stations. Your local council will provide details of your nearest station.

Your local waste facility or landfill operator might also handle some recycled products so it might be worthwhile giving them a call.

Recover

Recovering resources involves making energy from waste (EfW). This is where waste material is used in the production of heat or electricity, or in some instances gas recovery, to be used as a fuel.

This is not something that can occur on a building site, but is something that building materials may be used for. It must be noted that hazardous materials are not suitable without prior treatment.

Treat

Treatment of hazardous materials, which include chemicals and products containing chemicals, is a vital step prior to safely disposing of waste into landfill sites. It is important that the waste going into these sites will not potentially impact or harm the environment.

From the builder's perspective this is something that cannot be undertaken on the site itself, but separating potentially hazardous material makes it easier for the material to be processed quickly and easily.

TABLE 2.6 Common building materials that can be recycled

Material	Reuse or recycling method
Concrete	Crushed and used for future concrete works Crushed and used as road base or fill
Bricks and tiles	Cleaned and used on future projects Cleaned and sold on Crushed and used as backfill or gravel
Steel	Retained for use on future projects Sold on as second-hand product Recycled into new products
Aluminium products	Retained for use on future projects Sold on as second-hand product Recycled into new products
Gypsum plasterboard	Retain large sheets for use on future projects Recycled into new plasterboard product Used as a soil conditioner or for composting in gardens
Timber/green waste	Retain large beams/plywood sheets, etc. for use on future projects Reprocessed into other timber building and landscaping products Untreated materials can be used as firewood Chipped and used as mulch either on site or at other projects
Plastics	Recycled into new products
Clean fill/soil	Utilised in onsite landscaping Stockpiled for use on future projects Sold on as a landscaping material
Glass	Retain large sheets for use on future projects Crushed and used as aggregate in concrete Recycled into new products
Carpet	Retain large pieces for use on future projects Sold on as second-hand product Used onsite to prevent erosion, dust mobilisation and weed invasion Natural fibre carpets can be shredded and used as fill in garden beds or composted

Dispose

All waste must come to its final destination – that is landfill. It is vital that the amount of building and demolition waste going into landfill is reduced as much as possible. Reducing the waste builders are sending to landfill will have the following benefits:

- reducing energy consumption by using salvaged and recycled products in new construction
- reducing the amount of natural or raw resources having to be used
- creating energy when waste is used in the EfW process
- preserve space in limited landfill sites
- reduce material disposal costs.

Planning for waste minimisation

The steps in Table 2.7 will assist in the initiation of a resource recovery program on a typical construction site.

It is recommended that readers become familiar with their relevant state or territory Environmental Protection Authority (EPA), by either contacting them directly or by visiting their respective website as detailed in Table 2.8.

Erosion and sediment control

The final area to be addressed in this section is erosion and sediment control during the construction phase of the building which prevents sand, soil, cement slurry

TABLE 2.7 Planning for waste minimisation

Step 1: Commit to responsible waste management	• Develop and implement a business waste minimisation policy • Involve all staff members in this process • Incorporate waste minimisation into position descriptions • Request subcontractors sign project waste minimisation plans • Incorporate waste minimisation into site induction programs • Provide positive feedback to staff successfully minimising waste
Step 2: Identify resource pathways	• Review materials utilised during construction and demolition activities • Assess volumes of resources currently going to waste • Assess material avoidance, reduction, re-use and recycling options • Determine avoidance, reduction, re-use and recycling methods of each material
Step 3: Develop a project waste minimisation plan	There are several keys to implementing an achievable project waste minimisation plan, including: **To ensure important field staff will follow the plan:** • *ensure that all staff understand what is to be achieved* • *keep staff informed of progress* • *ensure provision of up-to-date training.* **To plan specific projects:** • conduct a site assessment • prepare a project waste minimisation plan • determine resources for the project and pathways for excess and/or waste materials • determine the location of both waste and materials recovery stations • set targets and objectives for each project • require subcontractors to recycle and dispose of their own waste. **To understand plan options and limitations:** • identify the collection, sorting and resource utilisation options available • determine the suitability of each option for each job. **Establish a monitoring and reporting program to:** • quantify results and identify shortfalls • provide a record for comparison across work sites and methods • set targets and objectives • determine financial outcomes. **Focus on high potential materials and practices that:** • are high volume, can be readily separated and collected • have a viable economic value for recovery.
Step 4: Educate staff about waste minimisation plan	• Communicate with staff • Inform staff of the methods and objectives of maximising resource recovery • Provide staff with copies of the 'No Waste Project Plan' • Involve staff during development and review of the project waste minimisation plan
Step 5: Implement waste avoidance strategies	• Limit the types of resources being consumed on the work site • Design works to avoid waste generation • Use modular/prefabricated frames and fit-outs when possible • Request minimal packaging from material suppliers • Ensure that materials that will generate minimal waste are used

TABLE 2.8 Current state and territory regulating environmental protection authorities

State / territory	Regulating authority	Current EPA Act	Website and contact numbers
WA	Environmental Protection Authority Western Australia	*Environmental Protection Act 1986*	www.epa.wa.gov.au Contact: 08 6364 7000
VIC	Environment Protection Authority Victoria	*Environment Protection Act 1970*	www.epa.vic.gov.au Contact: 1300 372 842 (1300 EPA VIC)
QLD	Department of Environment and Science	*Environment Protection Act 1994*	https://www.ehp.qld.gov.au Contact: 13 74 68 (13 QGOV)
NSW	Office of Environment and Heritage	*Protection of the Environment Operations Act 1997*	www.environment.nsw.gov.au Contact: 131 555
SA	Environment Protection Authority South Australia	*Environment Protection Act 1993*	www.epa.sa.gov.au Contact: 1800 623 445
TAS	Environment Protection Authority Tasmania	*Environmental Management and Pollution Control Act 1994*	www.epa.tas.gov.au Contact: 03 6165 4599
ACT	ACT Commissioner for Sustainability and the Environment	*Commissioner for Sustainability and the Environment Act 1993*	www.environmentcommissioner.act.gov.au Contact: 02 6207 2626
NT	Northern Territory Environment Protection Authority	*Northern Territory Environment Protection Authority Act 2012*	www.ntepa.nt.gov.au/ Contact: 08 8924 4218
National	National Environment Protection Council	*National Environment Protection Council Act 1994*	www.environment.gov.au/protection/nepc

and other light-weight building materials from reaching stormwater drains, rivers and finally the ocean. By successfully containing this material on the site, builders and workers can achieve the following:

- reduced losses from material stockpiles
- reduced mud and dust being carried onto local roadways
- cleaner stormwater pipes (reducing localised flooding)
- cleaner waterways and healthier aquatic life
- reduced clean-up costs to the community
- fewer public complaints and less chance of fines.

To achieve these outcomes most local councils will ask the builder/developer to produce an Erosion and Sediment Control Plan (ESCP). This is a simple plan that outlines which control measures will be placed where on the site (see **Figure 2.21**).

Control measures

There are many control measures that may be used to prevent sand, soil and cement slurry from leaving the site. These include but are not limited to:

- sediment fences
- stabilised entry/exit points
- wash-out areas
- material stockpiles
- grass filter strips.

Sediment fence

The most efficient and widely used sediment control for domestic sites is a specially manufactured geotextile sediment fence.

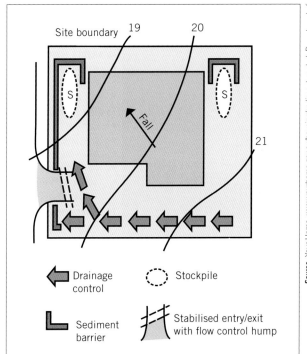

FIGURE 2.21 A simple erosion and sediment control plan

Sediment fences are placed on the low side of the block and have uphill returns at either end to prevent sediment flowing around them. These fences act like a sieve, trapping sediment but allowing water through. They are simple to construct and relatively inexpensive. It is important that these fences are regularly maintained and repaired as required (see **Figure 2.22**).

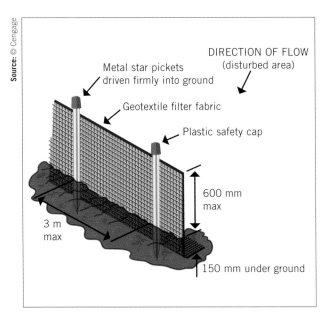

FIGURE 2.22 Sediment fence

allow mud and soil to be tracked onto the roadway. It is important that any soil that has inadvertently been tracked onto the road should be swept up immediately.

It is also important to contact the LGA to find out the requirements for length, width and acceptable material for the stabilised entry/exit point; please use Figure 2.23 as a guide only.

Wash out area

Another thing to remember is that no fine material should escape the site; this includes brick dust, paint and concrete slurry. Most builders set up a small contained area using either sediment fencing or hay bales to define the wash-out area for concrete trucks, wheelbarrows and the like, as well as containing the fine material while allowing clear water to drain away.

Material stockpiles

All loose material such as sand or soil should be stockpiled as per Figure 2.24.

Grass or vegetated filter strips

The final control method that can be used is grass filter strips. A 400-mm (minimum) wide strip of grass may be placed behind the kerb to act as a secondary measure outside the site. Strips of turf or vegetation are used to trap sediment, acting as a buffer zone between the site and the gutter. Refer to Figure 2.25 for more details.

Stabilised entry/exit point

It is important to protect the area where people and vehicles enter and leave the site, as all local government authorities (LGAs) have the power to hand out on-the-spot fines to builders and contractors who

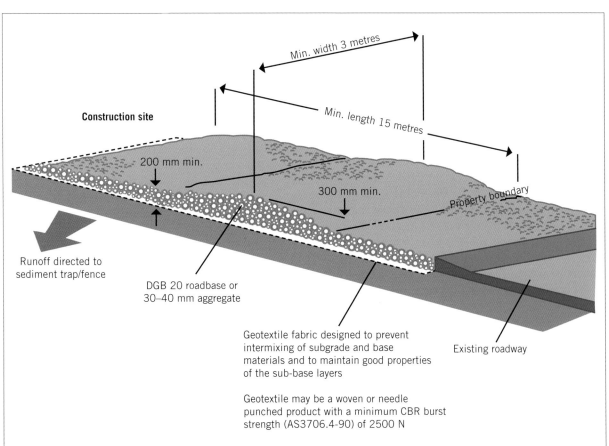

FIGURE 2.23 Stabilised entry/exit point

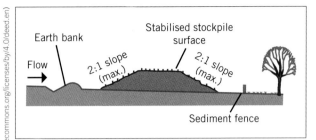

FIGURE 2.24 Stockpiled material

LEARNING TASK 2.5

PHOTOGRAPHY TASK – SEDIMENT AND EROSION CONTROL

Visit a job site that you are currently working on and photograph items on the site that are sediment and erosion control measures.

Use these as the basis for a short written report about the control measures.

3.3 Follow requirements to identify and report resource efficiency issues

To identify and report resource efficiency issues, it's important to understand who you need to report to and what available guidelines there are regarding using resources efficiently. When materials or tools are manufactured, installation instructions, user manuals or other technical documentation are produced, which will guide the user on the most efficient way to use that resource.

The business owner and the workers will need to pay attention to the daily activities on the job and identify where resources are being wasted or used inefficiently and then decide to change to a better way of working to gain more efficiency. There may be reporting requirements for the task that the organisation has in place, and these should be followed; however, it might just be as simple as having a discussion with the other workers in an informal manner to decide on a more efficient way of doing tasks and using materials. For example, it's noticed that excess water is being consumed when completing tasks because people are not turning taps off properly, or there are leaky tap washers. If this occurs in a large organisation there may be a process to report leaking tap washers. The same could occur for electricity use and lighting on a job. Always try and solve the resource efficiency issue at the lowest management level and as quickly as possible. The aim is to contribute to the conservation of resources and reduce our environmental impact to create a more sustainable future.

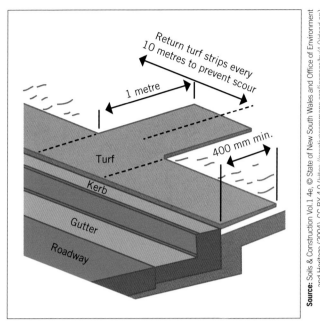

FIGURE 2.25 Grass filter strip

All new Australian housing needs to meet overall energy efficiency requirements as specified in the National Construction Code 2022. The Nationwide House Energy Rating Scheme (NatHERS) provides energy ratings for new dwellings to help create more energy efficient, resilient and comfortable homes for the future that cost less to run. The scheme can rate the energy performance for the whole home including major appliances, solar panels and batteries, in addition to the star rating for the building shell. All new homes across Australia will be required to meet the minimum energy efficiency rating of seven stars under the NatHERS. Achieving seven stars can make a big difference by considering things such as:

- involving a NatHERS assessor early in the planning of the construction
- orienting the house to suit the local climate
- prioritising window glazing and consider double or triple glazing
- zoning within the dwelling, including more internal doors
- prioritising insulation
- planning for fans
- checking and perhaps adjusting the colours of roofs and external walls
- obtaining advice on lighting options.

Source: nathers.gov.au ; https://www.nathers.gov.au/sites/default/files/2022-09/22726_Nathers_Newsletter.pdf

 COMPLETE WORKSHEET 3

SUMMARY

In Chapter 2 you have learnt how to work effectively and sustainably in the construction industry. We have covered the topics:

- work effectively in a team
- investigate construction industry employment pathways
- identify and follow environmental and resource efficiency requirements.
  You have also learnt that:
- The construction industry plays a significant role in the Australian economy and has strong links and influences with other parts of the Australian economy. The construction industry may be divided into parts or sectors, including residential construction, non-residential construction, engineering construction or public and private sectors.
- Employment conditions are the legal rights and obligations of both the employer and the worker and cover a range of topics such as job roles within the industry, industry trends, industrial relations, legal and ethical obligations as well as WHS issues.
- Individuals working in the building industry need to be responsible and be able to plan their work activities in a sequenced and structured manner.
- Workers need to be aware of the organisations that can assist them including unions (CFMMEU) and employer organisations, such as MBA and HIA.
- All workers undertaking work as a tradesperson must obtain a licence.
- Being part of a team and working within a team is important. Knowing the structure, roles and responsibilities for all team members and fitting within that organisational structure is important, as is working together with other team members to complete tasks regardless of any differences or disharmony that may arise.

- Individuals need to be aware of the different types of meetings common to the construction industry – from the simple to the complex and recognise the importance of these meetings.
- Identifying where you want to go within the building industry and finding a pathway forward is not always easy and talking with professionals may help.
- Training is critical – both onsite and offsite. Practical learning on the job and formal training are both necessary. Workers should gain an understanding of what training packages, qualifications and units of competency are available and understand how assessment is undertaken.
- Environmentally sustainable work practices are broken down into three parts: economic, environmental, and social sustainability. Sustainable housing development takes all these aspects into account.
- Waste minimisation is important and requires planning and implementation strategies to be effective.
- Workers need to understand how the work they are performing can affect the environment (e.g. erosion and sediment control) and learn how to minimise the impact on the environment including waterways.

REFERENCES AND FURTHER READING

Acknowledgement

Reproduction of the following Resource List references from DET, TAFE NSW C&T Division (Karl Dunkel, Program Manager, Housing and Furniture) and the Product Advisory Committee is acknowledged and appreciated.

Texts

Australian Bureau of Statistics (2012), *Year Book Australia, 2012 – Introduction*, retrieved from **www.abs.gov.au/ausstats/abs@.nsf/Lookup/1301.0Main+Features322012**

EPA South Australia (2004), *Handbook for pollution avoidance on commercial and residential building sites* (2nd edn), EPA South Australia, Adelaide

Fair Work Ombudsman (n.d.), *Unfair dismissal*, Australian Government, Canberra, **www.fairwork.gov.au search for 'unfair dismissal'**

Graff, D.M. & Molloy, C.J.S. (1986), *Tapping group power: a practical guide to working with groups in commerce and industry*, Synergy Systems, Dromana, Victoria

National Centre for Vocational Education Research (2001), *Skill trends in the building and construction trades*, National Centre for Vocational Education Research, Melbourne

NSW Department of Industrial Relations (1998), *Building and construction industry handbook*, NSW Department of Industrial Relations, Sydney

Mackay Regional Council (2012), Best practice guidelines for controlling stormwater pollution from building sites, O2 Environment + Engineering, Brisbane

PricewaterhouseCoopers (2013), *Productivity scorecard. Productivity in the construction industry*, PricewaterhouseCoopers, Sydney

Safe Work Australia (2013), *Guide for Preventing and Responding to Workplace Bullying*, search for this title on the internet

TAFE Commission/DET (1999 / 2000), *Certificate III in General Construction (Carpentry) Housing*, course notes (CARP series)

Towers, K. (2015), 'Docklands tower blaze exposes "dirty secret" of cheap cladding', *The Australian*, 29 April

Safework NSW (2022), *Pocket guide to construction: Safework NSW*, Sydney

Audio
visual resources
Title: 'Bullying and Harassment for Employees Training Video'

Running time: 14 minutes
Author: Channel 1 Creative Media
Available from: https://channel1.com.au

Web-based resources
Regulations/codes/laws

https://www.fairwork.gov.au, search for the Building and Construction General On-Site Award MA000020

www.australianapprenticeships.gov.au, Australian Apprenticeships

www.environment.gov.au, Department of Environment

https://www.fairwork.gov.au/ArticleDocuments/723/Introduction-to-the-national-employment-standards.pdf.aspx, Fair Work – Introduction to the National Employment Standards (PDF)

www.fairwork.gov.au, search for unfair dismissal

www.fairwork.gov.au, search for discrimination

www.safeworkaustralia.gov.au, search for bullying

www.yourhome.gov.au, Your Home – Australia's Guide to Environmentally Sustainable Homes

www.sustainablelivingguide.com.au/waste, Sustainable Living Guide

https://labourmarketinsights.gov.au, Job Outlook – An Australian Government Initiative

https://www.standards.org.au, Standards Australia

Industry organisation sites

www.ausleave.com.au, Australian Construction Industry – Long Service Leave

www.fairwork.gov.au, Fair Work Ombudsman

https://www.fwc.gov.au, Fair Work Commission

https://cg.cfmeu.org.au, CFMMEU – Construction, Forestry, Maritime, Mining and Energy Union

https://hia.com.au, Housing Industry Association

www.masterbuilders.com.au, Master Builders Australia

www.austehc.unimelb.edu.au, Technology in Australia 1788–1988

www.myskills.gov.au, MySkills – Australia's Directory of Training

ecobuy.org.au, Eco-Buy

www.aatinfo.com.au, Australian Apprenticeships Training Information Service

GET IT RIGHT

In the photo below, the workers are not working effectively.

Identify how the workers could be more productive. Provide reasoning for your answer.

Source: Richard Moran

_____

_____

_____

_____

_____

_____

 **WORKSHEET 1**

Student name: _____

Enrolment year: _____

Class code: _____

Competency name/Number: _____

To be completed by teachers

Student competent ☐

Student not yet competent ☐

Task: Read through the sections beginning at *Work effectively in a team* then complete the following questions.

1   What are some advantages to working affectively in a team?

_____

_____

2   What does it mean to participate as a team?

_____

_____

_____

3   There are three (3) main ways of dealing with the situation on a site. List them in your own words and identify which way is the best.

_____

_____

_____

_____

4   When working on a problem, what are some advantages of having a good discussion about the problem?

_____

_____

5   What number of employees on a work site do you need before you consider having a health and safety representative appointed on that work site to assist with safety?

_____

6   If disharmony occurs on a work site and there is conflict, when should it be dealt with?

_____

7   What are the keys to successful teamwork?

_____

_____

8 Is it a good idea to regularly rotate roles within a team? (Circle the correct answer)

Yes   No

9 Who is responsible for providing and maintaining equipment and safety in relation to work health and safety requirements? (Circle the correct answer)

PCBU/employer   Worker/employer

10 Who is responsible for *not* interfering with or misusing any item provided for the health, safety or welfare of people on a work site? (Circle the correct answer.)

PCBU/employer   Worker/employer

11 Who is responsible for providing information, instruction, training and supervision in relation to work health and safety? (Circle the correct answer.)

PCBU/employer   Worker/employer

 **WORKSHEET 2**

Student name: _____

Enrolment year: _____

Class code: _____

Competency name/Number: _____

Task: Read through the sections beginning at *Investigate construction industry employment pathways* then complete the following questions.

1   What is the difference (in simple terms) between *direct* and *indirect* labour?

Direct labour is _____

_____

Indirect labour is _____

_____

2   What four (4) parts of the Australian economy are strongly linked to the construction industry?

1   _____

2   _____

3   _____

4   _____

3   What are the three (3) main industry sectors?

1   _____

2   _____

3   _____

4   List seven (7) types of structure that fit into residential construction.

1   _____

2   _____

3   _____

4   _____

5   _____

6   _____

7   _____

5 List four (4) types of structure that fit into the non-residential or commercial sector.

1 _____

2 _____

3 _____

4 _____

6 List four (4) types of structure that fit into the engineering sector.

1 _____

2 _____

3 _____

4 _____

7 List two (2) types of buildings that would be found in the public or government sector.

1 _____

2 _____

8 List three (3) different trade categories or construction job roles.

1 _____

2 _____

3 _____

9 What is the meaning of the term 'industrial relations'?

Industrial relations is _____

_____

_____

10 What is the name of the national Act introduced in 2009 that allows all private sector employees to be covered by one national workplace relations system?

_____

11 Name three (3) areas of change covered by this new Act.

1 _____

2 _____

3 _____

12 Name four (4) of the 10 National Employment Standards (NES). Refer to Learning Task 2.2.

1 _____

2 _____

3 _____

4 _____

13 What is an enterprise agreement?

An enterprise agreement is _____

_____

_____

14 What is the legal definition of unfair dismissal?

Unfair dismissal is _____

_____

_____

15 If someone believes that they have been unfairly dismissed, within how many days must they notify the Fair Work Commission?

16 What is the full name of the program through which corporations and businesses provide employees with counselling and psychological support?

_____

17 List six (6) items that are considered *adverse actions* under the *Fair Work Act 2009*.

1 _____

2 _____

3 _____

4 _____

5 _____

6 _____

18 What does the acronym EEO stand for?

E _____ E _____ O _____

19 In what way is bullying a risk to health and safety?

_____

_____

20 What is harassment?

Harassment is _____

_____

_____

WORKSHEET 3

Student name: _____

Enrolment year: _____

Class code: _____

Competency name/Number: _____

Task: Read through the sections beginning at *Identify and follow environmental and resource efficiency requirements* then complete the following questions.

1   List four (4) technological changes that have occurred in the construction industry.

1   _____

2   _____

3   _____

4   _____

2   List three (3) materials that have been affected by changing technology.

1   _____

2   _____

3   _____

3   When did timber-framed cottages clad in asbestos cement start to be built in Australia? (Circle the correct answer.)

1   about 1945   2   about 1960   3   about 1975   4   about 1950

5   about 1965   6   about 1955   7   about 1970

4   What do the following building computer software acronyms mean?

1   CAD

C _____   A _____   D _____

2   BIM

B _____   I _____   M _____

5   What does quality control mean?

_____

_____

6   What does quality assurance mean?

_____

_____

7   Which Australian and New Zealand Standard covers quality assurance?

_____

8   Name the eight (8) key principles of quality control.

1   _____

2   _____

3   _____

4   _____

5   _____

6   _____

7   _____

8   _____

9   What are the six (6) key terms of quality control?

1   _____

2   _____

3   _____

4   _____

5   _____

6   _____

10   Which Australian Standard should be used by carpenters who are erecting wall frames?

AS   _____

11   List three (3) items that are included as part of site amenities.

1   _____

2   _____

3   _____

12   List three (3) items that are included as part of emergency procedures.

1   _____

2   _____

3   _____

PERFORMANCE OF SKILLS EVIDENCE

Student name: _____

Enrolment year: _____

Class code: _____

Competency name/Number: _____

Task: Working with your teacher, prepare to demonstrate your skills by completing the following task. To demonstrate competency, a candidate must satisfy all the elements, performance criteria and foundation skills of this unit.

Instructions:

1   Work effectively as a member of a team to plan and perform a construction task chosen by your teacher.

2   Work with members of a team to review the team's purpose, roles, responsibilities, goals, plans and objectives.

3   List own existing skills and the additional skills required for a selected tradesperson or skilled operator role in the construction industry.

4   Identify environmental and resource efficiency requirements that apply to entry level roles in the construction industry.

5   Prepare basic reports on:
    − an environmental hazard
    − a resource efficiency issue.

3 PLAN AND ORGANISE WORK

This chapter covers the following topics from the competency CPCCOM1013 Plan and organise work:

- determine and plan basic work task activities
- organise performance of basic work task.

The first element of working effectively on any construction site is being able to work to your full potential; that is, maximum productivity. To do this you need to plan all phases or steps of the work task to be undertaken.

This chapter addresses the skills required to plan work tasks and to maximise personal productivity on a construction site.

1. Determine and plan basic work task activities

Your ability to plan and organise is an extremely important life skill. Whether you are planning and organising a holiday, a party, a shopping trip or a task at work, your life will probably be more satisfying, more fulfilling, more successful and far less stressful if things turn out the way you want them to.

In your private life, if you are not a good planner and organiser, it probably does not matter much if you take the wrong clothes on holiday or that you forgot to buy milk at the supermarket. If you are a poor planner and organiser in your private life you mainly cause problems for yourself.

But, if you are a poor planner and organiser at work you will cause problems for yourself and others (see Figure 3.1) and if you cause enough problems for others at work, you may find it difficult to keep a job.

Source: Shutterstock.com/Juice Flair

FIGURE 3.1 Result of poor planning

1.1 Determine work task outcomes and other requirements

In the construction industry it is important that everybody is working efficiently. This means that all workers need to be good at planning and organising. The key to understanding the outcomes

and requirements for a work task will be identifying the task and understanding the purpose for that task based on the available information or documentation. You may need to identify the stakeholders, consult, or collaborate with others, taking into account all available information, codes of practice, standard operating procedures, plans or schedules and then ultimately you will need to monitor and evaluate their effectiveness as you work to achieve your outcomes. This is all part of being organised.

Good work organisation will:

- prevent accidents and injuries
- avoid wasted time and energy and unnecessary work
- avoid wasted materials, tools and equipment
- result in skilful and productive work
- result in better profits for the building company.

Prevent accidents and injuries

Working safely as an individual or part of a team involves taking responsibility for your own actions and taking time to check for hazards in the work environment, as well as identifying hazards in the work process. Many accidents and injuries may have easily been prevented if someone had thought through what was needed before starting the task.

It is common for a construction business to have policies in place that are used to describe the minimum standard in relation to planning and organisation for works.

 Look at **Figure 3.1** – if the person had only taken the time to think through what was needed, he would not be doing this task in a way that is dangerous.

Working safely is also, ultimately, working efficiently. Work injuries lead to lost work time, higher costs and higher prices quoted to complete jobs. Furthermore, working unsafely for an imaginary increase in your pay packet will never make up for your or another worker's damaged back, lost sight or hearing, damaged lungs or hands. More money in your pay packet cannot make up for you having to leave the building industry to work in a lower-paid job because your injuries make you unfit for building work.

Avoid wasted time and energy

Work in the building industry can be physically demanding, and wasting time and energy means less work is completed in a set timeframe. A good example of this is the age-old saying that you may often hear on the worksite: 'measure twice, and cut once!' Less work completed in a set timeframe means less money being earned. Building workers do not like having to do unnecessary work because someone else has not done their job properly. No one likes to have to try to fix or work around other people's mistakes.

All materials, power and energy come from somewhere and if you are careful to avoid unnecessary waste you will save money and help the environment.

Avoid wasted materials, tools or equipment

Wasted materials, tools and equipment are a cost that must come out of the price of the job. The more money spent on replacing materials, and replacing or repairing lost or damaged tools and equipment, means either less money for wages and company profits or higher prices quoted to do jobs. The higher the prices quoted to do jobs, the fewer jobs will be won and the less work will be available.

Skilful and productive workmanship

Good planning and preparation will enable you to work skilfully and productively. Confident planning helps you produce a high-quality product that is good value for money for your customers. It also increases the reputation and success of your business, which will ideally generate more work and contribute to financial success.

Alternatively, poor planning may result in poor quality of the finished product, that may affect the appearance as well as the structural quality of the building.

Better profits for the building company

By having a team of people who can plan and organise work correctly, the building company will be able to make good profits without having to quote higher prices to do the work. Ultimately, when the builder is happy the workers and the clients will be happy.

Mistakes can be expensive to fix and can cause delays on a job. Too many mistakes may send a builder broke, which could lead to staff being out of work and contractors having less work available.

In summary, by planning and organising a task you have been given on a construction site, you can:

- eliminate or reduce hazards to your safety
- reduce future problems by making important decisions in advance
- get the cooperation of other workers when you inform them about the task
- save time, materials, and tools
- have materials, tools, and equipment ready when and where you need them
- easily adapt to overcome problems you did not foresee.

By planning your tasks, you can make your work more satisfying, more fulfilling, more successful and less stressful.

1.2 Break the work task into its component tasks

Finding the component tasks within a major task involves an understanding of the overall objectives of your specific trade work and how that fits into the overall project. Tasks are often broken down into trade areas similar to how

a schedule for a construction job contents page is listed in trade areas. From the broad task list of things to be done to achieve a construction job it's important to break items down into manageable chunks so the job plan and timings can be managed in a more controllable way.

Identify the main objective and any specific requirements and then break down the tasks into a logical progression of how the work could be performed by identifying key steps or milestones. These sub-tasks could be broken down into different skill sets, or different material requirements. These should be sequenced in the overall project so they can be monitored and evaluated to keep the work flowing in a logical progression without having to go back and undo things because they are out of sequence.

The different resources that are required to achieve completion of the sub-tasks and then the main task need to be planned in such a way that as much as possible of that resource can be used while it is available. For example, when concreters are onsite it may be possible that while they're waiting for concrete to cure they can prepare paths or driveways in another location on the site rather than losing them to someone else's job while they wait for their concrete to cure.

Before you begin any task, it is essential that you make sure that you know exactly what you are required to do. Your supervisor may give you spoken instructions, written instructions, drawings, or a combination of all three.

While your supervisor is doing this, it is important for you to be patient and to check that you have clearly understood exactly what is required. Use your communication skills to check that you know, not just think you know, what the task is that you need to carry out. Always seek and offer feedback. You will learn more about developing your communication skills in Chapter 4.

It is very easy, through eagerness, enthusiasm or impatience to get on with the task, to not listen properly, to ignore what is being said, or to dismiss information that does not seem important to you. Many tasks have been botched right from the start because crucial information has not been fully understood. Because it is difficult to remember a lot of spoken instructions, a number of different sizes or dimensions, or a number of different steps in a procedure, always be prepared to write down what is required.

For simple tasks, it is a good idea to jot down key points (Figure 3.2) or crucial information to keep you focused on what is required and to stop you making mistakes or errors. For highly dangerous or complex tasks, a work method statement (WMS) will be required. If you are confused about a simple task, ask your supervisor specific questions, and jot down the answers so that you don't forget. Keep asking questions until you understand, otherwise you might make a mistake. Repeating your instructions back to your supervisor will help reduce any new misunderstandings.

FIGURE 3.2 Writing down key points

For more complex tasks, you may need to consider breaking up the task into the following categories:

- work area or work environment
- materials required
- tools and equipment required
- carrying out the task
- cleaning up, packing up and waste disposal
- reporting and reviewing.

1.3 Assess the component tasks to determine what needs to be done and how it is to be done

As stated before, when you are carrying out a task make sure that you have a clear idea in mind about the sequence of actions you need to follow.

This sequence may be best set out in a SWMS, a simple list of steps or a chart.

In the building industry, construction follows a logical order. For example, the construction of an entire building can be broken down into:

- **stages:** each of these stages can be broken down into a series of steps
- **steps:** each of these steps can be broken down into a series of tasks
- **tasks:** each of these tasks can be broken down into a series of actions
- **actions**.

Some stages, steps and tasks must be completed in preparation before the next stage, step or task can begin. These critical items must be done in the correct sequence and timeframe otherwise the entire construction process will stop.

The three examples that follow illustrate how important it is to follow the correct sequence and timeframe for a job.

Example 1

A building company needs to start the *first-floor stage* of construction, which involves placing the concrete

floor slab. The concreters on the job must follow the sequential steps:

- building the formwork
- tying the steel reinforcement
- pouring the concrete
- stripping and removing the formwork.

Building the formwork has a number of tasks, which in turn are composed of a number of actions that must be done before the concreters can begin the next step of *tying the steel reinforcement*. If the formwork is not built properly before the steel is tied in position there will be a serious problem when the concrete is being poured (see Figures 3.3 and 3.4).

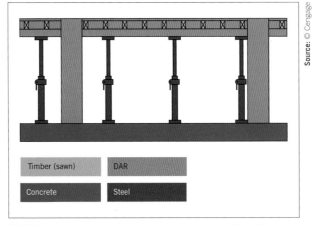

FIGURE 3.3 Good and accurate formwork – resulting in good and accurate concrete finish

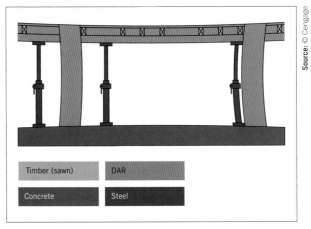

FIGURE 3.4 Poor formwork – resulting formwork collapses, concrete is wasted

So to complete the step of *building the formwork*, the concreters need to follow a sequence of carefully laid out tasks.

Complete TASK sequence:

- **Task 1:** Set out falsework and bearers and joists accurately.
- **Task 2:** Completely form deck.
- **Task 3:** Completely prop and brace the deck.
- **Task 4:** Fit edge forms to form deck

- Task 5: Brace the edge forms.
- Task 6: Inspect and check formwork before starting next step.
  Incomplete TASK sequence:
- Task 1: Completely form deck.
- Task 2: Temporarily prop the deck.
- Task 3: Fit edge forms.

Example 2

Bricklayers install damp-proof courses (action), flashings and brick ties as they build a wall (the task), because these things are impossible to install after the wall is finished (see Figure 3.5). If this is not done, the wall will have to be demolished and built again.

Source: Ed Hawkins

FIGURE 3.5 Preparing the wall frame ready for the brick veneer

Example 3

When a house is being built, the plumber and electrician must run their pipes and cables through the walls before the plasterboard is put on (referred to as rough-in stage) (see Figure 3.6). If the plasterboard (lining and fix-out stage) goes on first it will have to be removed, the pipes and cables installed and then the plasterboard replaced.

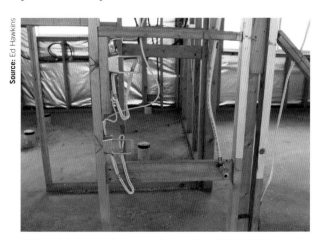

Source: Ed Hawkins

FIGURE 3.6 Rough-in the services before fitting linings

1.4 Estimate the time and the number of personnel required to complete each component task

Construction processes rarely go according to plan. This does not mean that a plan is therefore unnecessary. A plan is important because it tells when we must start and when we must finish a task. It also identifies key points along the way. A plan gives you:

- a predicted timeframe for the job to be completed
- the best chance of success and minimise any unexpected outcomes
- increases the safety aspects of the job
- a reference point for workers
- the number of people that should be engaged to carry out the job.

A good planner is someone who has a time schedule set up; this may be a computer-based *Gantt chart* (see Figure 2.9), *flow chart* or *critical path chart*. These charts allow a good planner to be flexible enough to change when circumstances require it, but still keep the project on track.

These examples show that planning the work steps is of vital importance to the correct completion of the whole project. No one wants to waste their time or their money or be unemployed with no money in their pocket; so, it is important that each person on a construction site plans and organises the job prior to starting work.

Estimating the time and number of personnel resources required to complete a task requires some skill and knowledge and a competent person will need to check the timings and numbers before decisions are made that affect a project. Prior experience is invaluable. However, there are resources available, and specialist advisers who can give good estimates of times and resources based on the type of job and the location of the job. Performing this task on a really large job may require a specialist like a quantity surveyor or project supervisor. Industry organisations publish different job task typical completion times and different job task resource requirements from time to time and these can assist you as you work to estimate the time and number of personnel required for your job.

 COMPLETE WORKSHEET 1

1.5 Identify the tools and equipment required, including personal protective equipment (PPE) for each stage of the task

After determining the materials required, you will have an idea of the tools and equipment you will require to undertake the task. You should list the tools and equipment necessary to do the task in the order in which you will use them (see Table 3.1).

Source: Shutterstock.com/Ground Picture

FIGURE 3.7 Transporting materials to and around the site

TABLE 3.1 Checklist for slab placement

TOOL CHECKLIST	
Project: Construct a concrete slab for a small shed	
Check	**Prepare foundation material**
✓	shovels
✓	pick
✓	mattock
✓	spud bar
✓	spade
✓	wheelbarrow
✓	8-metre tape measure
✓	dumpy or laser level
	Form up for the slab
✓	claw hammer
✓	nail bag
✓	8-metre tape measure
✓	four-fold rule
✓	pencil
✓	string lines
✓	hand saw
✓	circular saw
✓	lump hammer
✓	sledge hammer
✓	75 mm nails
✓	dumpy or laser level
	Place the reinforcement
✓	bolt cutters
✓	wire snips
	Pour and finish the concrete
✓	short-handled square mouth shovel
✓	immersion vibrator
✓	aluminium screed
✓	wheelbarrows
✓	wooden float
✓	steel trowel
✓	edging tool
✓	hose
✓	cleaning brush

Some specialist equipment may need to be hired; therefore, you must remember to order special equipment in time for it to be delivered to the site so that it is there when you need it.

After you have prepared your list of tools and equipment, you must plan how you are going to transport them to the place where you are going to use them.

First, you will probably need to carry the tools from the truck or from another part of the site. To do this you can use a tool carry-all, toolbox or wheelbarrow (Figure 3.7).

Second, you may have to bring the tools from another site or from home. Most tradespeople store their equipment at home in a shed or garage and take what is required to the job only when it is required. It would be impractical for most tradespeople to carry every tool they own to every job they attend. They would need a huge truck to cart around tools and equipment they did not need at the time, which would be not only a waste of time and effort but also a waste of money.

Therefore, it is important to plan what equipment you will need each day. If you do not take a tool to the job when you need it, you will either have to go and buy it or do another task that you do have the tools for.

As with materials, do not make the mistake of simply picking up the tools and equipment and heading for the job location. You should check first that the tools are fitted with the correct bits, drills or blades as well as they are working properly. Make sure that the appropriate personal protective equipment (PPE) is also at hand.

If time has elapsed since you delivered the materials to the location where you will be working, make sure that the path is still clear for you to carry the tools, equipment and PPE safely. Building sites are busy places and other workers may shift gear and move things around. By doing this they may unwittingly put things in your way.

In all cases PPE should be the last line of defence for safety on a job. If you can eliminate a hazard any way other than having to use PPE then do so. However, if you need to use PPE then make sure you follow the guidelines in any one of a number of resource documents such as codes of practice, user manuals, Australian Standards or even advice from competent technical advisers. Many tasks already have standard operating procedures that are available based on expert evidence from people doing similar tasks and it's good to learn from this information. Most state/territory work safety regulators will produce regular newsletters that discuss accidents or near misses that occur each month and give advice on what to do if you are doing similar tasks so that nothing goes wrong. Make sure that as you identify tools and equipment required for your task, you have also planned for any safety equipment or methods needed to complete the tasks.

LEARNING TASK 3.1

GROUP ACTIVITY – MATERIALS LISTING

In your group, find a plan for a simple carport on the internet, or ask your teacher to provide you with a simple shed or pergola plan. Write a materials list for the plans and then complete a tools and equipment list.

1.6 Plan the sequence of the component tasks in a logical order and to maximise efficient use of resources

When planning the sequence of the tasks in a logical order to maximise efficient use of resources, it's important to identify any tasks that require other tasks to be done beforehand. You need to know the logical order of how work can proceed. For example, you can't put the linings on the wall inside the building until all the framing, electrical and plumbing rough-in work is completed. Knowing these dependencies will assist you in identifying which ones are the most critical. Keeping the most critical items in mind you also need to consider availability of resources so that you don't hold up the overall workflow of the job. If possible, you may have a number of tasks occurring simultaneously and this may help the overall time flow of the job. However, there may be times where not many people are onsite, for example, during asbestos removal tasks, in which case the whole job might have to be given over to this one task. Once these important aspects are known the job can be planned for much more easily.

There needs to be room in your plan for flexibility and adjustment as unforeseen situations occur and in all cases you need to know which tasks are the most important critical tasks that will extend the final finish timeline of the job if they are delayed. There are several critical items in a construction that can occur at the same time and a small delay in any one of those items may not delay the overall completion date of the project. For example, the concreter may be pouring paths on the job while other trades are doing work inside. There are easy-to-use computer programs that make planning and sequencing of the tasks for construction quite easy and allow for easy altering of tasks as the construction proceeds. One type of chart that gives a good visual representation of a schedule of work on a job is a Gantt chart. You can create a simple Gantt chart by:

1 listing the tasks that need to be done to complete a project
2 list next to each task how long each task is expected to take

3 list which task is critical and that must be done before another task can start
4 list the resources you need for each of the tasks.

With a Gantt chart the columns (vertical columns of information) on the left contain each of the four items listed above. In the first column the tasks are listed on each row. Subsequent columns could contain timeframe for task, preceding task number, and resources for each task. These last three items may or may not be visible on the chart that you stick on the work site wall. Refer to Figure 2.9 for an example. The complexity of the overall project will determine how many rows of tasks, or steps, there will be. Across the top of the chart from column five onwards is where the dates are listed. You can enter these four pieces of information onto a chart using columns on the left and dates across the top and then simply block out on the dates for the item that is planned on the date that it is to occur.

LEARNING TASK 3.2

GROUP EXERCISE – GANTT CHART

Reuse either the plans for the carport or the small shed or pergola plans that you used in Learning Task 3.1. In your group, and with the assistance of your teacher, create a simple Gantt chart for building the carport, small shed or pergola. Refer to blank Gantt chart in Worksheet 2.

 COMPLETE WORKSHEET 2

1.7 Prepare a written work plan and a list of resources required to complete the overall work task

It is important to list the materials necessary to perform a task and the order in which you will use them.

A good motto to remember is 'just in time'. All materials need to be on the site 'just in time' to be incorporated into the building. The materials must be on the site when you need them.

This motto also demonstrates that while it is important to have materials delivered just before they are needed, it is also important not to have materials delivered too early to the job. This is because the site can become overcrowded with materials that are not needed and that get in the way. These materials may be damaged, lost, stolen, vandalised or simply slow the entire job down because they may need to be moved all the time so other people can get on with their work.

Therefore, you must have your materials on the site, ready to be used, in the sequence in which you will need to use them, *just in time* for you to use them.

An example of this is the list of materials to construct a small shed (see Figures 3.8 and 3.9).

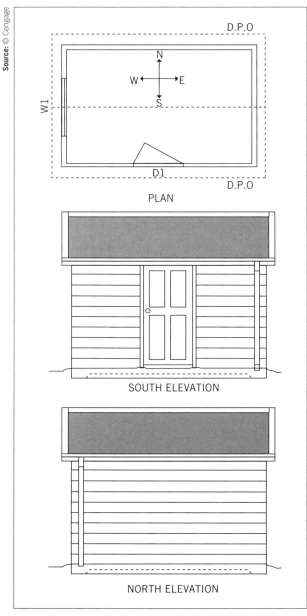

FIGURE 3.8 Plan, south elevation and north elevation of a small shed

Materials list for a small shed

Materials are required in the following order:
- formwork for the slab
- fill sand for slab preparation
- steel reinforcement
- concrete for the slab
- timber framing for the walls, ceiling and roof
- Colorbond® gutters and downpipes
- corrugated Colorbond® roofing
- Colorbond® ridge and barge cappings
- fibre-cement sheeting for eaves

- solid timber entry door and furniture
- aluminium sliding window
- external wall cladding
- external mouldings
- internal wall and ceiling lining
- internal fixings
- exterior and interior paint.

Once the materials are delivered to the site, you will need to consider where the materials are to be stored and used (Figure 3.10). Prior to moving the materials into location, inspect your route to make sure that the path is clear. Extra care may be needed if the material is large, heavy or bulky, so check the route you will be using. There may be stairs, long narrow corridors, doorways, ramps, planks, scaffolding, tight spaces, trenches or even mud. These obstructions or site conditions can place unnecessary strain on your body that may cause injury to yourself or others.

When you reach the place where you will be carrying out the task, make sure you locate your materials in a safe and secure position.

It is extremely important to stack or store the materials in the most appropriate way possible; the best way to stack each type of material is the way each material was delivered from the manufacturer. The manufacturer's stacking or storage method is intended to maintain the material in the best possible condition in the safest possible way, so when stacking or storing the materials, simply repeat the manufacturer's stacking or storage methods (see also Chapter 7 for further discussion of stacking and storage of materials).

Again, a Gantt chart can be used to show when resources are required to fit in with your work plan. This may be completed for a simple job in a simple way or you may have a very complex chart with colour coding and symbols for very large jobs. There are some paid computer programs or free programs that create Gantt charts. It's important to remember that any planning needs to work for your job and your organisation and must aid and assist you in planning your resources so that the overall work tasks can be kept on schedule.

LEARNING TASK 3.3

GROUP EXERCISE – GANTT CHART 2

Continuing on from Learning Task 3.2 with your group, use the Gantt chart that you created to develop a list of the *stages in construction* required to complete the job.

Select one of the stages and write down all of the steps for that particular stage.

FIGURE 3.9 East elevation and west elevation of a small shed

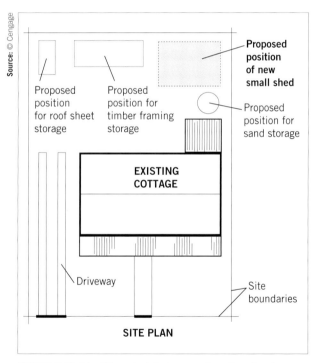

FIGURE 3.10 Identify material placement on the site plan

2. Organise performance of basic work task

2.1 Work with team members to review the work plan, schedule the work, allocate roles and responsibilities, and review work health and safety requirements

Now that you have your materials, tools, equipment and PPE on the spot where you will be working, it is time to take care to follow your instructions.

To complete the task properly you must follow a safe and efficient sequence of work. Make sure that you have your *safe work method statement* (SWMS) if one is required for the task.

If your instructions were spoken, think back to when you received them and be clear about what exactly you were instructed to do.

If your instructions were written, or came in the form of a drawing, make sure you read or view them again. Do not make the mistake of relying on your memory. Misremembering instructions, and not asking for clarification, can lead to expensive errors. Never be too lazy to double-check (Figure 3.11). Double-checking may help you to keep your job, or save a lot of money.

FIGURE 3.11 Double-check details before you start

Finally, it is time to do the task; you must follow a logical sequence of work to complete the task efficiently. If you have a SWMS, follow the steps outlined in the procedure. It is important that you do everything in the correct order and that you do not rush the task. By rushing the task you are likely to make errors. Speed is good but only if the task is done correctly the first time. Having to go back and redo tasks, or demolish, is more costly than taking a bit more time in doing the job. As mentioned earlier, a good rule to remember when setting out materials for cutting is 'measure twice, cut once'.

Periodically take the time to keep your work area clean and clear of debris. Taking a few moments to get rid of waste and to sweep up will save you or another worker from having an accident.

While you are working, you will not be using all your tools and equipment all the time. Therefore:

- make sure that the tools and equipment not being used are safely located
- do not block walkways
- never stand up, stack or lean tools or equipment in such a way that they can slip, slide or fall over and injure another worker, or be damaged.

Problem solving

Working on a construction site is not like working in a factory. Construction workers are not just assembling parts that have been manufactured somewhere else. A building is like a huge jigsaw puzzle, in which the people who are trying to put the puzzle together are also making the parts as they go along (Figure 3.12).

Source: Shutterstock.com/zhengzaishuru

FIGURE 3.12 A building is like a huge jigsaw puzzle

To an observer it may look as though all the workers are part of a mass production process, but this is not the case. The building that the observer is watching being constructed is probably of an original design, which has never been constructed before and is being erected on that particular site for the first time.

Therefore, although the workers have a great deal of experience and may be repeating tasks they have done many times before, they are doing that particular task on that particular spot on that particular site for the first time. In doing so they are dealing with different people and different materials in different situations. This means that challenges and difficulties can arise. How the workers deal with these challenges and difficulties will depend on their individual talent, training, knowledge and experience. In general, the

most successful construction workers have highly developed problem-solving skills.

In the construction process, it is important to solve each construction problem as it arises and to not allow problems to accumulate at the end of the job when it is too late to do anything about them. The accumulation of errors on a job can lead to structural failure, faulty finishes on the building and accidents and injuries to other workers.

Expect difficulties to arise and be prepared to overcome them. Common problems that arise can be broadly classed as follows.

- *Problems that arise with the procedure you had intended to follow.* Because of work going on nearby you may not be able to follow the procedure you had initially planned. You will have to change your sequence of work to accommodate what is going on near you or around you.
- *Problems that arise with the materials you had intended to use.* The materials may not be available in the order you need or at the time you need them. Or some of the materials may be incorrect or faulty and you will have to arrange for replacement material but still keep the work progressing. This also will require you to change your sequence of work.
- *Problems that arise with the weather.* This may affect your task sequence and your materials. This would require a radical replanning and re-organisation of the sequence of the task that you are doing.

Another problem to consider is if the environment you are working in is safe for you and the people working with you? Is it secure from other workers and the general public?

By looking at the work area and asking questions it may become apparent that the area needs to be secured in a way that protects others from danger. Some methods of protecting others are as follows.

Barricades

Barricades include any physical barriers placed to prevent entry, and to signify that a danger exists. They may be as strong as a timber or steel hoarding, or as light-weight as a coloured plastic strip or tape.

Hoardings

There are two main types of **hoardings**:

- type 'A' (fence-type hoardings)
- type 'B' (overhead-type hoardings).

Fence-type hoardings are used for residential, commercial or industrial sites where there is a minimal risk of objects falling on the heads of passers-by.

Fence-type hoardings may be constructed of timber and sheeted with plywood or galvanised steel posts and chain wire (Figure 3.13). They range in height from 1.8 m for most residential sites up to 2.1 m for commercial or industrial sites. Overhead-type

FIGURE 3.13 Typical chain wire fence type – Type A hoarding

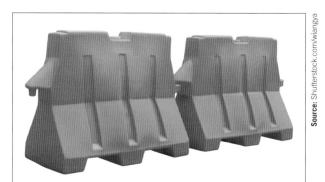

FIGURE 3.14 Plastic water-filled barrier

FIGURE 3.15 Board and trestle barricade

hoardings are primarily used on commercial or multi-storey construction sites where there is a high risk of objects falling on passers-by.

Barricade tapes

Barricade tape is a roll of coloured polyethylene plastic tape printed with a warning message that can be quickly run out around large areas. The tape can be easily nailed, stapled, tied or wrapped around posts, poles, stakes, railings or any convenient support, then rolled up again at the end of the job for later reuse. An example is the blue and white tape used by police to close off a crime scene.

Tube and fittings

These may be placed around open trenches or pits to prevent people from falling in.

Plastic water-filled barriers

These are hollow plastic interlocking units; once placed they are filled with water to create an interlocking barrier to prevent the entry of vehicles (Figure 3.14).

Board and trestles

Board and trestles used in building and construction are similar to those found on the side of the road where roadworks are underway – i.e. a horizontal timber board painted with black and yellow angled stripes held up at either end by a metal A-frame trestle (Figure 3.15).

Bollards

Bollards are solid vertical barriers less than 1 m high that visually or physically deter the entry of vehicles into an area of free pedestrian movement (similar to ram-raid posts).

Signage

Signage placed in significant locations around the perimeter of the work area informs others of the hazards, dangers and the PPE required to be worn in the work area. For more information about signage refer to Chapter 1 and Chapter 4.

Housekeeping

It is important that the work area is clean and clear of any hazards and obstructions even before starting to work. In some instances the work area may need to be cleared, with all rubbish and debris removed and work material neatly stacked or stored in another location to allow for easy access to the work space.

Cleaning up correctly is just as important a part of doing the task as planning and organising, but cleaning up does not just mean sweeping the floor. Cleaning up involves:

- collecting debris and waste materials and moving them to the waste storage area
- placing recyclable materials at the correct storage and collection area
- returning unused materials to the storage area and stacking or storing them neatly.

Packing up

Packing up means leaving the work area clean, safe and secure on completion of the task (Figure 3.16). This includes the following operations:

- Tools and equipment must be cleaned, maintained and stored correctly (Figure 3.17).
- Hire equipment items must be counted, accessories noted and then all equipment located where it can be picked up by the hire company with any faulty parts noted.
- Company-owned tools and equipment must be checked and maintained.
- Timber handles on tools should be wiped with raw linseed oil.
- Metal blades on tools should be wiped with mineral oil.
- Equipment with their own storage cases must be returned to their cases together with blades, attachments and accessories. Any damaged or faulty equipment must be reported so that it can be repaired.
- Safety signs and barricades must be removed and stored for later use.

FIGURE 3.16 Clean the work area daily

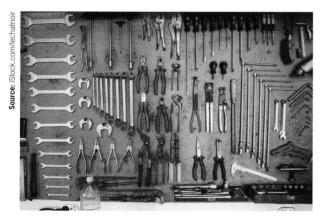

FIGURE 3.17 Clean, maintain and store tools and equipment

GREEN TIP

Remember that if you sort your waste into the different categories that your local council will charge you for, then you will not be charged the highest fee for waste. You will also be helping the environment by having your waste sorted appropriately for further processing.

Waste disposal

Waste disposal for many on the job site simply involves getting rid of the waste: some dispose of it correctly, while others are dodgy and dump the waste material on the sides of roads, in parks or even waterways, simply because they consider it too costly or are too lazy to dispose of it correctly.

Waste is defined as 'anything unused, unproductive, or not properly utilised' (*Macquarie Dictionary*).

A 2012 report from the Australian Government Department of Sustainability, Environment, Water, Population and Communities (DSEWPaC) titled *Australian Waste Definitions – Defining waste related terms by jurisdiction in Australia* defines waste as:

1 a substance or objects that:
   a is proposed to be disposed of; or
   b is disposed of; or
   c is required by a law of the Commonwealth, a State or a Territory to be disposed of
2 waste, in relation to a product, means waste associated with the product after it is disposed of.

Source: Department of Sustainability, Environment, Water, Population and Communities, 2012, © Commonwealth of Australia 2015. CC BY 4.0 (https://creativecommons.org/licenses/by/4.0/legalcode)

Note: this is the Australian Government's definition of waste; as well as the federal definition, the report also records each state and territory's definition.

Construction waste in Australia

In 2010–11, 18.2 million tonnes of construction and demolition waste was generated nationally, of which:

- 6.25 million tonnes (34 per cent) were disposed of to landfill
- 11.89 million tonnes (65 per cent) were recycled
- 0.06 million tonnes (0.3 per cent) were used in energy recovery (DSEWPaC, 2013/ © Commonwealth of Australia 2015 (CC BY 4.0)).

The graph in Figure 3.18 shows that waste from the construction and demolition (C&D) stream in Australia from 2010 to 2011 represented both the greatest amount of waste generated and the highest rate of recycling of the three main waste streams (Municipal Solid Waste, Commercial & Industrial, and C&D). The higher recycling rate (66%) is related to factors including:

- weight-based landfill levies which create a disincentive to sending large quantities of often heavy material, such as concrete, to landfill

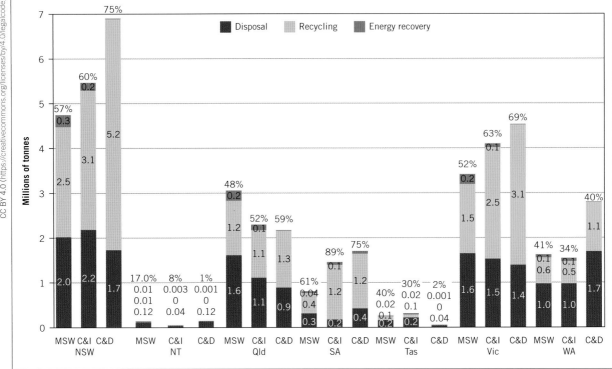

FIGURE 3.18 Australia total waste generation by waste stream, management and jurisdiction (excluding ACT), 2010–11

- established markets for recovered resources, particularly high-value materials such as metals
- effective recovery rate targets in New South Wales, South Australia, Victoria and Western Australia.

Because of increasing landfill levies and environmental impacts it is critical that builders develop a waste management plan to reduce the amount of waste going to landfill by separating waste materials and reusing or recycling waste materials as much as possible. This will reduce the amount of waste being used as landfill, and reduce building costs overall.

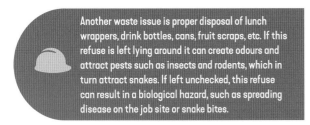

Another waste issue is proper disposal of lunch wrappers, drink bottles, cans, fruit scraps, etc. If this refuse is left lying around it can create odours and attract pests such as insects and rodents, which in turn attract snakes. If left unchecked, this refuse can result in a biological hazard, such as spreading disease on the job site or snake bites.

Therefore, it is every worker's responsibility to clean up as they go, to prevent accidents and hazardous situations, to provide a safe, hygienic site for all and to minimise the amount of waste taken to landfill.

2.2 Confirm availability of required tools and equipment, including PPE

Before and after you do a work task, it is important to report to your supervisor about the availability of tools and equipment, including any PPE needed. A checklist is a useful tool when performing complex tasks because competent people can make sure that the most important items are on the checklist and as people go through that checklist all they need to do is refer to the items listed. This helps to avoid missing things (see Table 3.2).

TABLE 3.2 Create a checklist for reporting

✓	Make a report for the supervisor.
✓	List any problems or difficulties you experienced.
✓	Report any damaged or faulty equipment.
✓	Report any unsafe or dangerous act or item.
✓	Report any uncooperative workers, who may have hindered your progress.
✓	Any other matters you feel are relevant.

1 Make sure that you inform your supervisor that you have completed the set task, so that it can be checked for accuracy and so that you can be allotted the next task to be done.
2 Let your supervisor know if you experienced any problems or difficulties while completing the task. This enables your supervisor to change any faulty or poor work processes.

3   Report any damaged or faulty equipment so that it can be repaired before it is required again. It is very frustrating when you find that the tools or equipment you need are not working or not working properly.

4   Advise your supervisor of any unsafe or dangerous behaviour or procedures you have experienced or witnessed as these can place everybody's health, safety and welfare at risk. Your supervisor needs to know about these WHS breaches as the worker, supervisor, manager or company can all be heavily fined for failing to follow the WHS Act and Regulations.

In addition, you should always report any lack of cooperation or assistance from other workers. On building sites many tasks are carried out by subcontractors who are paid to come onto the site, perform a task and then leave.

Some subcontractors feel pressured to get the task done as quickly as possible so that they can get on with the next task. This can lead to a lack of consideration, cooperation and assistance for other workers on the site, which may generate hostility and lead to arguments and fights. These actions can disrupt the smooth running of a job, slow the job down, lead to faulty work, unsafe work practices and waste time, materials and money. Your supervisor needs to know anything that can affect the efficient running of the job.

When you confirm the availability of tools and equipment, make sure you also have on hand any user manuals or safety data sheets for items involved in completing the task. These documents can ensure that you use the tools and equipment safely.

 COMPLETE WORKSHEET 3

SUMMARY

In Chapter 3 you have learnt how to plan and organise work by being able to:

- determine and plan basic work task activities
- organise performance of basic work tasks.

This unit of competency specifies the skills and knowledge required to plan and organise basic work tasks on a construction site.

The unit is suitable for those with basic skills and knowledge undertaking routine work tasks under the direction of more experienced workers.

REFERENCES AND FURTHER READING

Acknowledgement

Reproduction of the following Resource List references from DET, TAFE NSW C&T Division (Karl Dunkel – Program Manager – Housing and Furniture) and the Product Advisory Committee, is acknowledged and appreciated.

Texts

Department of Climate Change, Energy, the Environment and Water. Search for: *Australian waste definitions*

Department of Climate Change, Energy, the Environment and Water. Search for: *National waste report*

Web-based resources

European Construction Institute (1996), *Implementing TQ in the construction industry: A practical guide*, Thomas Telford Ltd, London, England

National Centre for Vocational Education Research (2001), *Skill trends in the building and construction trades*, National Centre for Vocational Education Research, Melbourne

Damian Oliver (2011), *Skill shortages in the trades during economic downturns*, National Centre for Vocational Education Research, Melbourne. Visit voced.edu.au to search for this document.

Safe Work Australia website. Search for: Construction induction training card (white card)

WorkSafe Victoria (2008), A handbook for the construction regulations – Working safely in the general construction industry, WorkSafe Victoria, Melbourne

Resource tools and VET links

http://training.gov.au, Skills training.gov.au

http://www.environment.gov.au, search for Australian Waste definitions – Defining waste related terms by jurisdiction in Australia (PDF)

https://www.epa.nsw.gov.au, search for: Report into the Construction and Demolition Waste Stream Audit 2000–2005 (PDF)

http://www.environment.gov.au, Department of the Environment

GET IT RIGHT

In the photo below, the person is working unsafely.

Identify the unsafe practices and provide reasoning for your answer.

Source: Richard Moran

_____

_____

_____

_____

_____

WORKSHEET 1

Student name: _____

Enrolment year: _____

Class code: _____

Competency name/Number: _____

Task: Read through the sections beginning at *Determine and plan basic work task activities* then complete the following questions.

1   What may happen if you are a poor planner and organiser at work and cause problems for other workers?

_____

_____

_____

2   List the five (5) benefits that good work organisation achieves.

1 _____

2 _____

3 _____

4 _____

5 _____

3   What does 'working safely' mean?

Working safely means _____

_____

_____

4   Work injuries can lead to _____

_____

_____

5   What must be done before you begin any task?

_____

_____

WORKSHEET 2

Student name: _____

Enrolment year: _____

Class code: _____

Competency name/Number: _____

Task: Read through the sections beginning at *Identify the tools and equipment required, including personal protective equipment (PPE) for each stage of the task* then complete the following questions.

1   What planning needs to be done when specialist equipment might be needed for a particular job?

_____

2   When collecting tools for use for the day's jobs what are some important things to check?

_____

3   When planning the sequence of tasks to achieve an overall job, does it matter in which order those tasks are completed?

_____

Source: © Cengage

Simple pergola

Project
Location:
Schedule by:
Date:

ACTIVITY	Week 1		Week 2			Week 3			Week 4			Week 5			Week 6			Week 7			Week 8		
	2	4	6	8	10	12	14	16	18	20	22	24	26	28	30	32	34	36	38	40		
1. Receive plans																						
2. Approval to proceed																						
3. Site establishment																						
4. Setting out pergola																						
5. Excavate for footings																						
6. Pour footings																						
7. Install posts																						
8. Install beams																						
9. Install rafters																						
10. Price structure																						
11. Paint structure																						
12. Hand over																						

 **WORKSHEET 3**

Student name: _____

Enrolment year: _____

Class code: _____

Competency name/Number: _____

To be completed by teachers

Student competent ☐

Student not yet competent ☐

Task: Read through the sections beginning at *Prepare a written work plan and a list of resources required to complete the overall work task* then complete the following questions.

1   What are the three (3) main methods of protecting people when setting up the work area or work environment?

    1 _____

    2 _____

    3 _____

2   What type of hoarding is a temporary wire fence around a building site?

    _____

3   Where should signage be placed before any work is started?

    _____

4   Why is it important to list the materials required and the order in which they are required?

    _____

    _____

    _____

5   What are the possible consequences of having materials delivered to the construction site earlier than they are required?

    _____

    _____

    _____

6   What is the most appropriate way to stack any material when stacking or storing materials on the job site?

    _____

    _____

7   What motto is used in regards to material being delivered to a site?

    _____

8   Why is it important to periodically clean up the work site?

_____

9   List three (3) types of problems that may arise on a construction site.

1   _____

2   _____

3   _____

10 Waste is defined as _____

_____

11 What are the three (3) steps involved in cleaning up?

1   _____

2   _____

3   _____

12 When packing up, what must be done with all tools and equipment?

_____

_____

13 What must happen with safety signs when packing up after completing the task?

_____

14 In 2010–11, how many millions of tonnes of waste did the construction and demolition industry in Australia produce?

_____

15 In 2010–11, how many millions of tonnes of waste did the construction and demolition industry in Australia send for recycling?

_____

16 According to the *National Waste Report 2013* from the Department of Sustainability, Environment, Water, Population and Communities (DSEWPaC, 2013), what were the three (3) contributing factors that produced this high rate of recycling?

1   _____

2   _____

3   _____

17 What sort of hazard is produced on the job site by not disposing of lunch wrappers and drink bottles, etc. correctly?

_____

18 What are the reasons given in the text for making a report to the supervisor?

1 _____

2 _____

PERFORMANCE OF SKILLS EVIDENCE

Student name: _____

Enrolment year: _____

Class code: _____

Competency name/Number: _____

Task: Working with your teacher, prepare to demonstrate your skills by completing the following task. To demonstrate competency, a candidate must satisfy all the elements, performance criteria and foundation skills of this unit by planning and organising a basic work task on a construction site that includes a minimum of ten component tasks and a team of at least three people.

All work must be performed to the standard required in the workplace and must comply with appropriate work health and safety (WHS) and environmental requirements, workplace requirements, drawings and specifications.

Instructions:

1   Choose a basic work task in consultation with your teacher, ensuring at least ten (10) activities to get the task completed. The task could include constructing a timber frame floor, or constructing the formwork for a rebated edge slab on ground.

2   Note what needs to be done for each task and how it is going to be done, estimating the time and number of personnel required to complete the task.

3   Identify the tools and equipment required, including any PPE to complete the task safely.

4   Sequence the activities in logical order for the efficient use of resources and prepare a written plan listing the resources required.

5   Work with team members to review the plan, allocating roles and responsibilities and reviewing work health and safety requirements.

6   Refer to any plans or specifications, user manuals, Australian Standards or codes of practice that may be helpful in performing this activity.

4 CONDUCT WORKPLACE COMMUNICATION

This chapter covers the following topics from the competency CPCCOM1014 Conduct workplace communication:

- convey and receive information and instructions
- access, interpret and present information
- participate in simple meeting processes.

This unit of competency specifies the skills and knowledge required to communicate effectively with other workers in a construction workplace environment.

The unit includes gathering, conveying and receiving information through verbal and written forms of communication.

The unit is suitable for those with basic skills and knowledge undertaking routine work tasks under the direction of more experienced workers.

1. Convey and receive information and instructions

The term communication can be applied in three ways.

1.1 Receive information and instructions from others using effective listening, questioning and speaking skills to confirm understanding

First, it can be applied to the *act* of communicating. Talking (Figure 4.1), singing, writing, miming, gesturing, signalling, drawing, sketching, kissing, hugging, hitting, punching or kicking are examples of communication acts. They all carry a message of some kind.

Source: iStock.com/Photo_Concepts

FIGURE 4.1 Communicating by talking

The message being communicated

Second, communication can also be applied to the *message* that is being communicated. These messages may be about emotions, feelings, wants, needs, thoughts, ideas, opinions, facts, knowledge, information, warnings, or any one of the many things people need to impart to others. For example, the communicating act of talking enthusiastically about a new house that your construction company is going to build, in addition to the information you are conveying, imparts a message that you are excited about the upcoming project. The communicating act of hugging someone imparts the message that you like that person. The communicating act of chastising someone imparts the message that you are upset. The communicating act of extending your arm forward with your hand flexed upward and palm toward the viewer imparts the message that you want that person to halt or stop.

The means of communicating

Third, communication can be applied to the *means* of communicating. For example, an MP3 file recording of you talking about your new building project, which you will send to an interstate friend, is a means of communicating. The email that you attach the MP3 file to is also a means of communicating. A poster that you

have developed to advertise the new house to prospective buyers is yet another means of communicating. In summary, communication can be understood as:

■ the act
■ the message
■ the means of communicating with another person or group of people.

Communication can only be successful if the person or group towards whom the message has been directed (often called the receiver) has understood the message exactly as the person who imparted or sent the message (often called the sender) intended it to be understood.

Communication audiences

In the construction industry, it is very likely that you will need to communicate with:

■ supervisors
■ other construction workers/tradespeople
■ customers/clients
■ suppliers.

To communicate effectively you will need to tailor your methods to suit each audience. For example, when speaking with other construction workers it is appropriate to speak informally and use industry jargon since they will understand what you mean. Whereas, when you are speaking with a client, more formal language and Plain English words are likely to be more appropriate.

Communication barriers

Each person has a different life experience, and these different life experiences shape the character, personality and social views of each individual. This will affect the way an individual understands a message from someone else.

Feedback

To be sure the correct message is received, the sender will require *feedback*. For the sender, feedback may range from a simple request for the receiver to repeat what was said, to an invitation for the receiver to discuss the message. For the receiver, restating the message back to the sender is a good check. Communication is a circular procedure that requires checking and rechecking to determine that the message received matches the message sent. Without feedback, failure to communicate is highly likely.

Unfortunately, successful communication is not as simple as Figure 4.2 implies. In reality, there are many obstructions to effective communication. These obstructions, sometimes called barriers or blocks, can exist in the:

■ environment in which the people are communicating
■ system by which they are communicating
■ people who are trying to communicate.

Barriers to communication can be conveniently categorised as physical, emotional, psychological or cultural.

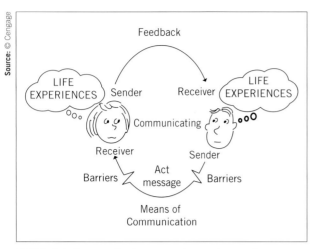

FIGURE 4.2 Communicating = creating understanding

Physical barriers

Physical barriers may occur in a number of forms, such as:

- the *environment* or the workplace surroundings, e.g. loud machinery noise or obscured safety signage, which can affect the act of communication
- electronic interference or static can cause disruption when using a telephone or two-way radio communication system
- physical disabilities such as deafness or failing hearing, and blindness or failing eyesight, can create a barrier to communication.

Emotional barriers

At different times people may be angry, resentful, frustrated and even disliked. These strong negative emotions can get in the way of good communication. The presence of these emotions in the sender can override the content of the message. The presence of these emotions in the receiver can distort the perception of the message.

Psychological barriers

Psychological barriers also occur, but are different to emotional barriers. People's psychological characteristics can influence the communication process. For example, people who as senders are continually aggressive, arrogant, judgemental, manipulative, confrontational or guilt-inducing in their communication behaviour are unlikely to encourage the feedback that is essential to successful communication. Similarly, people who as receivers avoid, make light of, refuse to acknowledge or divert communication messages are avoiding communicating frankly about an issue.

Cultural barriers

People can also have *cultural barriers* which may affect good communication. Today, there are many nationalities and educational levels found on the job site. These differences may have a great impact on the communication process at the work site. A person's

lack of knowledge can limit his or her effective use of complicated communication systems such as computers. Not understanding the language of signs or instructions might jeopardise a person's safety. Similarly, the lack of a shared language could cause serious misunderstanding between sender and receiver. In the workplace, the use of language that is too technical for beginners, or the use of jargon, or even the inability to read technical plans correctly, can lead to problems as well as expensive mistakes.

How do we overcome these barriers?

First, we must not make the assumption that it is the receiver's responsibility to understand the message we are trying to send. If it's our message, it is our responsibility to get it across clearly.

Second, we must anticipate the possibility of physical, emotional, psychological or cultural barriers to communication and, by imagining ourselves in the shoes of the receiver, try to make sure that our communication avoids or cuts through these barriers.

The communication process

The ability to communicate well is an important skill in the building industry. On building sites, poor communication leads to injuries, faulty work and wastage of time and materials. Good communication benefits everyone. Nearly everyone acknowledges the need for good communication, but very few people ever think of themselves as being poor communicators.

This is because most people think that just because they can talk, read and write they can communicate clearly.

While talking, reading, and writing are important, there are other aspects to successful communication that should be considered.

Creating understanding

Some people who are not expert talkers, readers and writers are excellent communicators. Why? Because they realise that to communicate they must create understanding in the mind of the person with whom they are communicating.

Good communicators create understanding. When we communicate we send a message to someone. As the sender of the message we need to make the information as comprehensible as possible, and carefully check that the person receiving the message (the receiver) has understood it. If the receiver does not understand the message as the sender intended it to be understood, communication has failed.

Both the sender and receiver have responsibility for successful communication. They work together to create understanding. They do this by checking that the message received is the same as the message sent. They check by asking for and giving feedback (Figure 4.3). The sender and receiver can check the message has been received by following the simple process in Table 4.1.

FIGURE 4.3 The feedback process

TABLE 4.1 Good communication process

The sender checks by:	The receiver checks by:
asking for the message to be repeated	repeating the message
asking questions	asking questions
asking for the message to be restated.	restating the message.

To make sure that you have overcome the barriers to communication, always seek and give feedback.

Six simple steps for clear communication
Finally, remember these six simple steps for clear communication:

- state the overall goal you require
- describe the main steps of any task in a logical order
- explain the details of each step slowly
- emphasise the critical points
- seek and offer feedback to check the other person's understanding
- summarise the main steps in the task in a logical order.
   Be aware that on a building site you may have to combine spoken language, written language and diagrams to make your instructions clear and understandable to the other person.

1.2 Convey information and instructions to others using effective listening, questioning and speaking skills to confirm understanding

Verbal communication is the act and means of sending and receiving information using spoken language. There are several different ways that this may occur including:

- direct or face-to-face
- landline or mobile telephone
- computer, using a specialised program such as Skype
- two-way radio.

Verbal communication methods
Many people believe that verbal or spoken communication is simple, that it simply involves the message being spoken to the other person and that, when this is done, communication has happened. This unfortunately is not the case – there are many factors to consider, such as background noise or interference, hearing disability and the emotional state of both parties involved in the communication process. It is important that spoken communication is, therefore, treated like a sword fight or fencing match – the first person will try to get the point (of his sword) in contact with the opponent (receiver); the receiver will wait and then respond, so the contest (communication) continues until an understanding is reached. The diagram in **Figure 4.4** shows that the sender and receiver change roles throughout the process, so that the sender becomes the receiver and so forth. This is in fact how good verbal communication should occur.

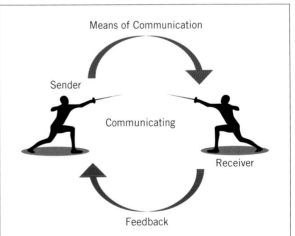

FIGURE 4.4 The communication process

Direct or face-to-face communication
Of all the communication processes used, face to face is the most complex process. This has many associated parts or components that are not exactly verbal communication, but without which face-to-face communication would be very difficult.

The components that are significant and that give additional meaning to the face-to-face communication process include:

- facial expressions
- eye movement
- body language
- voice inflection.

Facial expressions

Facial expressions add to verbal communication by allowing the sender and the receiver to see how the other person is processing the message (see Figure 4.5). From birth we have been trained to interpret facial expressions; even if the words being used are not known to us, we can still recognise facial expressions and the meaning they convey. Some of the facial expressions that may be encountered during verbal communication include:

- concentrating on the task at hand
- bothered, distracted or worried
- confused
- happy
- angry
- frightened
- in pain
- ill
- embarrassed.

FIGURE 4.5 Communication using facial expressions

Eye movement

Eye movement can also add to verbal communication; it allows each participant in the communication process to judge whether the other person is engaging in the conversation or if they are distracted. A person who cannot look you in the eye may have low self-esteem or be ashamed or embarrassed. A person whose eyes are constantly on the move might be distracted by activity around them.

Eye contact in Western culture is usually expected as a component to good communication. Avoiding eye contact can be perceived to communicate a lack of respect, or the person avoiding eye contact might be considered suspicious. Be aware, however, that in other cultures prolonged eye contact can be considered offensive or as a challenge to their authority. Remember to be culturally sensitive when working with people from different cultures, and avoid making assumptions.

Body language

Body language is similar to facial expression in that it allows the sender and receiver to see how the other person is processing the message as well as sending out other underlying messages (see Figure 4.6). An example of this is when two men are having a heated discussion and one starts to clench and unclench his fists by his side. This clearly indicates his anger and the possibility of him acting on it by hitting out.

Some of the more commonly recognised body language indicators are:

- hands – movement, signals and gestures
- feet – tapping, movement

FIGURE 4.6 Communication through body posture

- legs – movement, crossing the legs
- posture – leaning towards or away from; tense or relaxed.

Voice inflection
Voice inflection is not *what* is being said but the *way* in which it is being said. Consider the following two examples, and think about what they might mean:

- Person A says they agree with what is being said, but their voice inflections are under-confident or hesitant.
- Person B loudly and quickly agrees with what is being said, but their response is sarcastic and casual.

Even though both people are voicing agreement, their voice inflections might actually be communicating totally different things. Person A's might be a first-year apprentice who doesn't understand the message, but they may be worried they'll get in trouble if they don't say that they do. Person B might be uninterested and not listening properly, or they might be joking sarcastically because they actually don't agree with the message being sent.

Landline or mobile telephone
Today, due to mobile phones, tradespeople are always contactable. Where once the builder or tradesperson had to wait until they got home or to the office, today much of the day-to-day running of the business can be done at any time and in any location (see Figure 4.7).

Communicating using the telephone still has certain requirements or etiquette that should be followed, such as:

- give a polite greeting
- give your company name
- give your name
- offer assistance.

Unlike face-to-face communication, telephone conversation is *voice only*, so there are no visual cues such as facial expressions or body language, just the words being spoken and the voice inflections.

It is, therefore, important to remember to speak clearly and at a moderate pace. The caller may never have spoken to you before and has no visual cues; this means the person will need time to tune into your way of speaking. Rudeness and lack of consideration have crept into our telephone practices.

Some points to remember when making or receiving a phone call are:

- announce your full name when answering the phone or mobile
- be aware of your speaking volume. Talk loud enough, but do not shout. If you are in a noisy environment, either move quickly and safely to another location or call the person back at a quieter time
- try not to answer your call when meeting with others. If you really have to answer the call, politely excuse yourself and keep the call brief

- let the other person know when you have them on speaker phone, such as when in the car with other colleagues
- be sure to thank the other person if they have been helping you
- take notes during your call. Repeat back your notes to make sure you have understood them, especially the information you have asked for to ensure it is correct
- let the caller hang up first.

FIGURE 4.7 Mobile phone in use

Source: iStock.com/shotbydave

Computer-based communication – video conferencing
In today's fast-paced world, many people find it difficult to attend face-to-face meetings at a specific location or given time. An example of this would be an architect based in Melbourne who needs to meet with the building contractor who is working on a project in Surfer's Paradise. The meeting could be done over the phone, but if there are specific visual details that need to be presented then *video conferencing* is the best option available. There are many existing programs that can be used for video conferencing, including:

- Google Hangouts
- Apple FaceTime
- Skype
- Zoom
- Microsoft Teams
- Adobe Connect
- Citrix GoToMeeting.

The negative aspect of video conferencing compared to face-to-face meetings is that each attendee can generally only see the face of the other person, so very little communication through body language may be possible. There are also issues relating to delays or communication drop-outs as well as time-zone differences; for example, a teleconference at 9 a.m. in Sydney means it will be 12 midnight in London.

Two-way radio
The two-way radio or walkie talkie is another tool for communicating vocally. This is similar to the mobile

phone except that it has a limited range; once you get to the outer limits of the device range there is the added difficulty of static or electronic interference. When using this system of communication, follow these basic rules:

- use an individual call sign to identify yourself
- say 'over' to indicate you have finished speaking so the other person can reply
- turn your microphone off after saying 'over', otherwise you cannot hear the other person speaking
- spell out important words using the international alphabet; for example, (a)lpha, (b)ravo, (c)harlie, (d)elta.

Note: Speak clearly and at a moderate pace. Remember that radio frequencies are public, so be careful about what you say.

Delivering a verbal report

Speaking in front of others can occur in both formal and informal meetings. Collect all relevant information on the topic. Organise it into main points, subpoints and supporting details. If necessary, create an outline to visualise the structure of your report. Consider the audience and the format that suits the presentation or meeting setting.

Begin with a concise and engaging introduction. State the topic and the purpose of your report. Provide a brief overview of what you'll be discussing.

Present your main points in a logical order. Each main point should be a significant aspect of your topic. Make sure the order of these points follows a logical progression, such as chronological, cause-and-effect, or hierarchical. Under each main point, provide supporting details, evidence, examples or data. These details help reinforce your main points and make your report more informative and credible.

Use transition statements or phrases between main points and supporting details to create a smooth flow. This helps guide your audience through your presentation.

Summarise the key points you've discussed. Restate the purpose of your report and the main takeaways. Avoid introducing new information in the conclusion. If appropriate, ask for feedback or questions.

Practise the presentation ahead of time to ensure familiarity with the content, to improve the delivery and to identify areas that need refinement.

LEARNING TASK 4.2

ORDERING MATERIALS

Obtain a timber-cutting list containing typical section sizes, stress grades, species, etc. Contact a timber yard or hardware store to obtain a price for each of the materials. This will allow you to practise your telephone technique, as well as receiving and conveying information.

Gather, convey and receive through written communication

Non-verbal communication can also be used on the building site. Information may be conveyed and received in different ways. These include:

- written language
- diagrams, drawings and pictures
- sounds
- lights
- touch sensations
- odours.

All of these channels of information are important, but this text will deal only with written communication.

Examples of written communication

There are many situations where written communication can be used. Some of the many written documents that are used on a building site are listed below:

- workplace signage/dangerous goods labels
- sketches/drawings/plans/diagrams
- instructions issued by authorised personnel
- manufacturer specifications and instructions
- safety data sheets (SDSs)
- safe work procedures/work method statements (WMS)/safety tags
- regulatory and legal documents
- relevant Australian Standards/National Code of Construction (NCC)
- company procedures and regulations
- work bulletins/notices/work orders
- work schedules/job specifications
- memos/letters/work instructions
- emails/text messages/faxes/site diary
- contracts
- delivery dockets.

This is not a full and comprehensive list; there are many more documents that are used by builders and tradespeople each and every day on the job site. We will look at two of these written document types in more detail in the following sections:

- workplace signage/dangerous goods labels
- sketches/drawings/plans/diagrams.

Workplace signage

In any workplace in Australia, there are signs that convey information using colours, simple diagrams and text. There are seven different categories of workplace signage. They are outlined in Table 4.2.

Note: For greater detail on workplace signage refer to the section in Chapter 1, 'Safety signs and tags'.

Workplace signage is used to convey information. The message communicated by signage offers no possibility of feedback or questioning so it is important that people reading the signs are familiar with their purposes and meanings.

TABLE 4.2 The seven categories of workplace signs

1. Prohibition (must not do) signs These signs give instructions for things that the viewer *must not do*		**5. Mandatory (must do) signs** These signs give instructions for things that the viewer *must do*	
2. Restriction signs These signs allow the viewer to *do certain activities, but with limitations*		**6. Hazard warning signs** These signs give *warning of a non-life-threatening hazard*	
3. Danger (hazard) signs These signs give *warning of a potential life-threatening hazard*		**7. Emergency information signs** These signs point to the *location of, or directions to, emergency amenities*	
4. Fire-fighting equipment signs These signs indicate the location of *fire-fighting equipment*			

For workplace signage to be used as an effective communication tool, the signs should be placed or erected just above eye level (approx. 1500 mm from floor level).

Placement or erection of signs

For workplace signage to be used as an effective communication tool, the signs should be placed or erected so they are clearly visible and unlikely to be covered or moved (e.g. by stacks of material or machines or by opening a door). Other heights may be more suitable in some circumstances.

Important as it is to place or erect signs correctly to communicate a hazard or warning, it is equally important to remove signs that are no longer needed or appropriate. As soon as the warning or information set out on a sign is no longer relevant, the sign should be removed. This is particularly important where a sign has been used to warn of a temporary danger – such as demolition work in progress, explosive power tool in use, or people working above – because failure to keep signage up to date may encourage a tendency to disregard signs.

Dangerous goods labels

Dangerous goods labels/signs or placards are similar to workplace signage.

These signs MUST BE used to identify the particular dangerous good being transported or stored within a vehicle or contained area. There are many materials stored and used on building sites that are classified as dangerous; each class of dangerous goods is identifiable by specific standard labels.

These labels only identify the hazard associated with each product, they DO NOT explain how the dangerous goods must be stored, handled and used safely. For this information it is necessary to view or read the safety data sheets (SDSs), which are prepared by product manufacturers. Safety data sheets give information about the product's constituents, possible health effects, first aid instructions, precautions for use, and safe handling and storage.

By law, PCBU/employers are required to provide employees/workers with information about the safe use, handling and storage of dangerous substances at work. Having SDSs available for use on the job is an effective way of achieving this.

Note: For greater detail on this topic, refer to the section in Chapter 1, 'Dangerous goods'.

Table 4.3 provides a summary of the nine classes of dangerous goods.

Sketches/drawings/plans/diagrams

Instructions are often transmitted or transferred (passed on) by pictorial or drawn information. Examples of pictorial instructions include:

- quick sketches: onsite, supervisors, leading hands and other tradespeople may draw a quick sketch or hand drawing on a piece of timber or off-cut

TABLE 4.3 The nine classes of dangerous goods

1. iStock.com/alexandernative; 2. Shutterstock.com/Nicola Renna; 3. iStock.com/alexandernative; 4. Shutterstock.com/Charles Brutlag; 5. Shutterstock.com/infinetsoft; 6. Shutterstock.com/Nicola Renna; 7. Shutterstock.com/Soru Epotok; 8. Shutterstock.com/Soru Epotok; 9. Shutterstock.com/Benjamin Marin Rubio

Class 1 – explosives		Class 2 – gases	
Class 3 – flammable liquids		Class 4 – flammable substances	
Class 5 – oxidising substances		Class 6 – poisonous or infectious materials	
Class 7 – radioactive materials		Class 8 – corrosive materials	
Class 9 – miscellaneous dangerous materials			

board to show what needs to be done because it is quicker, easier, more accurate and more easily understood than a spoken description.

- plans: a set of building plans drawn by an architect contain information that would be impossible to understand if expressed in words
- product drawings or diagrams: a supplier of a product will often include drawings or diagrams of the product and how to correctly install it.

Technical information is often better expressed in pictures than by long and involved verbal explanations or lengthy instructions written in a book or manual.

The ability to draw and to interpret drawings of buildings and components is a special skill that must be learned and developed. But once learned, making a quick sketch to highlight or emphasise verbal or written instructions, will allow those instructions to be more easily understood.

In the building industry, the old adage 'a picture is worth a thousand words' has much greater meaning. Some things can be described only in diagrams (Figure 4.8).

Written documents

Written documents convey meaning through text printed on a sheet of paper; this may be printed or hand-written. Because there are many different documents and applications where written text is used,

FIGURE 4.8 Communicating by sketching a detail

Source: iStock.com/ez_thug

this book will focus on the basics of writing relevant documents, and not the different types of text that may be used to communicate ideas and thoughts.

When you write, it is important to keep in mind that you are writing to be understood. Your objective must be to get your message across as clearly and quickly as possible. One way of doing this effectively is to write in Plain English. Plain English does not mean just using simple words, it also means ensuring that your message is as clear, concise and as free of jargon as possible. The following section gives some tips on how write an effective document.

Writing a message or document

When writing a message or document you should ensure that you express your ideas:

- in a logical order; for example, from the first step of the task or job to the final step according to the construction sequence
- in the active voice; for example, 'I need more nails', not in the passive voice, 'More nails are needed'
- in short sentences, with one main idea to a paragraph.

Avoid expressions that may obscure your meaning, such as:

- **ambiguities**: including sentences that can be misread or misinterpreted; for example, 'I put the saw in the truck that had been damaged'. What was damaged, the saw or the truck?
- **tautologies**: using extra and unneeded words to express an idea; for example, 'lineal metres' when you mean 'metres', or 'climb up the ladder' when you mean 'climb the ladder', or 'reverse back the truck' when you mean 'reverse the truck'
- **clichés**: expressions that have been overused and lost their true meaning; for example, saying 'hit the road' when you mean 'we need to go'
- **jargon**: words or expressions that other people outside the industry may not understand; for example, 'optimise output by maximising labour input' when you mean 'achieve more by working harder'
- **slang**: common terms used by a specific culture, group or nation; for example, saying 'mud' when you mean 'mortar', 'helicopter' when you mean 'trowelling machine', 'bubble stick' when you mean 'spirit level', or 'chippy' when you mean 'carpenter'
- **technical terms**: when you are writing for non-technical readers; for example, saying 'Pour a 100 mm, 20 Mpa slab with SL72 mesh' when you just mean 'Lay a concrete path'.

Note: Jargon, slang and technical terms may also be referred to as 'metalanguage'; if used with persons new to the industry or not familiar with them, these terms may be confusing and their meaning may be lost.

Once you have written what you want to communicate, read it through (aloud) or, even better, get someone you trust to read it aloud to you. While you are reading it ask the question: Does it make sense?

Reading a message or document

To read a message or document correctly, there are a number of items to consider. These include:

- What type of information is being sent? Is the information technical, personal or entertaining?
- To whom is the information being sent? Is this document meant for me or someone else?
- Why is this information being sent? What is the reason for me receiving this communication?
- Is it relevant to me? Do I really need to read this and make a response?

Once you have determined that the communication is important or of value to you, you will need to obtain the information that is contained in the document. Reading a document takes time; however, there are methods that you can use to reduce the time, but will still allow you to obtain the essential information.

When you are seeking information from a document you can:

- predict the content from the title or pictures
- skim quickly through the document to get a broad idea of the content
- scan the document to find a specific piece of information
- read for the main ideas
- read in detail for deep understanding.

How you read will depend on the purpose of the message, how relevant it is to you, and how urgently you need to digest the content of the message. It is not usually necessary to read every document in detail. Choose a way of reading to suit your purpose and the document.

Different types of reports are used to communicate information related to safety, project details and milestones, and for compliance. For example, incident reports, progress reports, daily logs, safety reports, quality control reports, and cost and variation reports. They all have slightly different ways of presenting information. Check your workplace procedures and their standard document formats before submitting a report.

Finally, be mindful that written documentation is communication that is recorded and in some applications is required for legal purposes.

LEARNING TASK 4.3

UNDERSTANDING SIGNAGE

Identify six different examples of dangerous goods and the associated label or sign appropriate for them. Then write a short report listing these dangerous goods and categorising them using the Class 1 to Class 9 categories from Table 4.3. Submit your report to the group for discussion.

 COMPLETE WORKSHEET 1

---

2. Access, interpret and present information

2.1 Access and interpret basic information from a range of sources

Accessing and interpreting basic information from a range of sources is very important in a construction context. The following are steps that will help as you gather information:

- Narrow down the actual information you need. Do you need information on materials? Do you need information on a process? Do you need information on a regulation or a requirement? Your task here is to try and narrow down the information that you need so that you can proceed further with your accessing and interpreting of information.
- Identify where you might access a source of information. If you need information on materials, you could go to the manufacturer or supplier directly or visit their website. If you need information on a process you could go to a competent person, an engineer, a code of practice, or even an Australian standard. If you need information on a regulation or a requirement you may go to the regulating authority or the governing body website. The key here is to make sure that your source of information is a credible and accurate source of information and does not contain biases that may compromise the information's accuracy.
- Gather and collect the information from your sources and interpret it into a well-rounded understanding of the topic for your needs. As you interpret the information you have sourced, seek to identify the key ideas and main concepts, and work out how you will implement your findings, as well as how you will communicate the actions that need to be done with the information you find.

This information should help you to access and interpret information from a range of sources.

2.2 Select and sequence information to prepare a basic written report

A basic written report involves selecting and sequencing information in such a way that you communicate the main points and conclusions of your research ready for presentation to your audience. The following are some steps that you can take:

- Determine the purpose of the report and who is the audience you will be presenting the report to. Who might read this report? What are the expected outcomes of this report? What level of language, literacy and numeracy skills do the readers of the report have?

- Gather all your information and review it and place it in order of topics, making sure you keep track of your sources. Organise your ideas identifying your main ideas and points that may support your findings and then create an outline of your report. At this point you could sequence your ideas and decide on the most effective way to create a logical and coherent report that leads to the conclusions you may be recommending regarding the information you are communicating. Use techniques such as headings, subheadings and bullet points to sort and organise your information and the sequence of your ideas.
- Write your first draft, by filling in the work you have done so far, and then edit and revise your work. Try to use clear and concise language that's easy to understand. As you go through this process, take advantage of any technology that may assist you while writing your report. You could consider the review of spelling and grammar functions within a word processing program. You may even choose writing assistance tools.
- Ask another competent person to review your report and then finalise it and publish it in whatever medium you have chosen to present it to your audience. This may involve printing and binding, publishing on your company's internal website or displaying electronically in a meeting.

By checking with a competent person and doing a little more research, you should be able to produce a basic written report.

2.3 Select and sequence information to prepare and present a basic verbal report

The process of selecting and sequencing information for a verbal report is much the same as for a written report. The difference is you are presenting your report verbally. The main aim is to select and sequence information in such a way that we effectively communicate the main points and conclusions of the research. The following are some steps you can take:

- Identify the purpose of the report and who is going to receive the report. What are the main objectives and what actions might be expected from this report? What is the situation that the report is going to be presented in? Is it an onsite meeting with lots of distractions and noise? Is it in a quiet air-conditioned meeting room? Is it an online presentation? Is it a recorded presentation?
- Gather your information and take detailed notes, keeping track of your sources, organising the main ideas and arguments to support the purpose of the report. Identify the main points, and any subheadings, and any graphics or illustrations that may be required to support your points. Prepare all this information for easy access as you move on to the next point.

- Sequence your ideas in a way that provides a logical path by using headings, subheadings and any other media you've decided on that can guide the listener through your report. Consider the language literacy and numeracy levels of your audience and, if needed, ask a competent or technical person to review the information.
- Practise your presentation several times by yourself and in front of a constructive audience. Make any changes that may be necessary.
- Deliver your presentation to your audience, considering any feedback you gathered as you practised. Present your verbal report with good eye contact and in a confident manner, relating to your audience's level of understanding and using questioning techniques to check their comprehension. Speak clearly and audibly and use appropriate facial expressions and gestures. Be prepared to answer questions and use this as an opportunity to ensure your audience understands what it is you're trying to convey.

Of course, having researched and found all the relevant information, you need to prepare and present it in a logical way, using clear and understandable industry standard methods of introducing and closing your presentation. For example, if you are writing an email or a letter, then you will need to introduce yourself to the people you are presenting the information to and introduce the subject. You then close off by noting who wrote the report and where more information can be obtained and/or a contact for more information. Always include pictures, graphics or charts if it will help in the presentation because this is a good visual way of getting a point across to your audience.

By considering the above steps and seeking advice from a competent person, you should be able to select and sequence information to prepare and present a basic verbal report to an audience in your construction business.

2.4 Enter information into basic workplace records and documents

Entering information into basic workplace records and documents involves some detail and attention to accuracy as well as an understanding of the purpose of that information for the short and long term for your business. Information can guide your business into being more efficient, more productive, to meet business goals and compliance requirements, and to determine the profitability and so on. The types of information that you may want to, or be required to, enter into workplace records and documents may include:
- contact details for persons involved in the business transactions
- important information about jobs your business is doing
- regulatory information for your business
- safety records for your business

- financial records for your business
- communication
- plans and ideas for the future.

Each piece of information for each record may be different but, in all cases, entering information is an important task and must be done efficiently. The following are some ideas for a flow process that you might find useful:
- Consider the document or record carefully to understand what information is required and how it should be entered as well as how information has been entered in the past. Consider any requirements from regulatory organisations or industry bodies who may have recommendations on how to enter the information or store the information. Talk to other people who are competent and have experience.
- Gather all the information that you need to enter into the document or record and have it in a usable format. This may include gathering all the receipts or paperwork for a particular activity that are relevant to that record or document and finding a way to put them into the record or document in a format that is efficient and effective. You may consider scanning, photographing, downloading records or receipts, etc. If you are gathering the information in a hard copy, you will need an efficient filing system so you can interpret and review that information later on. If you are collecting the information electronically you probably need to be aware of a few different ways that information may be collected. Information may be stored or transferred as a PDF file. This requires an Adobe program to read the file. If your file involves lots of data that can be downloaded as a spreadsheet then this is normally done as a CSV file. This is a method of downloading files for spreadsheets that are 'comma separated value' files. For large amounts of data, it can be very helpful to have it in a spreadsheet so you can sort, filter and draw information from the data.
- Check the accuracy of your information before entering it into the document. This might involve checking calculations, confirming contact details, checking supporting documentation, etc.
- Enter the information in a neat and accurate way and keep in mind that the end user of the document needs to be able to draw information out accurately, clearly and efficiently. You might need to learn some new skills to enter information, especially if it involves use of new technology. Entering important business information is an excellent skill to improve your business efficiency. Many tradespeople now enter receipts and transaction records immediately after a purchase in such a way that the transaction goes directly to their business records, their bank records, and their accountant for some of the mandatory reporting that is required for your business; for example, GST reporting.

- Make sure you save the information you input, whether it be in a physical storage system or a digital storage system. Consider a back-up policy for your business so information is kept secure and only the information that is required is kept. There are laws on what information you can keep, and you need to have a privacy policy for your business that meets the requirements of these laws. Talk to your legal business adviser for assistance in this.
- Review, proofread and practise drawing information from the document so that you know your system works. Put your policy into practice for your business by providing a flow chart that outlines to your staff what your business policy is regarding how to enter information into business records and documents.

By following these steps, you can efficiently and effectively enter information into basic workplace records and documents.

 COMPLETE WORKSHEET 2

3. Participate in simple meeting processes

Onsite meetings occur on all building sites. Some commercial sites will have many meetings, whereas small residential sites will only have a few. Meetings involve both verbal and written communication processes.

3.1 Describe and follow simple processes and procedures for meetings

A number of different types of meetings may be held on a building site. These are:
- general staff meetings
- union meetings
- WHS meetings
- special-purpose committee meetings
- team meetings
- social club meetings
- special interest group meetings.

General staff meetings

General staff meetings are used to inform the staff about the company's performance, direction, future, policies, etc. All employees are usually expected to attend.

Union meetings

These meetings may be called by either union members or union organisers. At these meetings members may discuss pay and conditions, unresolved safety problems, disputes with employers, problems on the site, support for other employees or support for victims of tragedies. These meetings are usually attended by members of the particular union.

WHS meetings

WHS committees are generally formed only on sites with 20 or more employees. The members of the committee regularly inspect the site and hold meetings where they make policies and recommendations about health and safety on the job site. Committee members represent both management and general staff.

Toolbox talks and induction meetings provide WHS requirements and work instructions for a particular site. These can include specialist PPE or equipment being used. Along with safety factors, work activities or daily work tasks may also be communicated to team members along with their roles and responsibilities in performing these tasks. At times, numerous trades will attend these meetings and, in some cases, attendees must sign and report their attendance and understood the information.

Special-purpose committee meetings

Special-purpose committees are formed to undertake particular tasks or responsibilities. Their meetings will usually be conducted according to a program or timetable they have established in order to achieve their objective. When this has been done the committee will generally be disbanded. The committee may be composed of specialists, experts or simply volunteers who want to get something done.

Team meetings

Team meetings are held by work teams that have been formed to increase efficiency on the job. The teams meet regularly with the objective of planning, organising and carrying out the work as successfully as possible (Figure 4.9).

FIGURE 4.9 Typical team meeting

Social club meetings

These meetings are usually held by volunteers to arrange social functions after work for interested staff. Activities can range from barbecues to concerts, restaurants or visits to the snow. The success of the social club depends on the enthusiasm and shared interests of the staff.

Special interest group meetings

These meetings are similar to social club meetings but are focused on specific interests. In a large organisation there may be enough people with a common interest to form a special interest group and hold meetings to arrange opportunities for them to explore their shared interest.

Conducting meetings

Broadly speaking, there are two ways of conducting meetings: they can be formal or informal.

Formal meetings

Formal meetings are run by elected office bearers called a chairperson, secretary and treasurer.

Each meeting will follow an **agenda** (Figure 4.10). The agenda is prepared by the secretary and sent to the people attending the meeting before the day of the meeting. It will include the time and location of the meeting, as well as what will be discussed and the order in which it will be discussed.

Source: © Cengage

FLYINGFOX CONSTRUCTIONS

PROJECT: Renovation of committee room

PROJECT MEETING

Venue: Site office
Date: Monday 7 October, 2024
Time: 9 am

AGENDA
1. Welcome, introduction & apologies

2. Project update:
   – carpentry work
   – floor tiling and carpet laying
   – painting and decorating

3. Proposed variations to plans

4. General Business

Next Meeting:

FIGURE 4.10 Typical example of a meeting agenda

The meeting will be controlled by the chairperson. When each item is raised by the chairperson it will be discussed, and the meeting will follow a set procedure for making decisions using movers, seconders, speakers for and against the motion, then finally a vote by the meeting for or against the motion. During the meeting the secretary will write the **minutes** of the meeting, which are summaries of the discussions and the decisions agreed on (Figure 4.11). The minutes then become the formal record of the meeting.

Note: These procedures are usually followed only when the items to be discussed affect a lot of people, have legal implications, are required by law, or need to be accurately recorded for future reference.

Company meetings, union meetings, work health and safety meetings and special-purpose committee meetings are usually conducted as formal meetings.

Informal meetings

Informal meetings are usually conducted with simple procedures and without the election of office bearers. A group of people will simply get together to discuss an area of interest (Figure 4.12). There may not be a formal agenda, but everyone will have a general idea of the focus of the meeting.

Someone will probably lead the meeting, and if there are to be a number of meetings, group members may take turns at being leader.

There will be no set procedures for conducting discussions and making decisions. The success of the meeting will depend on the reasonableness, fairness, self-discipline and depth of the desire to achieve a successful outcome of each of the members of the group.

All members will probably take their own notes about issues that affect them directly. During the meeting, individuals may volunteer or be coaxed into undertaking tasks that have been decided on by the group.

Informal meetings are much more dependent than formal meetings on the ability of the members to work as a team.

Team meetings, social club meetings and special-interest group meetings are often conducted as informal meetings. Of course, many meetings are neither formal nor informal. In the workplace, depending on the situation, meetings may be hybrids or combinations of both extremes.

3.2 Provide constructive contributions to meeting discussions

Whether a meeting is formal or informal, it is important to participate. When you attend a meeting, make sure that you:
- understand the purpose of the meeting
- contribute to relevant discussion
- are prepared to listen
- offer only constructive criticism
- deal with issues or problems, not with people or personalities
- are prepared to resolve problems
- are prepared to accept and carry out the decisions reached by the group, because the group's decision applies to everyone.

Source: © Cengage

(i)

WORK HEALTH AND SAFETY COMMITTEE MEETING

Date: Friday 11 October, 2024
Venue: Hut 18
Time: 10 am

Agenda

1. Welcome and apologies
2. Minutes of previous meeting
3. Business arising from minutes
4. General business
   4.1 Non tagging of electrical leads
   4.2 No hand rails on stair well
   4.3 Bricks falling from fourth floor
5. Other business
6. Next meeting
7. Close

(ii)

WORK HEALTH AND SAFETY COMMITTEE MEETING

Date: Friday 11 October, 2024
Venue: Hut 18
Time: 10 am

Minutes

Present: A Adams, W Calper (chair), N Couri, G Boon, F Jones, J Melendez, P Nguyen (secretary)

1. Apologies: B Grune, N Price
2. Minutes of the previous meeting were read and accepted.
3. Business arising from minutes:
   3.1 Missing safety signs have been replaced on entrance gate.
4. General Business:
   4.1 Untagged leads. N Couri reported that all untagged leads have been removed from the site.
   4.2 No hand rails on stair wells. J Melendez is to organise a team to install hand rails on all stair wells.
   4.3 Bricks fallling from fourth floor. A Adams reported that toe boards had been removed from the bricklayer's scaffold. These have been replaced, which should prevent any further incidents.
5. Other Business:
6. Next meeting was set down for Friday 16 December at 10 am.
7. The meeting closed at 11 am.

FIGURE 4.11 Typical examples of meeting documents – (i) Formal agenda, (ii) Minutes of the meeting

Source: Shutterstock.com/Phovoir

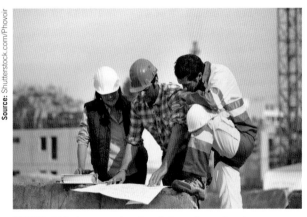

FIGURE 4.12 An example of an informal meeting in progress

During the meeting

During the meeting it is important to make sure that:

- everyone sits where they can see each other
- everyone has a place to put their agenda and take notes
- everyone has an opportunity to make a contribution
- no-one is allowed to dominate the meeting and control the outcomes.

Because meetings are often called to resolve problems or deal with difficult situations, you are likely to be required to help resolve a **conflict**. In such a situation, the most common ways of responding are to:

- withdraw from the situation, which allows others to win and, because the conflict is not resolved, may allow it to grow out of control

- suppress your feelings and refuse to acknowledge the problem, which does not allow others to recognise your feelings and have the opportunity to behave differently
- compromise, which can lead to dishonesty and possibly degenerate either into haggling or exaggerated ambit claims
- confront the other person, which can lead to win/lose ego-fired battles of will, which have nothing to do with the pros or cons of the issue at hand.

All of these ways of responding can lead to a win/lose situation. In time, successive win/lose situations can produce a culture of tit-for-tat responses, where it is more important to win and get even than to solve a problem. This leads to a breakdown of harmonious and cooperative work relationships and an unhappy, unsatisfactory and unproductive workplace.

The ideal response to conflict is the win/win response.

A win/win solution is a solution that meets everyone's needs. Not only do we get what we want but the other people also get what they want. We give up trying to persuade or convince the other party that we are right, and we give up trying to destroy the argument that they are using to try to convince us that they are right. Instead we set out to cooperate and to find a solution that will benefit everyone.

LEARNING TASK 4.4

GROUP ACTIVITY – SITE MEETING

Split into groups of four students. In your groups, conduct a 'mock' site meeting in which each group member participates. Rotate the members and vary the scenarios to allow each person to act to perform a main role, such as chairperson, treasurer, etc.

 COMPLETE WORKSHEET 3

SUMMARY

In Chapter 4 you have learnt about the skills and knowledge required to communicate effectively with other workers in a construction workplace environment.

You have also learnt about gathering, conveying and receiving information through verbal and written forms of communication.

You have also learnt to:

- convey and receive information and instructions
- access, interpret and present information
- participate in simple meeting processes.

REFERENCES AND FURTHER READING

Texts

Basic Work Skills Training division, NSW TAFE Commission (1995), *Workplace Communication, NCS001, A Teaching/Learning Resource Package*, Basic Work Skills Training Division, South Western Sydney Institute of TAFE

Eagleson, Robert D. (1990) *Writing in Plain English*, Australian Government Publishing Service, Canberra

Elder, B. (1994), *Communication Skills*, Macmillan Education, Melbourne

Emmitt, S. & Gorse C. (2003), *Construction Communication*, Blackwell Publishing

Graff, D.M. & Molloy, C.J.S. (1986), Tapping group power: a practical guide to working with groups in commerce and industry, Synergy Systems, Dromana, Victoria

National Centre for Vocational Education Research (2001), *Skill trends in the building and construction trades*, National Centre for Vocational Education Research, Melbourne

NSW Department of Education and Training (1999), Construction industry: induction & training: workplace trainers' resources for work activity & site OH&S induction and training, NSW Department of Education and Training, Sydney

Web-based resources
Resource tools and VET links
training.gov.au, Skills training
www.worksafe.qld.gov.au, search for information about Placarding for storage of hazardous chemicals under the *Work Health and Safety Act 2011*

GET IT RIGHT

In the photo below, the worksite has some communication issues.

Identify these issues and provide reasoning for your answer.

Source: Richard Moran

_____

_____

_____

_____

_____

_____

 **WORKSHEET 1**

Student name: _____

Enrolment year: _____

Class code: _____

Competency name/Number: _____

Task: Read through the sections beginning at *Convey and receive information and instructions* then complete the following questions.

1   Name the two (2) parties involved in the communication process.

   1   _____

   2   _____

2   Name the four (4) barriers that may obstruct effective communication.

   1   _____

   2   _____

   3   _____

   4   _____

3   List the procedures for checking whether communication has been successful.

   The sender checks by:

   _____

   _____

   The receiver checks by:

   _____

   _____

4   List the four (4) specific ways that verbal communication may occur.

   1   _____

   2   _____

   3   _____

   4   _____

5 Face-to-face communication involves direct verbal communication as well as four other additional or associated components. List the four (4) additional components.

1 _____

2 _____

3 _____

4 _____

6 List the things you should do to comply with phone etiquette when you answer the phone.

_____

_____

_____

_____

7 Name three (3) examples of written documents that are directly related to safety on the work site.

1 _____

2 _____

3 _____

8 Explain in your own words why signs should not be located on the backs of doors.

_____

_____

_____

9 Sketches and drawings are a useful means of communicating information. State two (2) typical materials found onsite that could be used to sketch or draw on:

1 _____

2 _____

10 Complete the following sentence. In the building industry, the old adage '_____' has much greater meaning.

_____

11 Instructions are often transmitted or transferred through the use of drawings or pictures. Name three (3) different people who use pictorial instructions to convey information to workers on-site.

1 _____

2 _____

3 _____

12 List four (4) basic points to follow when writing a document in Plain English.

1 _____

2 _____

3 _____

4 _____

13 List six (6) types of language that can obscure the meaning of your written message and that you should avoid.

1 _____

2 _____

3 _____

4 _____

5 _____

6 _____

14 When reading a document there are a number of different methods that may be used to obtain the essential information in a reduced time. List four (4) methods that can be used.

1 _____

2 _____

3 _____

4 _____

 **WORKSHEET 2**

Student name: _____

Enrolment year: _____

Class code: _____

Competency name/Number: _____

Task: Read through the sections beginning at *Access, interpret and present information*, then complete the following questions.

1   List the four (4) reasons why clear and concise communication on any building site is important.

    1 _____

    2 _____

    3 _____

    4 _____

2   What are the two (2) main forms of communication used on a building site?

    1 _____

    2 _____

WORKSHEET 3

Student name: _____

Enrolment year: _____

Class code: _____

Competency name/Number: _____

Task: Read through the sections beginning at *Participate in simple meeting processes* then complete the following questions.

1   List three (3) different types of onsite meetings that may be held on a building site.

    1 _____

    2 _____

    3 _____

2   List two ways of conducting meetings.

    1 _____

    2 _____

3   List the types of meetings that are often conducted 'formally'.

_____

_____

_____

4   Describe an agenda for a meeting.

_____

_____

_____

5   Describe the minutes of a meeting.

_____

_____

_____

6   Regardless of whether a meeting is formal or informal, what must attendees be prepared to do?

_____

_____

_____

7   List four (4) important things to make sure of during a meeting.

1   _____

2   _____

3   _____

4   _____

8   List and describe the four (4) most common ways of responding in a conflict situation. Describe the kind of situation these ways of responding can lead to and the effect each situation can have.

1   _____

2   _____

3   _____

4   _____

9   Name and describe the ideal response to a conflict.

_____

_____

_____

PERFORMANCE OF SKILLS EVIDENCE

Student name: _____

Enrolment year: _____

Class code: _____

Competency name/Number: _____

Task: Working with your teacher, prepare to demonstrate your skills by completing the following task.

Materials needed:

Workplace documents such as plans, schedules, safety manuals, safe operating procedures

Pen and paper for note taking

Access to computer software or web pages for creating documents and/or filling in online forms

Sample meeting agendas or related documents

To demonstrate competency, a candidate must satisfy all the elements, performance criteria and foundation skills of this unit by:

1 Conveying and receiving information and instructions to and from others

   Role play with your teacher or a co-worker to practise sending and receiving information and instructions. Use house plans, safety manuals, safe operating procedures or codes of practice for this activity. Practise active listening and clarifying your statements and then summarising to check comprehension. Watch for verbal and non-verbal cues.

2 Accessing, interpreting and sequencing information

   Using workplace documents such as work schedules, safety manuals or site plans, practise logically writing down procedures in point form in order of operations. Check each other's interpretation of the information accessed and sequenced to see if it is a workable solution.

3 Presenting information in verbal and written reports

   Prepare a verbal and written report about the information you used to complete the previous activities. Report on how those activities progressed. Use language appropriate for your audience.

4 Entering information into workplace records and documents

   Use workplace software or tools to perform data entry for items such as your timesheet, an invoice or docket, or receipt from a supplier.

5 Participating in simple meeting processes

   Review a simple meeting agenda and any related documents prior to a simulated meeting.

   Practise listening and participation in the meeting by asking questions, providing feedback and contributing to the discussion.

5

IDENTIFY CONSTRUCTION WORK HAZARDS AND SELECT RISK CONTROL STRATEGIES

This chapter covers the following topics from the competency CPCWHS3001 Identify construction work hazards and select risk control strategies:
- plan and prepare
- prepare and implement job safety analysis (JSA)
- prepare and implement safe work method statement (SWMS) for high-risk work.

This unit of competency specifies the skills and knowledge required to participate in preparing a job safety analysis (JSA) for general hazards, and a safe work method statement (SWMS) for high-risk work hazards on construction sites as specified in work health and safety (WHS) legislation. It includes meeting all relevant requirements of the National Construction Code (NCC), Australian Standards and Commonwealth and state or territory legislation.

Not a prerequisite, but ...

If you have started in the construction field, you should have already completed a unit of competency CPCCWHS1001 Construction induction course and received a 'white card'. This is a mandatory requirement in every state and territory to work in any capacity in the construction field. It will also give you an insight into this chapter as basic information would have already been covered in the course. This chapter will be a good refresher for any participant. Figures 1.2–1.5 in Chapter 1 show samples of some white cards that are acceptable across state borders.

1. Plan and prepare

Before commencing any job task, you must first plan and prepare. Conduct a workplace assessment to identify risks and select appropriate personal protective equipment (PPE) or other measures. You must also determine the level of risk for each hazard you have identified. This section will take you through how to do this.

1.1 Review job task, work site and compliance requirements

Whatever job task you are being asked to do, you need to consider that task carefully:

1 Look at what can harm you, your work colleagues or members of the public.
2 Decide what you are going to do to complete the task.
3 Write down or record how you are going to complete the task.
4 Tell your colleagues what you have decided and ask for input.
5 Modify the process of completing the task and amend if necessary.
6 Complete the task.

Understanding the safety documents and legislation requirements

The main documents that affect all workers in the various states and territories are based on the federal *Work Health and Safety Act 2011* (see Table 2.1).

It is worth noting that Western Australia has not implemented the model *Work Health and Safety Act 2011* but Victoria's WHS legislation has changed as of 22 September 2021 and can be viewed on the Victorian Government website by searching for *Workplace Injury Rehabilitation and Compensation Act 2013* (legislation. vic.gov.au).

Extra risks in any workplace: COVID-19

All workplaces, whether they are in construction, hospitality, retail, health care or other industries, must have a COVID-19 safety plan. This is still a dynamic situation as different states will have different rules. However, the basics are:
- provide staff with information and training on COVID-19 which includes:
- where to get tested
- physical distancing
- wearing of masks
- hand sanitiser
- cleaning.

Contact your state or territory health department for the latest information.

Reviewing job tasks

You may have done a task many times and not given a second thought to reviewing how that task is done,

but that is when 'complacency will breed contempt'. A good example of this is driving forklifts. When you first train to be a forklift driver, the training requires you to do a preliminary check of the machine before starting it up. You may do this a few times because you are unsure of the consequences if you don't, but after a time you tend to cut corners as you know the machine, you only drove it half an hour ago and the boss wants you to unload a truck as soon as possible.

Consequence: you forgot to notice a pool of oil under the hydraulics – 'O' rings have burst.

You load up the forks and as you take the load off the truck the hydraulics completely fail and the load comes crashing down.

Possible outcomes:
- load is damaged
- machine is damaged
- injuries to the truck driver
- injuries to the forklift driver
- investigation by local authority
- fines and time off work for the injured parties
- larger fines for the company.

These outcomes could have been avoided if you had reviewed the job task and checked the machine.

Reviewing the work site

Any time you arrive on a work site, especially in construction, things have changed due to the nature of the work to meet the end goal of a finished product. For this reason, you must be involved with 'toolbox' meetings. Toolbox meetings must be attended by any worker who is working on the site that day. These meetings are held every morning to discuss what is happening on the site that could involve potential risks.

Things that could be discussed are:
- any new machinery coming on site and where it might be working
- what is happening with scaffolding
- any excavation work going on
- changes to storage of material
- site access changes
- any near misses that haven't been reported
- SafeWork inspections.

The talk shouldn't be onerous and there should be a time limit of 15 minutes. Most toolbox meetings take between 5 and 10 minutes.

Make sure everyone's attendance is recorded. The key factors are that the discussion is relevant to the work being carried out and that it covers the health and safety risks and controls relating to that work.

Reviewing compliance requirements

With legislation and compliance issues constantly changing, the construction manager, project manager or architect must keep their finger on the pulse. Changes to the NCC/BCA usually occur every two years but can be every 12 months. Australian Standards can be upgraded, codes of practice can be upgraded, council legislation can be amended and methods of construction can be changed.

For this reason, the people in charge must be diligent in reviewing compliance issues, having notifications sent to them or subscription services to the various items mentioned above. Another key resource would be through an association such as the Housing Industry Association (HIA) or the Master Builders Association (MBA), which keep their members well informed of compliance requirements.

1.2 Select and use personal protective equipment (PPE) for each part of the task

As can be seen in the hierarchy of controls (Figure 5.1), personal protective equipment is usually a last resort control measure to reduce risk. However, sometimes it is the only control measure, especially with hand or fixed power tools.

Types of PPE and their uses

PPE can be grouped according to the part of the body it will protect:

- head – safety helmets, sun hats
- eyes/face – safety spectacles, goggles, face shields
- hearing – ear muffs, ear plugs
- airways/lungs – dust masks, respirators
- hands – gloves, barrier creams
- feet – safety boots and shoes, rubber boots
- body – clothing to protect from sun, cuts, abrasions and burns; high-visibility safety garments and fall protection harnesses.

Safety helmets

Wearing safety helmets on construction sites may prevent or lessen a head injury from falling or swinging objects, or through striking a stationary object. Safety helmets must be worn on construction sites when:

- it is possible that a person may be struck on the head by a falling object

First, identify and assess the risks, then decide the best way to control them by applying the Hierarchy of Controls as follows:

LEVEL	CONTROL	DEFINITION
Level 1	Elimination	Controlling the hazard at source
Level 2	Substitution	Replacing one substance or activity with a less hazardous one
	Isolation	Separating the hazard from the person
	Engineering	Installing guards on machinery
Level 3	Administration	Implementing policies and procedures for safe work practices
	Personal Protective Equipment	Use of safety glasses, hard hats, protective clothing, etc.

Hierarchy of Controls

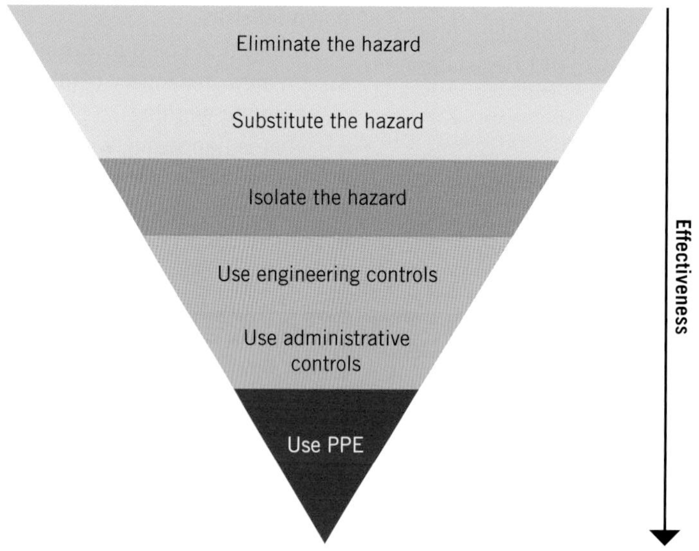

Eliminate the hazard

Substitute the hazard

Isolate the hazard

Use engineering controls

Use administrative controls

Use PPE

Effectiveness

FIGURE 5.1 Hierarchy of controls

- a person may strike his/her head against a fixed or protruding object
- accidental head contact may be made with electrical hazards
- carrying out demolition work
- instructed by the person in control of the workplace. Safety helmets must comply with AS/NZS 1801 Occupational protective helmets, and must carry the AS or AS/NZS label, and must be used in accordance with AS/NZS 1800 Occupational protective helmets – Selection, care and use.

Sun shade

The awareness of skin cancer for building workers is increasing. The neck, ears and face are particularly exposed. Workers should wear sun protection at all times when working outdoors (including in the winter). Sun shades include wide-brimmed hats and foreign legion-style sun shields fixed to the inner liner of safety helmets, or safety helmet 'foreign legion sun brims'.

Eyes/face protection

The design of eye and face protection is specific to the application. It must conform to AS/NZS 1337.1 Eye protectors for industrial applications. The hazards to the eyes are of three categories:

1 physical – dust, flying particles or objects, molten metals
2 chemicals – liquid splashes, gases and vapours, dusts
3 radiation – sun, laser, welding flash.

The selection of the correct eye protection to protect against multiple hazards on the job is important. Most eyewear is available with a tint for protection against the sun's UV rays, or may have radiation protection included.

Face shields

Face shields give full face protection, as well as eye protection. They are usually worn when carrying out grinding and chipping operations, when using power tools on timber. Shields are worn for full-face protection when welding. The shield may come complete with head harness or be designed for fitting to a safety helmet.

Hearing protection

You should always wear ear protection in areas where loud or high-frequency noise operations are being carried out, or where there is continuous noise. Always wear protection when you see a 'Hearing protection must be worn' sign, and when you are using or are near noisy power tools. The two main types of protection available for ears are:

- ear plugs – semi- and fully disposable
- ear muffs – available to fit on hard hats where required.

Choose the one that best suits you and conforms to AS/NZS 1270 Acoustics – Hearing protectors.

Disposable dust masks

Dust masks are available for different purposes and it is important to select the correct type. If the work that you are undertaking is mowing or general sweeping, a nuisance-dust mask is appropriate and is designed to filter out nuisance dusts only. A nuisance-dust mask is easily recognised as it has only one strap to hold it onto the face. If, on the other hand, you are working with toxic dusts (e.g. bonded asbestos), you will require greater protection. A mask with two straps and labelled with either P1 class particle dust filter (minimum protection) or P2 class particle dust filter (mid-range protection) will be required.

Respirators

Half-face and full-face respirators have cartridge type filters that are designed to keep out dusts, smoke, metal fumes, mists, fogs, organic vapours, solvent vapours, gases and acids, depending upon the combination of dust and gas filters fitted to the respirator. Cartridge type filters can be identified by the classification ratings from AS/NZS 1716 Respiratory protective devices.

Respirators fitted with P2 class particle dust filter are suitable for use with the general low-toxic dusts and welding fumes that are commonly found on construction sites. Some full-face or half-face respirators may also be connected to an air supply line or bottle that provides clean filtered air to the user. These are generally used for loose (friable) asbestos removal or in contaminated or low oxygen environments, where face and eye protection is also required. Further information on respirators and dust masks should be obtained from the manufacturers. It is very important to be trained in the correct methods of selecting, fitting, wearing and cleaning of the equipment in accordance with AS/NZS 1715 Selection, use and maintenance of respiratory protective equipment. When selecting, it is important that tight-fitting respirators and masks must have an effective seal on the face to ensure that all air entering your respiratory passages has been fully filtered.

Gloves

Gloves are used to protect hands and arms from both physical and chemical hazards. Stout leather gloves are required when handling sharp or hot materials. Rubberised chemical-resistant short or long gloves are used when handling hazardous chemical substances. Gloves should conform to AS/NZS 2161.1 Occupational protective gloves – Selection, use and maintenance.

Creams

Barrier creams may be used when gloves are too restrictive, to protect the hands from the effects of cement and similar low-toxic hazards.

Foot protection

It is mandatory to always wear protective footwear at the workplace. Thongs are not permitted at any time. Footwear should conform to AS/NZS 2210.1 Safety, protective and occupational footwear – Guide to selection, care and use. All safety footwear must have:

- stout oil-resistant, non-slip soles or steel midsoles to protect against sharp objects and protruding nails
- good uppers to protect against sharp tools and materials
- reinforced toecaps to protect against heavy falling objects.

Safety boots should be worn in preference to safety shoes on construction sites to give ankle support over the rough terrain. Safety joggers may be required when carrying out roof work or scaffold work; they must have reinforced toecaps. Rubber boots should be worn when working in wet conditions, in wet concrete, or when working with corrosive chemicals. They must have reinforced toecaps.

Clothing

Good quality, tough clothing is appropriate for construction work. It should be kept in good repair and cleaned regularly. If the clothing has been worn when working with hazardous substances, it should not be taken home to launder but sent to a commercial cleaning company; this will prevent the hazards from contaminating the home and the environment. A good fit is important, as loose-fitting clothing is easily caught in machine parts or on protruding objects. Work pants should not have cuffs or patch pockets, as hot materials can lodge in these when worn near welding or cutting operations. Clothing should give protection from the sun's UV rays, cuts, abrasions and burns.

Industrial clothing for use in hazardous situations should conform to AS/NZS 4501.2 Occupational protective clothing – General requirements.

GREEN TIP

Don't take personal protective equipment home with you that has been contaminated from working in a hazardous environment. Read the SDS for materials or products and study the important information about how the material, and its use, affects people and the environment.

Fall protection harnesses

In some instances when working at heights, it may be necessary to wear a harness. These are specialist items that need to be fitted to the individual. A harness must be correctly fitted or serious injury may result if or when the person falls. All harnesses must comply with AS 2626 Industrial safety belts and harnesses – Selection, use and maintenance.

Cleaning and maintenance

All PPE must be cleaned and maintained on a regular basis. This must be done by someone who has been trained in inspection and maintenance of such equipment. Remember, your life and wellbeing depends on this PPE; if it is faulty or damaged or simply not functioning properly, you are at risk.

Other PPE

Other than the items previously listed, you must wear high-visibility vests when working near machinery or near public roads, or it could be a condition of entry to the site. When working at heights, if no scaffolding or edge protection is provided, fall protection harnesses must be worn, the worker must undergo associated training and there must be a recovery plan if the worker does fall off the roof. For more detail on this topic, see G.P. Costin, *Construction Skills*, 4th edn, Cengage, 2024, Chapter 1.

1.3 Inspect work site and identify hazards relevant to job task

No matter what trade you are in, there will be hazards normally associated with those tasks. The most common types of hazards in any area are slips, trips and falls.

- Slips occur when there is little friction or traction between footwear and the walking surface. The type of footwear can be an issue, depending on where you are working. Most job sites would require steel cap boots with rubber non-slip soles. This type of footwear is fine for general building work. However, if you are working on roofs, it would be better to wear safety joggers with a non-slip sole and reinforced toecaps.
- Trips occur when the foot collides with an obstruction or there is foot contact with a highly tactile surface, resulting in the loss of balance and a likely fall.
- Falls occur because of either a slip, trip or loss of balance, or where the surface a person is standing on or slipping onto collapses or moves from underneath their footing. This causes the person to rapidly descend from a height or on the same level to the ground or lower level without control.

With any of these issues, the most common injuries are strains and sprains. View the YouTube video 'Sprains and Strains' from SafeWork NSW by searching for sprains and strains videos.

Everybody is responsible for safety on the work site. For some more WHS tips on legislation and what to do (and not to do) on the work site, view the YouTube video 'Funny Workplace Safety Training Video' from Channel 1 Creative Media at https://www.youtube.com/watch?v = Fcst9n5bgh4.

Musculoskeletal injuries to the body can have long-term implications on your ability to enjoy a full active lifestyle. By being careful and planning your work to avoid these types of injuries you can enjoy a long engagement with the construction industry and an active lifestyle when not at work. Take the time to research this type of injury with the work health safety regulator in your state or territory so you can avoid becoming injured.

Inspecting the work site

The process for inspecting the work site is as follows:

1 Collect information – Collecting relevant information from your local WorkSafe department will give you checklists, method of approach and how to record.

2 Consult outside groups – Interested parties such as associations (e.g. HIA and MBA) or unions (e.g. CFMMEU) will have many ideas on how to conduct these inspections.

3 Consult government bodies – Your local state or territory will have numerous publications on this issue and can guide you. Safe Work Australia has published a model code of practice on how to manage work health and safety risks. SafeWork NSW has published a pocket guide to construction safety. See below for links to these important resources:

- Safe Work Australia, *Code of Practice: How to manage work health and safety risks*, https://www.safeworkaustralia.gov.au/system/files/documents/1901/code_of_practice_-_how_to_manage_work_health_and_safety_risks_1.pdf
- SafeWork NSW, *The Pocket Guide to Construction Safety*, https://www.safework.nsw.gov.au/__data/assets/pdf_file/0004/386446/pocketguide-to-construction-safety.pdf

4 Conduct regular inspections on the work site, have your checklists ready and look for equipment becoming worn over time. Look for ergonomic issues, electrical hazards, equipment operation, fire protection, scaffolding issues, etc. Also, take photos or videos of hazardous conditions to keep a visual record that could be used in a toolbox meeting.

5 Investigate workplace incidents thoroughly to identify hazards.

6 Review any illnesses, near misses or close calls and injuries that have occurred and put in an action plan for prevention of any further occurrences.

1.4 Determine and record level of risk for each identified hazard

When you have identified the hazards that you have come across in the workplace, such as those in Figure 5.2, determine and record the level of risk for each identified hazard.

Several hazards can be seen in Figure 5.2:

1 messy floor – could cause slips and trips – clean up floor before starting work

2 power cord on ground near feet – another trip hazard. Does not comply with WHS of having power cords out of the way. Has it been tagged? There is a tag on the black cord, but is it in date?

3 no offcut bin – provide a bin for reuse of waste material

4 no vacuum system for dust – connect a vacuum system to the compound saw at the rear

FIGURE 5.2 What hazards can you identify in this image?

Source: Richard Moran

5 inappropriate footwear – choose appropriate safety footwear

6 no ear protection – anything with a decibel rating of 85 dB must have ear protection

7 no eye protection – chips of timber can easily fly into the worker's eye – wear eye protection

8 untidy workbench – if the worker cuts through the timber with the slide compound saw, it might hit the piece of pine lying across the bench

9 material not held firmly with clamps or similar – timber could slip

10 if the worker is working outside – no protective sunhat, or the work site might have a hard hat rule that is not being observed

11 clothing – no long sleeves for outdoors, but the worker could have on sunscreen.

Applying the risk rating

Now that you have determined that there are hazards associated with the picture in Figure 5.2, you must look at each hazard and give it a risk rating. This is usually done with the workers at the work site, with management involvement.

There are various models of risk ratings, but all are a matrix of the likelihood of something happening together with the consequence if it does happen (see Figure 5.3).

With all parties agreeing to the risk rating, they then sign off on the risk rating so everybody takes ownership of that process.

Under the *Work Health and Safety Act 2011*, you must put in controls of risk levels 1 and 2 or high and extreme, depending what matrix you are using. Using the hierarchy of controls (see Figure 5.1), you then re-evaluate the risks associated with that hazard.

Source: Small Business Development Corporation, Government of Western Australia

Likelihood / Consequence	Level	Rare The event may occur in exceptional circumstances. Less than once in 2 years 1	Unlikely The event could occur at some time. At least once per year 2	Moderate The event will probably occur at some time. At least once in 6 months 3	Likely The event will occur in most circumstances. At least once per month 4	Certain The event is expected to occur in all circumstances. At least once per week 5
Negligible No injuries, Low financial loss	0	0	0	0	0	0
Minor First-aid treatment Moderate financial loss	1	1	2	3	4	5
Serious Medical treatment required High financial loss, Moderate environmental implications, Moderate loss of reputation, Moderate business interruption	2	2	4	6	8	10
Major Excessive multiple long-term injuries, Major financial loss, High environmental implications, Major loss of reputation, Major business interruption	3	3	6	9	12	15
Fatality Single death	4	4	8	12	16	20
Multiple Multiple deaths and serious long-term injuries	5	5	10	15	20	25

Legend

Risk Rating	Risk Priority	Description
0	N	No Risk: The costs to treat the risk are disproportionately high compared to the negligible consequences.
1–3	L	Low Risk: May require consideration in any future changes to the work area or processes, or can be fixed immediately.
4–6	M	Moderate: May require corrective action through planning and budgeting process.
8–13	H	High: Requires immediate corrective action.
15–25	E	Extreme: Requires immediate prohibition of the work, process and immediate corrective action.

FIGURE 5.3 Risk matrix diagram

Hopefully, the risk has been reduced, resulting in a safer work environment.

To see how the risk matrix in **Figure 5.3** works, pick hazard 7 relating to **Figure 5.2** – no eye protection.

The likelihood of getting something in the operator's eye would be classed as likely and the consequence could be classed as serious. Now intersect the two and you will see a figure 8 in an orange square.

Now look at the risk priority; it comes up with a high factor which 'requires immediate corrective action'. In other words, wear eye protection or PPE as the control measure.

Review the hazard – it has now been removed, so the likelihood of it happening unless you have faulty eye protection is rare. The consequence is negligible, i.e. no risk for hazard 7.

COMPLETE WORKSHEET 1

2. Prepare and implement job safety analysis (JSA)

2.1 Review requirements of work health and safety legislation for preparation of job safety analysis (JSA) using template

Now that you have identified hazards and determined their associated risk, it is time to prepare and implement a job safety analysis (JSA). This will involve reviewing relevant legislation, and determining other factors associated with the recording and storing of a JSA.

A JSA is a form of risk assessment, which details step by step how a task is to be carried out safely. Sometimes the JSA is called a job safety and environmental analysis (JSEA) and includes environmental aspects as well. It is the employer's duty of care to ensure workers are trained and competent for their work. There are no specific legal requirements to have a JSA or any regulation prescribing the format or content for JSAs. Another definition is a systematic examination of a job through the identification of potential hazards, risk assessment and evaluation of practical measures to control the risk.

The main components of a JSA are:

- tasks – a step-by-step list of basic activities of the task, e.g. prestart checklist of machinery
- hazards – a list of potential hazards at each step of the task
- control measures – controlling each identified hazard using an approach that is logical, links up with the hierarchy of controls and is cost-effective.

Work health and safety legislation for preparing a JSA

The federal legislation is the *Work Health and Safety Act 2011*.

Every state has adopted this Act except Western Australia, with Victoria only adopting it in September 2021. Refer to Table 1.2 in Chapter 1 to see which state department is the regulating authority and is responsible for this Act.

In New South Wales, JSAs are not usually used and a safe work method statement (SWMS) is the usual document that covers any aspect, sometimes called a safety management sheet. For all other states, SWMSs are for prescribed high-risk construction work (HRCW).

Documents that should be accessed before attempting to prepare a JSA are:

- *Work Health and Safety Act 2011*, https://www.legislation.gov.au
- information sheets from the various government departments in each state or territory – see Table 1.2 for details
- codes of practice – practical ideas on how to achieve standards of health and safety required under the *Work Health and Safety Act 2011*; to have legal effect, it must be approved by the various states or territories
- standard operating procedures (SOPs) – a set of step-by-step instructions compiled by an organisation to help workers carry out routine operations.

JSA template and example

Look at the JSA template and example in Figure 5.4 and follow the tips on how to use it.

2.2 Break job task into logical steps, determine tools, equipment, plant and materials to be used for each step, and record on JSA

The second column, 'Job tasks', lists the tasks associated with building the wall frames. The sixth column, 'Tools, equipment, plant and material needed', lists tools and equipment needed to complete that task. You can get information on what types of tools, equipment, plant and materials that might be used for job tasks by looking at the material installation guidelines from the manufacturer of the material and also by using codes of practice and talking to a competent person about what might be required to complete a task.

Date:			Is this JSA:	⊠ New ☐ Reviewed	JSA Number:	1234-2021
Do personnel require a licence / ticket / permit / trade skill / other:				⊠ Yes ☐ No	Specify:	Carpentry trade

Exact location: 123 Nowhere Street, Somewhere
Task: Cut, assemble and erect timber wall frames for a new home on concrete slab
Reference materials: AS 1684, NCC book 2

Once developed, the JSA must be signed below by each JSA team member involved in the development of the JSA.

JSA team members

Name	Position	Signature
Bill Bloggs	Carpenter	
Sammy Sparrow	Leading hand	
Harry Houdini	Labourer	
Harry Potter	2nd year apprentice	

This JSA is not valid until signed by the site supervisor or nominee.
Each section of the work sheet must be completed before it can be validated and before work can start.

Authorisation

Site supervisor / nominee:	Mark Wyborn	Signature:		Date:	
JSA team leader:	Sammy Sparrow	Signature:		Date:	

Hazard / energy	Guidelines for assessing if a hazard is present	Recommended control measure – the better control is the first control measure listed in each section below
NOISE **Acoustic mechanical vibrations energy**	Would you have to shout to be heard less than a metre away from the person to whom you're speaking? YES ☐ NO ⊠	• Switch off the source of the noise where possible. • Move work to a quieter area. • Erect a sound-absorbing barrier between employee and source. • Work in rotating teams to reduce the employees' exposure time. • Specify the hearing protection required (plugs, muffs or both).
HEAT **Thermal energy**	Is there a risk of burns/scalds, cold burns, heat exhaustion, sunburn? YES ⊠ NO ☐	• Install a barrier between heat source and employee. • Work in rotating teams to minimise long exposure to heat or cold. • Supply personal cooling devices. • Ensure adequate cool drinking water is available. • Wear additional clothing, gloves, boots.
LIGHTING	Is the lighting good enough to see where you are and what you are doing clearly? YES ⊠ NO ☐	• Install additional and/or improve the permanent lighting (low voltage in confined spaces). • Move the current lighting to achieve best effect (out of shadows). • Move the job being undertaken to well-lit area. • Install temporary lighting.
AIRBORNE SUBSTANCES **Chemical energy**	Are there any airborne contaminants released or generated when performing this task? YES ☐ NO ⊠ If so, what sort are they? (e.g. welding fumes, dusts)	• Reduce the dust or fumes by wetting down. • Enclose the source of the dust. • Install permanent or temporary extraction ventilation to remove dust into drum for disposal. • Clean up all spills immediately, and vacuum if dry. • Provide and instruct in use and maintenance of respiratory protection.

FIGURE 5.4 JSA template and example

Hazard / energy	Guidelines for assessing if a hazard is present	Recommended control measure – the better control is the first control measure listed in each section below
CHEMICALS **Chemical energy**	Does the task involve the handling or the use of chemicals? YES ☐ NO ☒ If yes, please list the types of chemicals (e.g. sodium chlorate, diesel)	• Source a less hazardous chemical. • Install a temporary or permanent barrier between employee and chemical. • Reduce the volume of chemical stored or used. • Minimise the time the employee is exposed to the chemical. • Identify the need for specific permits and/or gas testing (e.g. confined spaces). • Refer to safety data sheet (SDS), and always specify the use of the appropriate personal protective equipment (PPE).
GASES **Chemical energy**	Are there any gases released or generated when performing this task? YES ☐ NO ☒ If so, what sort are they? (e.g. smells, SO_2)	• Dilute the gas by doing the job in open air or well-ventilated place. • Contain the gas by installing a permanent or temporary enclosure around the source. • Remove the gas by extraction ventilation or vacuum. • Instruct employees in the use and maintenance of appropriate personal protective equipment (PPE).
PLANT, MACHINES AND EQUIPMENT **Kinetic or potential energy**	Is plant conveyors and/or machine moving parts exposed which can be guarded? YES ☒ NO ☐ Are additional emergency stop mechanisms required to prevent risk of injury? YES ☒ NO ☐ Are there any potential electrical, mechanical or pneumatic hazards? YES ☒ NO ☐	• Specify the correct machine or piece of equipment to do the job. • Identify all the protective guards, grating, mesh which must be in place. • Ensure the correct signs are in place (e.g. this machine starts automatically). • Specify the signs and/or barricades required (e.g. bunting, no entry, authorised personnel only, restricted access). • Specify the type of permit required. • Specify the isolation required; electrical, high voltage, mechanical and pneumatic (air) or other energy sources.
HAND TOOLS **Biomechanical energy**	Will the task require the use of hand tools? YES ☒ NO ☐	• Specify the testing requirements for all electric hand tools and extension leads. • Specify any tools not to be used for the job. • Specify any personal protective equipment (PPE) related to using tools.

FIGURE 5.4 (*Continued*)

Hazard / energy	Guidelines for assessing if a hazard is present	Recommended control measure – the better control is the first control measure listed in each section below
HAZARDOUS MANUAL TASKS **Biomechanical energy**	Will you perform any of the following actions repeatedly? Bend down YES ☒ NO ☐ Reach above your head YES ☒ NO ☐ Reach forward YES ☐ NO ☒ Twist (at waistline) YES ☐ NO ☒ Maintain an awkward posture YES ☐ NO ☒ Are actions repeated frequently? YES ☒ NO ☐ Do you manually move loads over long distances? YES ☐ NO ☒ Does the task involve pushing, pulling or carrying loads? YES ☒ NO ☐	• Fix the item as part of a modular change-out rather than in situ. • Build or erect scaffolding to gain better access. • Specify need for scissor lift, cherry picker (or personnel cage) to gain better access. • Ensure that all employees are trained in correct lifting techniques. • Ensure that there are adequate numbers of employees to do the job. • Work in rotating teams to share the need to frequently lift or carry loads. • Limit the number of times the load must be moved by changing the drop-off or original storage/ destination point. • Use a mechanical lifting device where possible (e.g. crane, forklift, trolley). • Reduce the size or weight of the load to be carried or lifted (e.g. smaller bags, boxes, drums, containers).
SAFE WORKING AT HEIGHTS **Kinetic or potential energy**	Could an injury occur because of a person falling? YES ☐ NO ☐ Is a person required to work where there is a risk of falling from one level to another? YES ☐ NO ☐ Is a fall injury protection system the principal means of protection? YES ☐ NO ☐	• Use the safe work at height permit. • Have a safe working area by means of work platforms or scaffolds complete with floors, guardrails, kickboards, and a safe method of access and egress. • Use fall injury prevention systems to prevent falls and falling objects. • Wear protective helmets with chinstraps. • Use tool lanyards or tool belts. • Fit close-fitting floorboards and kick-rails and netting. • Practise good housekeeping, signage and drop-zone barricading to prevent injuries from falling objects.

FIGURE 5.4 (*Continued*)

Hazard / energy	Guidelines for assessing if a hazard is present	Recommended control measure – the better control is the first control measure listed in each section below
	Does a person need to exit from an elevated work platform (EWP) in the raised position? YES ☐　　NO ☐ Creating an open hole with edge protection, floor or walkway? YES ☐　　NO ☐ Scaffolding is to be erected or dismantled? YES ☐　　NO ☐ Is work to occur on or near the edge of a fragile surface? YES ☐　　NO ☐ Could an injury occur because of an object falling? YES ☐　　NO ☐	
CONCURRENT OPERATIONS	Are other jobs/tasks in progress which could pose an interaction risk to employees carrying out this task? YES ☐　　NO ☐ Are there other jobs/tasks in progress which could be put at risk by carrying out this task? YES ☐　　NO ☐ Is there a risk from accidental falling objects, spillage or other interactions, accidental or otherwise, between this task and any other concurrent tasks being carried out? YES ☐　　NO ☐	• Reschedule work. • Provide controls, such as area/vessel isolations, or drop-zone barricading and signage, to prevent injuries from falling objects, spillage or other interactions.
Cutting frame	Does the task/job require personnel, materials or equipment to be lifted such that a suspended load risk is created? YES ☐　　NO ☒	

FIGURE 5.4 (*Continued*)

Step	Job tasks *List the key job steps required to perform the task, in the sequence in which they are to be carried out.*	Hazards *List the identified hazards with each step.*	Existing controls *List the controls already in place (e.g. safe operating procedure, trained competent operator).*	Assess risk *1–5* *1 & 2 must be controlled*	Tools, equipment, plant and material needed	Items completed
1	Unload timber from truck and stack	Splinters, strains	Use of truck crane with trained operator, rigger's gloves, hi-vis	4	Crane unloading from truck, gluts evenly spaced to stack timber, hi-vis vest, hard hats	
2	Prepare and check tools	Electric shock, misfires	Electrical tested and tagged, EPT serviced regularly	7	Visual check of leads, safe tags in place	
3	Set up cutting bench	Slips, trips and falls	Clean site, erect stands for power leads	4	Piggy tail stands, roller cutting bench with placement for compound mitre saw	
4	Read plans and mark out plates	Misread dimensions, slips, trips and falls, sore back – bending over	Check measurements twice before cutting, clean area for marking out, use correct height saw stools for marking out	2	Correct height saw stools, tape measure, pencil, plans	
5	Create cutting lists	Slips, trips and falls	Clear site	2	Site diary, cutting list	
6	Cut all necessary parts	Kickback on saw, sawdust and noise	Clear site, use of PPE – ear and eye protection, clamping of material to cutting bench	10	PPE, compound mitre saw, cutting checklist, marker for pieces	
7	Nail frame together	Misuse of nailing gun, fingers jammed under frame, splinters, slips, trips and falls	Keep site clean, trained operators of nail guns, rigger's gloves, PPE	10	PPE, Paslode nail guns	
8	Stand and erect frame, brace frame	Slips, trips and falls, strains	Keep site clean, correct lifting techniques used	8	Ropes, PPE	
9	Shoot plate to line into concrete	Noise, misfires	Trained operator, PPE, SOPs for Ramset gun	10	Ramset gun with appropriate charges and fixings, chalk line	
10	Clean-up	Dust and debris, slips, trips and falls	Vacuum dust at rear of saw, appropriate dust mask, clear rubbish away as you go	5	Tool vacuum on saw, broom, shovel, bins, PPE	

FIGURE 5.4 (*Continued*)

All personnel working under this JSA must sign off below to acknowledge they have read, understood and accepted the conditions stipulated.

Name	Signature	Position	Date	Time

FIGURE 5.4 (*Continued*)

Changes to the JSA List new or changed job steps and identify the hazards and risks	The nominated person responsible for the job must consult with, communicate to and instruct all concerned with the work. All persons working on the job must be aware of the changed/new job steps, hazards and risk controls.

Step	Job tasks *List the key job steps required to perform the task, in the sequence in which they are to be carried out.*	Hazards *List the identified hazards with each step.*	Existing controls *List the controls already in place (e.g. safe operating procedure, trained competent operator).*	Assess risk *Initial rank*	Additional controls *List identified additional control measures that are to be implemented.*	Assess risk *Residual rank*
1	Stand, erect and brace frame; cypress pine, not radiata	Slips, trips, falls and sprains	Keep pathways clear, correct lifting techniques used	8	Use of HIAB crane to lift frames into position once assembled and braced for lifting	5
	I UNDERSTAND THE ABOVE CHANGES, INCLUDING THE HAZARDS, RISKS AND THE CONTROL MEASURES. TO BE SIGNED ALONG HERE BY ALL PERSONS WORKING ON THIS JOB.					
2.						
	I UNDERSTAND THE ABOVE CHANGES, INCLUDING THE HAZARDS, RISKS AND THE CONTROL MEASURES. TO BE SIGNED ALONG HERE BY ALL PERSONS WORKING ON THIS JOB.					

FIGURE 5.4 (*Continued*)

Rating	Safety	Health	Environment	Equipment and assets	Business continuity	Community and reputation	Liability
1 **Minor**	Single minor injury to one person. First aid or no treatment required. No lost time.	Reversible health effects of minor concern requiring first aid treatment at most.	Issues of non-continuous nature with promptly reversible impact or consequence (e.g. within shift). Low-level incident, site contained.	Below $5000 (or 0.01% of operational budget based at $50 000 000).	Loss of operations for > one day. Reduction in capacity, < 10% for up to one month.	Unsubstantiated, low media profile or no media attention. One-off complaint, which is resolved via existing procedures.	Below $50 000 (or 0.1% of operational budget based at $50 000 000).
2 **Moderate**	Medically treated injury. Reversible injury. Requires treatment but does not lead to restricted duties.	Reversible health effects of concern that result in medical treatment but not restricted duties.	Issues of a non-continuous nature and minor impact and consequence. Low-level incident, site contained. Short-term reversible (e.g. within days).	Between $5000 and $50 000 (or 0.01% to 0.1% of operational budget).	Loss of operations for > one da0y. Reduction in capacity, < per 20% for up to one month. Minor disruption to supply of services or technical support.	Substantiated, low impact, low media profile. Unresolved, low-level community dissatisfaction. Repeated community complaints.	Between $50 000 and $250 000 (or 0.1% to 0.5% of operational budget). Financial or accounting issue with ability to resolve with existing resources.
3 **Serious**	Reversible injury or moderate irreversible impairment. Less than 10 days lost time.	Severe but reversible health effects. Results in a lost time illness of less than 10 days.	Issues of a continuous nature – limited impact and consequence. Incident resulting in some site contamination. Medium-term recovery impact.	Between $50 000 and $500 000 (or 0.1% to 1% of operational budget). Threat to property by known extreme organisations.	Loss of operations for one day to one week. Reduction in capacity, < 30% for up to one month. Increased government interest.	Substantiated, public embarrassment, moderate media profile (front page, one day). Repeated community complaints. Community demonstration. Impact on share price.	Between $250 000 and $1 750 000 (or 0.5% to 3.5% of operational budget). Financial or accounting issue requiring chief financial officer (CFO) resolution.
4 **Major**	Severe irreversible damage to one or more persons. Lost time injury greater than 10 days.	Severe and irreversible health effects or disabling illness.	Compliance issue with large fine, media attention. Serious harm not immediately recovered. Significant site contamination or off-site impact. Long-term recovery.	Between $500 000 and $1 000 000 (or 1% to 2% of operational budget). Confirmed threats, without actions.	Loss of operations for one week to one month. Reduction in capacity, < 50% for up to one month. Regulatory inquiry.	Substantiated, public embarrassment, high impact, major media attention. Local or state media interest. Severe community dissent. Criticism from a non-government organisation (NGO) and/or government.	Between $1 750 000 and $5 000 000 (or 3.5% to 10% of operational budget).
5 **Catastrophic**	Single fatality. Permanent disabling injuries.	Life-threatening or permanently disabling illness.	Issues of a continuous nature with major long-term impact and potentially serious consequences.	Above $1 000 000 (or more than 2% of operational budget). Escalating threats or actions.	Loss of operations for > 1–3 months. Loss of permit to operate. Total loss of production for more than one month.	Substantiated, public embarrassment, multiple impacts, long-lasting widespread media coverage. Severe, prolonged community dissent.	Above $5 000 000 (or more than 10% of operational budget).

FIGURE 5.4 (Continued)

Risk matrix

		Consequence				
		1 **Minor**	**2** **Moderate**	**3** **Serious**	**4** **Major**	**5** **Catastrophic**
Likelihood	**A** **Almost certain**	10	16	20	23	25
	B **Likely**	7	12	17	21	24
	C **Possible**	4	8	13	19	22
	D **Unlikely**	2	5	9	14	18
	E **Rare**	1	3	6	11	15

Risk result	Rating	Definition	Level of involvement
Note when a potential consequence is classified as catastrophic, immediate and ongoing intervention is required from the CEO to ensure control measures are adequate.			
19–25	Critical	Imperative to eliminate or reduce risk to a lower level by the introduction of controls. Formal risk assessment required.	CEO needs to review.
11–18	High	Corrective action required. Normally permits required to perform work. Safe work procedure or job hazard analysis mandatory.	Site manager review required.
6–10	Moderate	Corrective action required. Safe work procedure or job hazard analysis required.	Supervisor review required.
1–5	Low	Corrective action where practical. Take 5 risk assessment required.	Manage by routine procedures at operational level.

Rating	Descriptor	Description	Suggested frequency
A	Almost certain	The event is expected to occur.	Recurring event during the lifetime of a project/operation (e.g. more than once per month)
B	Likely	The event will probably occur.	Event that may occur frequently during the lifetime of a project/operation (e.g. at least once per year)
C	Possible	The event should occur.	Event that may occur during the lifetime of a project/operation (e.g. once in three years)
D	Unlikely	The event could occur.	Event that is unlikely to occur during the lifetime of a project/operation (e.g. once in 10 years)
E	Rare	The event may occur only in exceptional circumstances.	Event that is very unlikely to occur during the lifetime of a project/operation (e.g. once in 15 years)

FIGURE 5.4 (*Continued*)

Breaking the job into logical steps helps you to gain a thorough understanding of what you will be doing to complete the job.

2.3 Identify work site and task-related hazards and levels of risk relating to each step, and record on JSA

The third column, 'Hazards', lists the hazards associated with each task. You will complete this section when making your own assessment. The fifth column, 'Assess risk', is used to record the risk assessment, which is determined using the risk matrix (see Figure 5.3). You can identify work site and task-related hazards and risk levels by referring to codes of practice, discussions with competent people, reading instruction manuals for tools, and studying installation guidelines and warnings for different products and materials. Safety data sheets also hold a lot of hazard information for products.

2.4 Apply hierarchy of controls to determine risk control strategies for each hazard in each step of the job task, discuss and confirm with relevant personnel, and record on JSA

You will see that certain control measures have been included in the fourth column, 'Existing controls'.

These were determined using the risk matrix (see Figure 5.3) and have been colour-coded as per the risk matrix. For each hazard in each step of the job task, discuss and confirm control measures with relevant personnel and record them in the JSA.

Always seek to avoid hazards by using other people's experience, documents based on research on how to do tasks safely, such as codes of practice and user instruction manuals for tools and materials. Researching and applying correct procedures is the best way to avoid becoming injured from hazards in the construction industry.

2.5 Review work site and job task immediately before starting work and discuss JSA with relevant personnel to confirm as still applicable, or to amend as required

Before starting work, discuss the JSA with relevant personnel to confirm it is still applicable or to amend as required. You can see in the example JSA in Figure 5.4 that after discussing with the relevant staff it has been noted that the species of timber being used, which is cypress pine, is heavier than the usual radiata pine. So an amended section of the JSA has been completed (see Figure 5.4). This can happen on a job site, which throws out the JSA from the normal run of events. In this scenario, cypress pine has turned up instead of radiata pine due to low supply of radiata and has been delivered at no price increase. The issue is that it is much heavier than radiata and an amended JSA had to be completed.

2.6 Store JSA securely onsite in accordance with compliance requirements

Digital or paper documents – that is the question. In the never-ending effort to comply with the requirements of the various interested government departments and authorities, you must keep accurate records of the work that is being done. You must have a system to show that you are working safely and that every person on the job site complies with the requirements. If you have paper documents, they should be kept in a folder stored in the onsite office. There are many construction job management software programs that save time once they are set up. They also keep track of all your jobs, so if a similar job comes up the form is already pre-filled and only minor changes are needed.

Some examples of JSA software and safety management resources are:

- Job Safety Analysis Software, https://www.vectorsolutions.com/solutions/vector-ehs-management-software/job-safety-analysis/
- Smart JSA Software, https://sitemate.com/safety/jsa-software/

- Managing your safety, https://hia.com.au/resources-and-advice/managing-your-safety.

In all states other than New South Wales, you will use a JSA for all non-high-risk work. An effective JSA should:

- be short
- encourage honest conversation
- include a review of the emergency plan for the site
- provide evidence of the process
- be accessible by various stakeholders
- be developed by the people who are working onsite.

Whether they are stored digitally or are paper-based will be the user's choice. Whatever method is used, JSAs should be stored with other WHS forms, such as induction forms, audit forms, incident reports, staff training records, maintenance or asset logs and any safety notes such as conversations, toolbox talks, feedback or suggestions.

Every person working on site should have their own file that lists their qualifications and any licence requirements.

All of these documents must be readily available for inspectors from WorkSafe or any other authorities.

LEARNING TASK 5.2

CLASS/GROUP ACTIVITY: COMPARING THE COST OF ONE CONTROL FOR A HAZARD OVER ANOTHER

Pick a hazard from Learning Task 5.1 and, considering the controls that you've identified based on research documentation, make a note of two different controls that could cost different amounts of money to implement. Estimate the risk rating by considering the **consequences** of the hazard occurring, and the **likelihood** of the hazard occurring for the two different control options. Now compare and present the two risk ratings and the two costs and discuss this with your classmates or teacher.

 COMPLETE WORKSHEET 2

3. Prepare and implement safe work method statement (SWMS) for high risk work

A safe work method statement (SWMS) is a document that states the high-risk work activities that will be carried out on the work site, the hazards associated with these activities and the measures that will be put in place to control the risks.

3.1 Review requirements of work health and safety legislation for preparation of safe work method statements (SWMS)

Except for Western Australia (who use *Work Health and Safety Act 2020*), all states and territories now have adopted the *Work Health and Safety Act 2011*, which is administered by various government departments. Refer to Table 1.2, which lists the names and contact details of the regulating authorities. Safe Work Australia is the overarching lead authority responsible for harmonising WHS laws throughout Australia. It creates draft legislation as and when required.

No matter what state you live in, it is mandatory to conduct a SWMS for the following high-risk areas:

- risk of a person falling more than 2 metres
- likely to involve disturbing asbestos
- work in or near a shaft or trench deeper than 1.5 metres or a tunnel
- work on or near chemical, fuel or refrigerant lines
- tilt-up or precast concrete elements
- work in areas with artificial extremes of temperature
- work on a communication tower
- temporary load-bearing support for structural alterations or repairs
- work on, in or adjacent to a road, railway, shipping lane or other traffic corridor in use by traffic other than pedestrians
- work in or near water or other liquid that involves the risk of drowning
- demolition of a load-bearing structure
- work in or near a confined space
- work on or near pressurised gas mains or piping
- work in an area that may have a contaminated or flammable atmosphere
- work in an area with movement of powered mobile plant
- diving work.

In Western Australia high-risk areas also include:

- work involving removal of hazardous substances, such as lead paint or PCBs
- work in areas of contaminated soils
- exposure to excessive noise or vibration
- exposure to excessive dust emissions from work activities, plant or traffic.

In New South Wales, if a task is considered high-risk work outside the nominated high-risk construction work, then a SWMS must be prepared using a risk assessment.

How to conduct the risk assessment

SafeWork NSW has a checklist to be used when conducting a risk assessment. This outlines the correct steps to keep you and others safe. See the following link from the SafeWork NSW website and go to template 4, 'Your site-specific risk assessment template': https://www.safework.nsw.gov.au/your-industry/construction/house-construction.

3.2 Determine work site conditions and job task requirements

If you are a new contractor or starting up the work site, each phase of the build will present itself with challenges of all kinds. Not just the obvious hazards of excavations, working at heights, confined spaces or demolition work (to name a few).

You also must be aware of the following hazards:

- Ergonomic hazards – which occur from repeated movement or performing heavy tasks without the proper devices. A good example of an ergonomic hazard is a bricklayer who repeats the same movement constantly using a product that can range from 3 kilograms to 8 kilograms every time they lay a new brick or block.
- Biological hazards – which may or may not be easy to determine, and include working on a contaminated site such as an old service station site.
- Chemical hazards – which may be highlighted on the site, but further investigation may be needed. Examples of work sites where chemical hazards might be present include building on farmland that was heavily doused with chemicals, or old building sites where dieldrin was used for termite treatment. Safety data sheets (SDSs), previously called material safety data sheets (MSDSs), are useful sources of information to enable you to handle safely the chemicals on site. Any SDS will have:
  - the product name, ingredients and properties
  - who made or imported it
  - how it could affect your health
  - how to use and store it safely.
- Work organisation hazards – which could include bullying, various types of harassment, lack of role clarity, inadequate reward and recognition, poor support in the workplace, poor organisation skills. These hazards would be hard to quantify in a SWMS, which is meant to be for defined high-risk areas, but they could certainly be brought to attention in a toolbox meeting or reported to HR or WHS personnel.

3.3 Determine and record high-risk work site and task hazards relevant to job task

As most domestic, commercial and industrial builds will require working at heights greater than 2 metres, the associated risk consequence could be as high as 'catastrophic' and have a likelihood of 'almost certain'. If no control measures are put in place, the work site could be classed in the critical risk rating.

Of the designated high-risk areas as determined by Safe Work Australia, the number one spot is falling from heights 2 metres and above. Statistics show that most falls that cause injuries are between 2 metres and 4 metres. For heights greater than 4 metres, people tend to put in the correct safety procedures. Falling from heights is the number one killer in New South Wales on construction sites.

From personal experience, I fell less than 1 metre off a stepladder and broke a rib, suffered severe lacerations to the hand and had concussion. Reasons for my injuries were not clearing the site of debris and working on unstable ground, which are easy things to avoid. I was even coming down the ladder!

Falling from heights is a very serious matter. Even from lower heights, injuries can occur – as proved by the author. There are many options available to prevent you from falling, such as:

- different types of scaffold:
  - modular
  - unit frame
  - tube and fittings (pole and coupling)
- specially designed trestles with handrails and other anti-fall devices
- edge protection devices
- harness systems
- elevated work platforms.

It is not just falling from roofs or similar situations; it is also falling through voids such as stair openings where stairs have not yet been put in place. Trenches can also pose problems for falls, and special methods and considerations must be put in place. Australian Standards that cover this area can be found in G.P. Costin, *Construction Skills*, 4th edn, Cengage, 2024, pages 43CS and 44CS.

The reason why some tasks are called 'high risk' is because there is much greater risk of injury happening from doing the task. Take extra precautions when considering doing high-risk activities because it has been found that there is a much greater chance of something going wrong and someone being injured or something getting damaged.

3.4 Break job task into logical steps, determine tools, equipment and materials to be used for each step, and record on SWMS

Breaking the whole job into 'bite-size chunks' can be a tricky task and sometimes it can go overboard. Keep the SWMS as short as possible. Focus on the logical sequence of events in the job task and don't break it down too much. Select the tools that would normally be used and are common to the task. Any specialised equipment will have to be taken into consideration in the working at heights as well as licensing situation. In the SWMS on scaffolding, the installers must have the appropriate licence for the type of scaffold they are building, and if you are supervising the site ensure they are erecting the scaffold safely and in accordance with the law.

The SWMS must be easily understood by all workers, including those from a non-English speaking background. Including pictures and diagrams is very useful and will assist in ensuring the message is better understood.

In the SWMS template and example in Figure 5.5, the second column, 'Tasks to be done', highlights the tasks, while the tools and equipment are listed in the seventh column, 'Plant and equipment, licences needed'.

3.5 Identify high-risk work site and task-related hazards and levels of risk relating to each step, and record on SWMS

A visit to the site as well as looking at the plans will help determine if it is a high-risk work site. Having contour lines on the site plan will also help in determining levels of scaffold needed.

As you have already broken down the tasks at hand, you now must determine the hazards associated with those tasks. Using the likelihood and consequence risk matrix, in whatever form you choose, you will come up with an intersecting square or cell to make a judgement call on what rating of risk you will put on that hazard.

This process should be completed by the people who will be doing the work. Management may also get involved as they might be more aware of the legislation and Australian Standards and codes of practice.

In the SWMS template and example in Figure 5.5, the third column, 'Hazards', highlights the hazards, while the risk rating for each hazard is in the fourth column, 'Risk rating'. The risk ratings have been colour-coded the same as the matrix for easy identification.

3.6 Apply hierarchy of controls to determine risk control strategies for each high-risk hazard in each step of the job task, discuss and confirm with relevant personnel, and record on SWMS

Using the hierarchy of controls (Figure 5.1), what type of control methods will you apply to the risks that have been identified? For a refresher, the hierarchy of controls from the most effective to least effective are:

1 Eliminate the hazard – highest and most effective level of control
2 Substitution – substitute the risk with a lesser risk
3 Isolation – isolate people from the risk
4 Engineering – change the risk through engineering (e.g. guards on machines, auto feeds on spindle moulders)
5 Administrative controls – in a practical sense, it is creating JSAs and SWMSs
6 PPE – lowest level of protection but sometimes the only level of protection, protecting the worker from harm.

It is important that each high-risk hazard associated with each step in the job task is discussed and confirmed with the relevant personnel and recorded on the SWMS.

WHS FORM 05: SAFE WORK METHOD STATEMENT			
(PCBU) Cengage Construction Pty LTD ABN 265596642 Ph 03 8263 5551, 0489 963 123		Principal contractor (PC)	Cengage Construction Pty LTD ABN 265596642 3 Shortland Street, Port Melbourne
Work activity	Installing modular scaffold for new dwelling, height is 3 metres off ground but can be up to 5 metres at places.	Work location	37 Somewhere Place, Nunawading
High-risk construction work	Risk of a person falling more than 2 metres	Works manager	Joe Bloggs
		Contact phone	0415 321 456
Have workers been consulted about the SWMS?	YES		
Person responsible for ensuring compliance with SWMS	Joe Bloggs	Date SWMS provided to PC	23/10/2024
Person(s) responsible for reviewing the SWMS	Sam Smith	Last SWMS review date	23/10/2024
Date received	26/10/2024	Signature	
Worker's name	Sammy Sparrow, Harry Hope, Bill Ford	Date received	All on same date 27/10/24
Worker's signature		SWMS Audit	Job 123456 – scaff

Task no.	Task to be done	Hazards	Risk rating (as per matrix, Figure 5.3)	Control measures	New risk rating	Plant and equipment, licences needed
1	Set up base level of scaffold	Public and worker access, allowable bearing pressure (ABP) of soil	17	Isolate area with tape/bollards. Confirm with engineer of ABP	8	Basic scaffolder's licence, engineer's report, torpedo level, hard hat, safety boots, danger tape, 'scaff tag' in place
2	First level working deck in	Slips, trips and sprains	10	Correct lifting techniques. Fully decked out working deck (5 planks) clean scaffold of components. Hand material to scaffolders building scaffold, no throwing	6	Tool belt with scaffold key, podger hammer and torpedo level, all with lanyards, PPE of rigger's gloves, hard hat and steel cap boots
3	Adding stairway bay to scaffold	Slips, trips, falls and sprains	14	Correct lifting techniques. Clean scaffold of components. Install handrails, mid-rails and toe boards. Hand material to scaffolders building scaffold, no throwing	6	Tool belt with scaffold key, podger hammer and torpedo level, all with lanyards, PPE of rigger's gloves, hard hat and steel cap boots

FIGURE 5.5 SWMS template and example

Task no.	Task to be done	Hazards	Risk rating (as per matrix, Figure 5.3)	Control measures	New risk rating	Plant and equipment, licences needed
4	Completing each sequence of lifts where needed	Slips, trips, falls and sprains. Overreaching. Overhead power lines or other structures	14	Correct lifting techniques. Clean scaffold of components. Install handrails, mid-rails and toe boards. Hand material to scaffolders building scaffold, no throwing. Energy authority notified and 'tiger tails' installed if needed. De-energise area. Use of movable working platforms to erect next lift. No reach greater than 1 metre	6	Tool belt with scaffold key, podger hammer and torpedo level, all with lanyards, PPE of rigger's gloves, hard hat and steel cap boots. Portable working platform See 'Erecting, altering and dismantling scaffolding – Part 1: Prefabricated steel modular scaffolding' on the SafeWork NSW website for details
5	Handover to builder	Scaffolding not complete. Unlicensed workers are altering, removing or adding to scaffold	25	'Scaff tag' showing completed when inspections will be carried out. Only licensed contractors to alter, remove or add to scaffold	10	Vigilant review of scaffold and seek feedback from workers on site. Safety talk at toolbox meeting.

Changes to the SWMS List new or changed job steps and identify the hazards and risks	The nominated person responsible for the job must consult with, communicate to and instruct all concerned with the work. All persons working on the job must be aware of the changed/new job steps, hazards and risk controls.

Step	Job tasks *List the key job steps required to perform the task, in the sequence in which they are to be carried out.*	Hazards *List the identified hazards with each step.*	Existing controls *List the controls already in place (e.g. safe operating procedure, trained competent operator).*	Assess risk *Initial rank*	Additional controls *List identified additional control measures that are to be implemented.*	Assess risk *Residual rank*
1	Review working on scaffold	Trench has been dug for drainage – scaffold collapse	See SWMS Audit 123456 – scaff	25	Seek engineer report, stop work, close scaffold	2

I UNDERSTAND THE ABOVE CHANGES, INCLUDING THE HAZARDS, RISKS AND THE CONTROL MEASURES. TO BE SIGNED ALONG HERE BY ALL PERSONS WORKING ON THIS JOB.

FIGURE 5.5 (*Continued*)

In the SWMS template and example in Figure 5.5, the fifth column, 'Control measures', highlights the control measures. The risk rating has been reassessed after putting the control measures in place to bring down those risks, as shown in the sixth column, 'New risk rating'.

3.7 Review work site and job task immediately before starting work and discuss SWMS with relevant personnel to confirm as still applicable, or to amend as required

There will be times when issues arise onsite that can radically change the effectiveness of the SWMS, such as trenches being dug near scaffolding, or cranes or heavy plant coming onsite that were not addressed in the current SWMS. Before work starts every day, the toolbox meeting is important to discuss any issues that have been raised from the previous day, or address any new items that will occur during the current day and modify the SWMS. Discuss those issues with the workers involved and work out a strategy to overcome the issues (see 'Changes to the SWMS' see Figure 5.5).

3.8 Store SWMS securely on site in accordance with compliance requirements

Storing the SWMS requires the same protocols as a JSA.

Whether they are stored digitally or are paper-based will be the user's choice. Whatever method is used, SWMSs should be stored with other WHS forms, such as induction forms, audit forms, incident reports, staff training records, maintenance or asset logs and any safety notes such as conversations, toolbox talks, feedback or suggestions.

Every person working onsite should have their own file that lists their qualifications and any licence requirements.

All of these documents must be readily available for inspectors from WorkSafe or any other authorities.

 COMPLETE WORKSHEET 3

SUMMARY

In Chapter 5 you have learnt how to identify construction work hazards and select risk control strategies.

This unit has covered the skills and knowledge required to participate in preparing a:

- job safety analysis (JSA) for general hazards
- safe work method statement (SWMS) for high-risk work hazards on construction sites.

JSAs and SWMS are prepared as specified in work health and safety (WHS) legislation. It includes meeting all relevant requirements of the National Construction Code (NCC), Australian Standards and Commonwealth and state or territory legislation.

REFERENCES AND FURTHER READING

Costin, G.P. (2024). *Construction Skills* (4th edn), Cengage Learning Australia, Melbourne, Vic

Useful weblinks

www.safeworkaustralia.gov.au, *Code of Practice: Construction work*
www.safeworkaustralia.gov.au, Safety data sheets
www.legislation.gov.au, *Work Health and Safety Act 2011*

Australian Standards
AS/NZS 1576.1–6 Scaffolding
AS/NZS 4576 Guidelines for scaffolding
AS/NZS 1801 Occupational protective helmets

GET IT RIGHT

In the photo below, the pallets on the job site are temporarily stored against the brick wall.

Identify why the materials are stored like this and consider if you can suggest a better way, and provide reasoning for your answer.

Source: Adrian Laws

_____

_____

_____

_____

_____

_____

 **WORKSHEET 1**

Student name: _____

Enrolment year: _____

Class code: _____

Competency name/Number: _____

Task: Read through the sections beginning at *Plan and prepare* then complete the following questions.

1 What are some things to consider when you are asked to perform a job task?

_____

_____

_____

2 List four (4) possible consequences of not following the planned safety procedures when completing a job:

1 _____

2 _____

3 _____

4 _____

3 Should everyone onsite attend a 'toolbox talk' that is dealing with overall site safety on a particular day?

TRUE   FALSE

4 List three (3) things that could be discussed at a 'toolbox talk'.

_____

5 Should everyone's attendance at a toolbox talk be recorded?

YES   NO

6 The National Construction Code rarely changes and you don't need to keep updated very often.

TRUE   FALSE

7 Should personal protective equipment be the first safety choice every time you do a hazardous task?

YES   NO

8 List some safety measures you might consider when planning to do a job that may be hazardous.

_____

_____

_____

9   List six (6) types of PPE and what they are used for in the table below. The first row has been completed for you.

PPE	USED FOR:
Hardhat	Head protection

10 List four (4) steps in the process of inspecting a work site:

1 _____

2 _____

3 _____

4 _____

11 What is the purpose of using a risk matrix or similar tool to be able to determine a level of risk rating?

_____

12 Name three (3) things that could be discussed at a toolbox meeting.

1 _____

2 _____

3 _____

13 What are two (2) areas that could be changed in compliance requirements?

1 _____

2 _____

14 What two (2) industry associations keep their members well informed of compliance requirements?

_____

 **WORKSHEET 2**

Student name: _____

Enrolment year: _____

Class code: _____

Competency name/Number: _____

Task: Read through the section beginning at *Prepare and implement job safety analysis (JSA)* then complete the following questions.

1 Whose responsibility is it to ensure workers are trained and competent for their work?

_____

2 What is a job safety analysis (JSA)? List four (4) features that help describe it.

_____

_____

_____

_____

_____

3 State the three (3) main components of JSA:

1 _____

2 _____

3 _____

4 List three (3) documents that should be accessed before attempting to prepare a JSA:

1 _____

2 _____

3 _____

5 Is it necessary to break the job into logical steps when planning for a JSA? If so, why or why not?

_____

_____

6 Once you have written up a JSA, do you need to review it or change it before you start the task?

_____

7 Should a JSA record be kept?

YES   NO

8   What is the current national legislation for WHS?

_____

9   Are JSAs used in New South Wales? If not, what is commonly used?

_____

10 When would you not use a JSA but use an SWMS?

_____

11 Name the four (4) features that an effective JSA should have.

_____

_____

_____

_____

 **WORKSHEET 3**

Student name: _____

Enrolment year: _____

Class code: _____

Competency name/Number: _____

Task: Read through the sections beginning at *Prepare and implement safe work method statement (SWMS) for high-risk work* then complete the following questions.

1   List five (5) high risk areas of work that a SWMS is required for.

  1   _____

  2   _____

  3   _____

  4   _____

  5   _____

2   List four (4) hazards that may not be obvious if you don't stop and think through your work processes.

  1   _____

  2   _____

  3   _____

  4   _____

3   Circle the correct response to the following statement: Injuries only occur when doing work at heights above 2 metres.

  TRUE   FALSE

4   When creating a SWMS, how long should the document be?

  _____

5   Does creating a SWMS need to involve the workers who will do the job?

  YES   NO

6   What two pieces of information are used to come up with a risk rating for an activity that has hazards?

  _____

7   List and describe the six (6) hierarchy of controls for the most effective control of hazards.

1   _____

2   _____

3   _____

4   _____

5   _____

6   _____

8   How should SWMS documents be stored and who should sign that they have read the document?

_____

PERFORMANCE OF SKILLS EVIDENCE

Student name: _____

Enrolment year: _____

Class code: _____

Competency name/Number: _____

Task: Working with your teacher, prepare to demonstrate your skills by completing the following task.
To demonstrate competency, a candidate must satisfy all the elements, performance criteria and
foundation skills of this unit by

1   Developing, in consultation with relevant personnel, two job safety analysis (JSAs) for general hazards.
    Each JSA should be for a different job task and on a different work site.

2   Revising a JSA prior to starting work, identifying changed conditions and, where appropriate,
    amending the JSA to reflect changed hazards and risk control strategies.

3   Developing, in consultation with relevant personnel, two safe work method statements (SWMS) for
    high-risk work. Each SWMS should be for a different job task and on a different work site.

4   Revising a SWMS prior to starting work, identifying changed conditions and, where appropriate,
    amending the SWMS to reflect changed hazards and risk control strategies.

PART 2

SETTING OUT

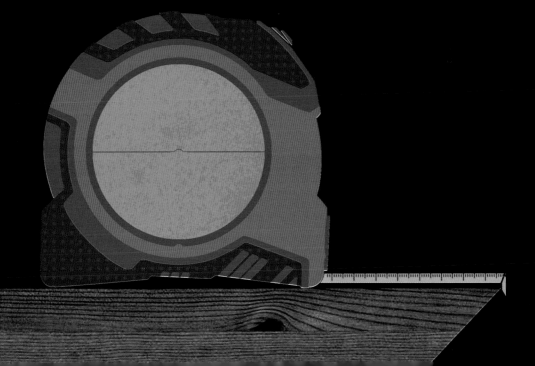

6

CARRY OUT MEASUREMENTS AND CALCULATIONS

This chapter covers the following topics from the competency CPCCOM1015 Carry out measurements and calculations:

- obtain measurements
- perform basic calculations.

This unit of competency specifies the skills and knowledge required to undertake basic measurements and calculations to determine task and material requirements in a construction work environment.

A person working at this level would be expected to complete tasks assigned to them, under supervision.

1. Obtain measurements

In the building industry, linear measurements are expressed in millimetres (mm) and metres (m). Working drawings of houses and buildings generally have their **dimensions** shown in **millimetres**, although some long dimensions on site plans may be in metres. Measuring tools used in house and building construction are marked in graduations of millimetres and metres.

Note: Centimetres are not recognised as true SI units and are therefore not used in the building industry.

Accurate measuring using an appropriate tool is critical to obtain true and accurate measurements (see **Figure 6.1**).

FIGURE 6.1 Accurate measuring and marking is critical

On a site, tradespeople refer to dimensions in either millimetres or metres. For example, a tradesperson may take a measurement of 5432 mm. The tradesperson may say that the measurement is either 'five thousand four hundred and thirty-two millimetres', or 'five metres four hundred and thirty-two millimetres'.

Most tradespeople carry three measuring tools on the job. These are the four-fold rule, 2 m folding rule and the retractable tape measure (see **Figures 6.2, 6.4** and **6.5**). To obtain accurate measurements you must first select the right tool for the task and then be able to use that tool correctly.

Before any tradesperson starts measuring, he or she will consider these four basic questions.

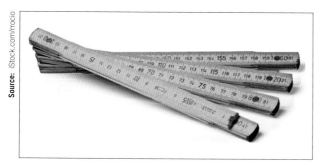

FIGURE 6.2 2 m folding rule

- *What am I to measure?*
  Are the measurements from an existing structure or from a set of plans?
- *What gear or tools do I need?*
  If the measurements are from an existing structure are there other items needed to undertake the work? For example, ladders or scaffoldings for measuring something at a height.
- Which measuring device is most appropriate?
  For example, should the tradesperson use a retractable tape measure or a long open-carriage tape measure?

 Is it safe to measure the item? There may be hazards associated with undertaking the measurements; for example, measuring at height could result in a fall.

- *Where and how do I record the measurements?*
  It is important to write down what the measurements are in a way that someone else can pick up your measurements and be able to read them.

1.1 Select most appropriate equipment and method for obtaining the measurement

To obtain correct measurements it is important to select the appropriate tool for the task. By selecting the most appropriate tool, the chances of incorrectly measuring are greatly reduced. It is also important to check the tool to make sure that it functions properly. Measuring tools that are damaged or broken may result in incorrect measures, which in turn result in the wrong **quantity** of materials being ordered.

Measuring tools that are used in the construction industry include:
- scale rule
- four-fold rule
- retractable tape
- long tape measures
- trundle wheel
- laser distance-measuring devices and range finders.

Scale rule

The scale rule is a plastic rule 150 or 300 millimetres in length that is used to scale off dimensions not given on a drawing (**Figure 6.3**).

Four-fold rule

The four-fold rule is 1 metre long and is made in four hinged sections so that it can be folded for convenience (**Figure 6.4**). This rule is best used for measuring and

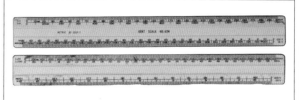

FIGURE 6.3 A typical scale rule

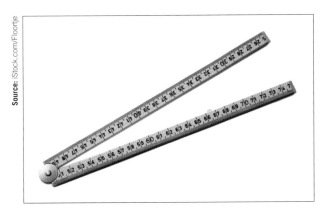

FIGURE 6.4 The four-fold rule

marking out dimensions of less than 1 metre. If the rule is used to measure or mark out a dimension greater than 1 metre an error is likely to compound each time the rule is moved.

Depending on whether the rule is made from boxwood or plastic, the rule blades are approximately 4 or 5 millimetres thick. Because of this thickness, a **parallax error** (when an object or point of measurement appears to shift or change position as a result of a change in position of the observer) is likely to occur when measuring or marking out with the rule laid flat on a surface. To overcome this problem, the rule should be used on its edge so that the graduations marked on the blades of the rule are in contact with the surface of the material being measured or marked out.

Tape measures

Retractable tape

The retractable tape is available in lengths from 1 metre up to 10 metres and is most suitable for measuring and/or marking out dimensions over 1 metre and up to 10 metres (Figure 6.5). Retractable tapes have a hook on the end, which adjusts depending on whether the measurement to be taken or marked out is internal or external.

FIGURE 6.5 A typical retractable tape

The hook slides a distance equal to its own thickness so that an internal measurement; for example between two walls, begins from the outside of the hook; an external measurement; for example, from the end of a piece of timber, begins from the inside of the hook. Care must be taken when returning the blade to the case so

that the hook does not slam against the case, which will stretch or distort the hook making the tape inaccurate. This can also tear the hook from the end of the tape, making it useless and needing replacement.

Long tape measures

For measuring distances greater than 10 metres there are long tape measures available. They are constructed from either steel or fibreglass and are manually wound back into a closed case or open carriage. These tape measures range in length from 30 metres up to 100 metres and are used on construction sites where measurement of longer distances is required. These tape measures generally have a specialist hook on the end so that they may be hooked onto a nail for simple one-person operation (see Figure 6.6).

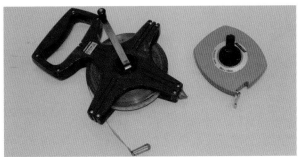

FIGURE 6.6 Open reel and closed-case long tapes

Construction workers regularly take measurements using a retractable tape measure. It is extremely important when taking measurements that the retractable tape measure is kept level (horizontal) or plumb (vertical). If the measuring device is not kept level or plumb, measurement errors *will* result.

Another technique that construction workers use to obtain accurate readings when using a retractable or long tape measure is to have one worker hold the tape so that the starting measurement is 100 millimetres or even 1 metre. This technique is used so that if the tape end is damaged or the hook is stretched no error will be recorded. It is important that the person who is reading the tape measure deducts the 100 millimetres or 1 metre from the measurement; if they forget to do this the recorded reading will be 100 millimetres or 1 metre too long.

Trundle wheel

A trundle wheel is a light-weight wheel constructed from timber or plastic which has a circumference of exactly 1 metre. It has a handle attached to allow a person to walk it along easily, and a counter so that every rotation (1 metre) is counted. These tools are not accurate and are mainly used to obtain a quick estimate of distances over 100 metres up to several kilometres (see Figure 6.7).

Laser distance-measuring devices and range finders

Laser distance measuring devices and range finders can be used to accurately measure distances up to 30–40

FIGURE 6.7 A trundle wheel

metres with an error margin of around 3 millimetres. They are simply placed against the object or surface you wish to measure from and pointed at the object you wish to measure to (see Figure 6.8).

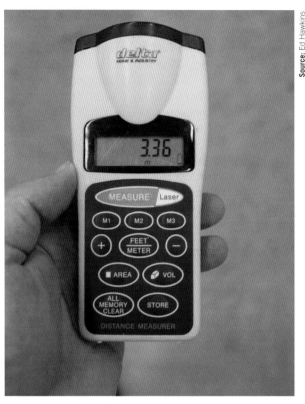

FIGURE 6.8 A laser distance-measuring device

LEARNING TASK 6.1 MEASUREMENT AND PLANNING

Task 1

Using a retractable tape measure and the plans for a building you have access to (ask your teacher for assistance in accessing a building plan if you don't have one), complete the following steps.

1   Select a window from the plan. Determine what size the window is from the plan.
    The window to be measured is in ....
    .... (name of room).
    According to the plans the window is ....
    metres high and .... metres wide.

2   Select what gear or tools you will need to measure the window and record them in the space below.
    ....
    ....

3   Explain how you intend to complete the task safely. Record your answer in the space below.
    I will complete the task safely by:
    ....
    ....

4   Now go and measure the actual window and record the window measurements in the space below.
    Height: .... metres
    Width: .... metres

Task 2

Using a four-fold rule and a retractable tape, accurately measure and record lengths of various materials and surfaces as directed by your teacher/supervisor.

Suggested materials for practice measuring could include the following:

- various lengths of timber
- section sizes of various timber
- spacings between exposed framing
- surface of walls
- surface of a concrete path or driveway
- height and width of nominated windows
- height and width of nominated doors
- length, width and thickness of standard MDF or plywood sheets, etc.

Note: Practice measuring should also take place during practical exercises.

Technological change for these devices moves at a fast pace and like all electronic items they need to be looked after, as moisture and rough handling will damage them.

Using the measuring device is very important, but what needs to be measured is of equal importance. An example of this is measuring up for a concrete slab. What needs to be measured so that the correct amount of formwork can be ordered? What measurements need to be used to determine the surface area of the slab so that the correct amount of paving paint can be ordered? And what measurements are needed to determine the **volume** of the slab, so that the correct amount of concrete can be supplied? All these questions relate to three measurements.

The final important factor to consider when measuring is that in Australia our measurements are all metric and not imperial – that is inches, feet and yards; a system that is used in the USA.

Metric scale (International System of Units)

Every day you will use a system of measurements for a range of activities. These activities include measuring temperature, weight and even the amount of petrol you put in your car. So, it is important to have a basic understanding of standard systems of measurement and the terms that relate to them. The units of measurement adopted for the building industry are metric units taken from the *Système International d'Unités* (SI), or International System of Units. The **SI units** commonly used in construction, together with examples of what they are typically used for, is discussed below.

When using metric units, small letters are used for all symbols except where the value is over one million, as in megapascals (MPa), or where the unit name is derived from a proper name, as in newtons (N) or pascals (Pa).

Linear measurements in the building industry are measured in millimetres (mm) and metres (m), with 1000 millimetres equal to 1 metre. Most plans and working drawings show dimensions in millimetres; however, where greater lengths are recorded, such as on a site plan, the length may be shown in metres. Where millimetres are used it is generally accepted that the symbol (mm) need not be used, and for metres a decimal point is used to separate the metres from the millimetres, again without the need for the symbol (m). For example, if a figure of 3600 is used it is accepted as millimetres, whereas 3.600 would indicate metres.

Surface area measurements must be accompanied by the symbol (m^2) and volume measurements must be accompanied by the symbol (m^3) to ensure correct interpretation of the calculation.

For example:

2.500 m × 5.000 m = 12.5 m^2 indicates surface area and

2.500 m × 5.000 m × 0.150 m = 1.875 m^3 indicates volume.

Measurement will be a major part of everyday activity in each section of the building industry. A carpenter is constantly measuring and cutting timber to length; a bricklayer will continually measure the length and height of brickwork and determine the position of openings; a concreter will measure and cut timber formwork; and a dry wall plasterer will measure the length and height of walls and the position of windows and doors.

COMPLETE WORKSHEET 1

1.2 Use a ruler or tape to obtain linear measurements accurate to 1 mm

Some of the first tools a construction worker purchases are a tape, a hammer and a nail bag. An onsite carpenter will need at least a 5- or an 8-m retractable tape that will hook onto the carpenter's nail bag. For shorter measuring tasks a four-fold rule will be useful because the markings are clear right to the end of the rule. In construction measuring you work to an accuracy of 1 mm and your tape or rule will have marks on it in 1 mm increments. It is important when measuring to make sure you are looking directly onto the surface that you are measuring and that the marks on your measuring instrument are as close as possible to what you're measuring. For example, if you use a thick measuring rule and you look at the marks from an angle down onto your material then you will encounter parallax error and your actual measurement may be incorrect. Therefore, it is important to view your measurements at 90 degrees to your material and have the marks on your measuring tool as close as possible to your material.

The following are some simple steps for using a rule or tape to obtain linear measurements:

1. Choose a rule or tape that has good clear markings and that is suitable for the length of the measurement you need to take.
2. Place the rule or tape along the material that you wish to measure by hooking the end of the tape onto the end of your material and rolling out the tape to the location that you want to mark, or the measurement that you want to mark. Ensure that the end of your tape is not worn and is working correctly to give you an accurate measurement. Tape measures are designed to have a loose end to allow for the thickness of the end of the tape if you are either pushing against something to measure or hooking over something to measure. Remember, the level of accuracy for construction in this instance is 1 mm. Ensure you are reading in a straight line from the start to the measurement reading location. Double-check the measurement that you have located so you do not make an error. If necessary

and suitable, round up the measurement rather than round down to ensure you have enough or more material than you need rather than being short of material.

3   Record the measurement, either by writing it down or putting a mark on your material so it is not forgotten. Check the measurements and if necessary, use another method to check your measurement to make sure that it is correct. For example, many smartphones these days have a measuring tool that uses the camera. The camera that will give a reasonable degree of accuracy and will allow you to estimate whether your initial measurement was close or not.

Refer to Figure 6.1 which shows accurate measuring and marking. By following these simple steps and discussing measurements with a competent person you can use a ruler or a tape to obtain linear measurements to the accuracy of the markings on your tool.

LEARNING TASK 6.2

MEASURING WITH A TAPE MEASURE

1   Using a retractable tape measure, set two short pegs at 5 m apart down a slope. Measure the distance down the slope from top to bottom.
2   Replace the bottom short peg with a long peg or stake, then measure 5 m horizontally from the top short peg to the long peg or stake.
3   Is there a difference between the 5-m mark measured down the slope and the 5-m mark measured horizontally? Discuss which measurement is correct with your teacher/ supervisor and other students.

COMPLETE WORKSHEET 2

2. Perform basic calculations

2.1 Take basic measurements and calculate quantities of materials in a construction environment, using basic formulas for each of: weight, area, volume, perimeter, circumference, ratio and percentage

Weight is the measure of the heaviness, or mass, of an object(s). In Australia, weight is in kilograms or tonnes in the construction industry. A tonne is 1000 kilograms. Depending on the weight of the object, it could be a number of kilograms, or it could be a number of tonnes. We don't want to say the concrete truck coming up the driveway weighs 36 000 kilograms and this needs to be designed for by the engineer. It's much better to say the truck coming up the driveway needs engineering to design for a load of 36 tonnes.

At the design stage of a construction the load-bearing capacity of the soil will be measured and determined. From there for the whole height of the building everything will be planned for so that all that load can sit stably on the soil foundation. To ensure stability in a building, all materials and forces within that building need to be planned for and that's when the Australian Standards and the National Construction Code are referred to by the designer/architect.

Formulas commonly used in the construction industry are listed in Table 6.1. Accurate measurement will save time, cut down waste and ultimately reduce costs.

GREEN TIP

Cutting down waste will save money and help reduce landfill in your local area. It's good for the environment.

TABLE 6.1 Formulas commonly used in the construction industry

Any figure/shape	Perimeter formula	Unit of measurement
© Cengage	$P = S_1 + S_2 + S_3 + S_4$	m or mm
Square ('L' shaped, or complex shaped area calculations are done by breaking the shapes down into simple shapes, and then adding the singular areas up to get a total) © Cengage	**Area formula** $A = S^2$ $A = S_1 \times S_2$ $A = L \times W$ $A = L \times H$	**Unit of measurement** m^2

>>

Rectangle ('L' shaped, or complex shaped area calculations are done by breaking the shapes down into simple shapes, and then adding the singular areas up to get a total)	Area formula $A = L \times W$ $A = L \times H$	Unit of measurement m^2
Triangle	Area formula $A = \dfrac{B \times H}{2}$ $A = \frac{1}{2} \times B \times H$	Unit of measurement m^2
Parallelogram	Area formula $A = B \times PH$ $A = L \times W$	Unit of measurement m^2
Trapezoid	Area formula $A = \dfrac{(B1+B2)}{2} \times PH$ $A = \frac{1}{2} \times (B1 + B2) \times PH$	Unit of measurement m^2
Circle	Area formula $A = \pi r^2$ $\pi = 3.142$	Unit of measurement m^2
Volume/circle-prisms	Volume formula $V = \pi r^2 h$ $\pi = 3.142$	Unit of measurement m^3
Volume	Volume formula $V = L \times W \times Th$ $V = L \times W \times H$	Unit of measurement m^3
Pythagoras' Theorem (3 – 4 – 5) s (3) d (5) s (4)	Pythagoras' Theorem formula $d = \sqrt{S_1^2 + S_2^2}$	Unit of measurement m or mm

Legend: A: Area; B: Base; C: Circumference; D: Diameter; L: Length; H: Height; Th: Thickness; P: Perimeter; r: Radius; S: Side; W: Width; PH: Perpendicular Height; π(pi) = 3.142

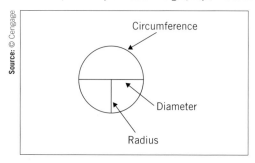

Source: © Cengage

Once measurements have been taken, a good tradesperson will then calculate the materials that are needed. Most of these calculations are worked out using a calculator (Figure 6.9).

FIGURE 6.9 A calculator

It is therefore important to know the system (formula) to use to obtain the correct calculation.

Simple measurement and calculations

The sections following will address:

- linear measurement
- area measurement
- volume measurement.

Linear measurement

In the building industry, linear measurement may include:

- the perimeter of a slab or property for a fence (see Example 1)
- the circumference of an edge board (screed board) for circular section of paving (see Example 2)
- the length of pipe or cable needed as it twists and turns from attachment point to end point
- the total length of timber for a job (see Example 3).

EXAMPLE 6.1

Find the total length of fencing required to enclose a site measuring 35.750 m × 23.500 m (see Figure 6.10).

Perimeter of fence $= S_1 + S_2 + S_3 + S_4$
$= 35.750 + 23.500 + 35.750 + 23.500$
$= 118.5$ m of fencing

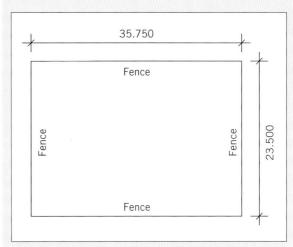

FIGURE 6.10 Example 1: Fencing an enclosed site

EXAMPLE 6.2

Find the total **linear metres** of screed boarding for a section of circular paving with a radius of 2.25 m (see Figure 6.11).

Formula for the circumference of a circle:
$= \pi \times$ diameter
$\pi \times 4.5 = 14.137$ m board required.

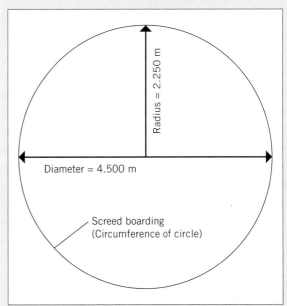

FIGURE 6.11 Example 2: Find the total linear metres of screed boarding for circular paving

Generally, timber is sold in linear metres, with lengths starting from 1.8 m and increasing in increments of 0.3 m up to 7.2 m.

When orders are placed with suppliers, the lengths required will be given within the standard range: 1.8 m,

2.1 m, 2.4 m, 2.7 m, etc. An example of a timber order would include:

- section size of the timber, e.g. 90 × 45, 50 × 25 (mm symbols are left out but understood)
- species of timber required, e.g. radiata pine, western red cedar
- shape, profile or finish of the timber, e.g. quadrant mould, colonial architrave, pencil round skirting, dressed all round (DAR), rough sawn (RS) or rougher header (RH)
- number of pieces, e.g. 24
- lengths required as supplied in the incremental lengths previously described, e.g. 2.7, 4.8, 5.4, 6.0.

The longest lengths are ordered first and progress down to the shortest.

Therefore, a typical timber order would look like: 90 × 45, radiata pine, RH, 25/3.6, 10/2.7, 16/2.1.

EXAMPLE 6.3

Find the total number of linear metres in 20 × 8 Tasmanian oak cover strips 21/4.8, 12/2.4, 10/2.1.

$$21 \times 4.8 = 100.8 \text{ m}$$
$$12 \times 2.4 = 28.8 \text{ m}$$
$$10 \times 2.1 = 21.0 \text{ m}$$
$$= 150.6 \text{ m}$$

Area measurement

Area measurements may be for regular or irregular shapes, and are expressed as the number of square metres that the shape contains. For land subdivision, because of the larger areas involved, hectares (ha) are used. But when measuring or describing building blocks the unit of measure will be square metres (m^2).

Area measurement in square metres (m^2) is used in building construction to determine:

- floor area of a building – used as a means of describing the size of the building
- floor area of individual rooms to determine the quantities of flooring or floor coverings
- wall and ceiling area for quantities of sheeting required to cover the walls and ceilings
- roof area to determine the number of roof tiles or the sheet roofing
- area of a building block to determine minimum and/or maximum coverage for building regulations.

Square measurement is also used for the calculation of the number of bricks required to construct a wall, or for the number of pavers for a path or driveway. This number is then multiplied by the number of bricks or pavers that will cover 1 square metre. For example, a

metric standard brick (230 mm long × 76 mm high) single-skin wall, say (5 m long × 2 m high) 10 m^2 will be multiplied by 50, as there are approximately 50 bricks needed per square metre. Therefore, the number of bricks will be 10 m^2 (wall) × 50 bricks/m^2 = 500 bricks.

The area of plane figures and rectangles is found by multiplying the length by its width or breadth.

Regular shapes

The following are examples of area measurement for regular shapes.

EXAMPLE 6.4

A building measures 14.000 m × 10.000 m. Calculate the area of sheet flooring.
Note: Area of the floor = area of the building.

$$\text{Area of floor} = L \times W$$
$$= 14.000 \text{ m} \times 10.000 \text{ m} = 140 \text{ m}^2$$

EXAMPLE 6.5

A wall 5.400 m long and 2.400 m high is to be covered with plasterboard. Calculate the *net* wall area if the wall has a window 2.400 m × 1.200 m and a door 2.100 m × 0.900 m.

$$\text{Area of wall} = L \times W$$
$$= 5.400 \text{ m} \times 2.400 \text{ m}$$
$$= 12.960 \text{ m}^2$$

Deductions:

Area of window	= 2.400 m × 1.200 m = 2.880 m^2
Area of door	= 2.100 m × 0.900 m = 1.890 m^2
Total deductions	= 4.770 m^2
Net wall area	= 12.960 m^2 – 4.770 m^2 = **8.190 m^2**

plasterboard will be required.

EXAMPLE 6.6

A building 16.750 m long is to be covered with roof tiles (see **Figure 6.12**). Calculate the area of a gable roof with a pitch length of 3.600 m.

$$\text{Area of roof} = L \times W$$
$$= 16.750 \text{ m} \times 3.600 \text{ m}$$
$$= 60.300 \text{ m}^2.$$

A gable roof will have two sloping sides. Therefore, the area calculated will be doubled to allow for tiles to cover both sides.

Total area of roof tiles will be
60.300 m^2 × 2 sides (60.300 m^2 + 60.300 m^2)
= 120.600 m^2

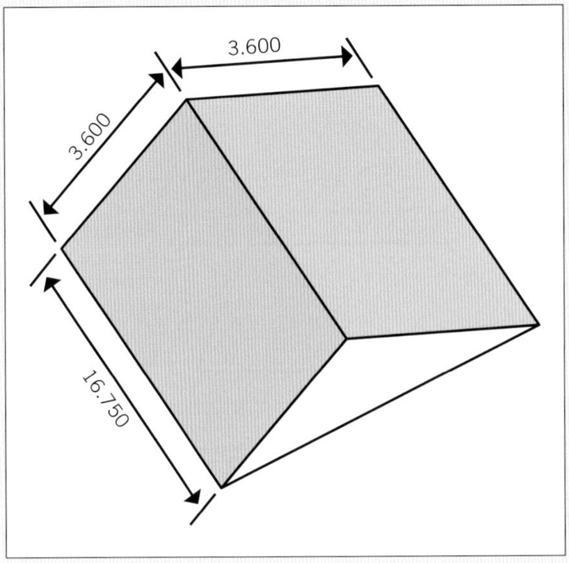

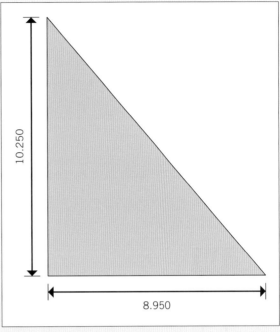

FIGURE 6.12 Example 6: Calculate the area of a gable roof

Source: © Cengage

EXAMPLE 6.7

Calculate the area of paving for a triangle-shaped courtyard with a base length of 8.950 m and a perpendicular length of 10.250 m (see Figure 6.13).

Area of a triangle = $\frac{1}{2}$ base × perpendicular length.

$$\left(\frac{8.950}{2}\right) \times 10.250 = 45.869\,\text{m}^2.$$

FIGURE 6.13 Example 7: Calculate the area of paving for a triangle-shaped courtyard

Source: © Cengage

EXAMPLE 6.8

A circular concrete slab with a radius of 4.886 m needs to be painted with paving sealer. Calculate the total area that needs to be painted (see Figure 6.14).

$$\begin{aligned}\text{Area to be painted} &= \pi r^2\\ &= \pi \times 4.886\text{ m} \times 4.886\text{ m}\\ &= 74.999\text{ m}^2\\ &= 75.000\text{ m}^2 \text{ (rounded)}\end{aligned}$$

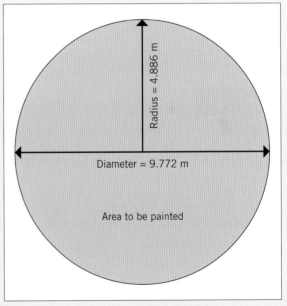

FIGURE 6.14 Example 8: Calculate the total area that needs to be painted

Source: © Cengage

Irregular shapes

The area measurements for irregular shapes in the following examples are for shapes with straight sides, such as irregular **quadrilaterals** (four sides) or irregular polygons (five or more sides). Irregular figures may be subdivided into triangles, parallelograms or rectangles. To find the area of the whole figure, add the areas of the parts into which it has been divided.

In some cases, areas may be determined by considering the figure as a rectangle from which parts have been removed. For example, if a quadrilateral has only two parallel sides, its area may be found provided the length of the two parallel sides and the perpendicular distance between them is known. The formula to calculate the area of this figure would be:

Half the sum of the parallel sides × perpendicular distance between parallel sides

Typical examples of calculations involving irregular shapes are:

EXAMPLE 6.9

Find the area of Figure 6.15.

$$\text{Area of trapezoid} = \left(\frac{B1+B2}{2}\right) \times PH$$

$$= \left(\frac{20.000 + 36.000}{2}\right) \times 10.000$$

$$= 28.000 \times 10.000$$

$$\text{Total area} = 280\,m^2$$

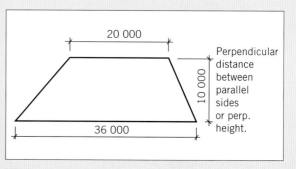

FIGURE 6.15 Trapezoid

EXAMPLE 6.10

Find the area of the block of land shown in Figure 6.16.

$$\text{Area of block} = \left(\frac{B1+B2}{2}\right) \times PH$$

$$= \left(\frac{90.000 + 96.000}{2}\right) \times 46.000$$

$$= 93.000 \times 46.000$$

Total area $= 4278\,m^2$ of land.

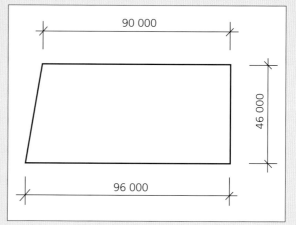

FIGURE 6.16 Quadrilateral

EXAMPLE 6.11

Calculate the area of the irregularly shaped concrete path in Figure 6.17. Divide the shape into separate sections and find the area for each section, then add the two areas together to find the total area.

Section one (parallelogram)

$3.600 \times 0.762 = 2.743\,m^2$

Section two (rectangle)

$2.134 \times 1.000 = 2.134\,m^2$

Total area

$2.743 + 2.134 = 4.877\,m^2$.

The areas of quadrilaterals with sides of different lengths, and irregular polygons, can be found by drawing in diagonals on the figure, finding the areas separately and then adding them together to find the total area.

Typical examples of this tape of calculation are shown in Example 12.

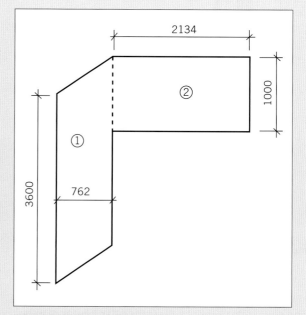

FIGURE 6.17 Irregular polygon

EXAMPLE 6.12

Find the area of the building block shown in **Figure 6.18**.

Area of block $= \frac{1}{2} \times B \times PH$

Area of triangle A $= \left(\dfrac{38.000}{2} \right) \times 10.000$

$= 190\,m^2$

Area of triangle B $= \left(\dfrac{38.000}{2} \right) \times 15.000$

$= 285\,m^2$

Total area of the block is $190 + 285 = 475\,m^2$.

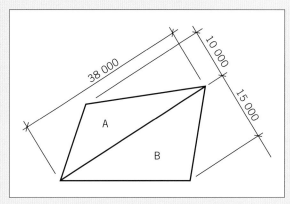

FIGURE 6.18 Polygon

EXAMPLE 6.13

Calculate the area of floor tiles required for a shop with the dimensions shown in **Figure 6.19**. Add 15 per cent to the total area calculated to allow for waste in cutting the tiles.

Area of floor tiles $= \frac{1}{2} \times B \times PH$

Area of triangle A $= \left(\dfrac{14.400}{2} \right) \times 4.600$

$= 33.120\,m^2$

Area of triangle B $= \left(\dfrac{14.400}{2} \right) \times 1.600$

$= 11.520\,m^2.$

Total area of floor tiles required:

$33.120\,m + 11.520\,m = 44.640\,m^2$

Add 15% for waste $= 44.640\,m^2 \times 1.15$

Total tiles required $= 51.335\,m^2$

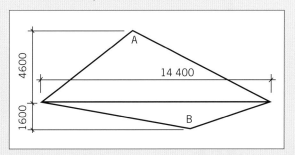

FIGURE 6.19 Polygon

GREEN TIP

Cutting down waste can save money because you only order enough material to do the job without having too much excess left over.

Volume measurement

Volume is measured in cubic metres (m³), and is found by multiplying the length by the width by the depth (or thickness).

Calculations of volume in the building industry are used to determine:

- volume of soil excavated from the foundation for footings
- volume of soil to be removed over an area of a sloping site to provide a level area
- amount of material required as filling, e.g. under floor slabs
- quantity of materials to be able to do a particular job, e.g. cubic metres of sand for use in bricklayer's mortar
- volume of concrete for strip footings, slabs, etc.
- cubic metres of timber.

Timber can also be priced in cubic metres for bulk orders. Therefore, the total cubic metres of the timber

order is calculated by multiplying the total linear metres by the end section size of the timber, e.g. 90 mm × 45 mm radiata pine 56/7.2 m.

Linear metres $= 56 \times 7.200 = 403.200\,m$

Cubic metres $= 403.200 \times 0.09 \times 0.045$

$= 1.633\,m^3.$

Typical calculations of cubic metres are given in the following examples.

Calculation of various solid shapes

In the building industry today, workers are dealing with three-dimensional objects. It is important to understand the differences between these shapes. In the section below you will be looking at:

- prisms
- cylinders
- cones
- pyramids.

Prisms

Prisms are solid objects with two ends formed by straight-sided figures, which are identical and parallel to one another. The sides of the prisms become *parallelograms*. The ends may be formed by common plane geometric shapes (e.g. the square, rectangle, triangle, pentagon, hexagon, octagon).

Cylinders

Cylinders have their ends formed by circles of equal diameter. The ends are parallel and joined by a uniformly curved surface.

EXAMPLE 6.14

A trench is to be excavated in soil on a level site for a footing to a depth of 600 mm from the top of the ground. The trench is to be 18.500 m long by 0.450 m in width. Calculate the cubic metres of soil to be excavated.

Volume of soil to be removed
$$= L \times W \times D$$
$$= 18.500 \times 0.450 \times 0.600$$
$$= 4.995 \, m^3 \text{ of soil removed.}$$

EXAMPLE 6.15

Calculate the volume of concrete for a strip footing for the trench in Example 14 if the depth of the footing is 450 mm.

Volume of concrete for strip footing
$$= L \times W \times D$$
$$= 18.500 \times 0.450 \times 0.450$$
$$= 3.746 \, m^3.$$

EXAMPLE 6.16

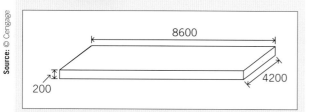

Calculate the volume of concrete for a slab-on-ground for a garage floor. The garage is 8.600 m long by 4.200 m wide and the floor is 200 mm thick.

Volume of concrete for slab
$$= L \times W \times D$$
$$= 8.600 \times 4.200 \times 0.200$$
$$= 7.224 \, m^3.$$

When calculating the volume of concrete, a **percentage** is added to the total to allow for irregular forms, varying thickness and losses caused by spilling when transporting and depositing the concrete. Typical waste allowances would be 10 per cent for concrete poured on the ground and 5 per cent for concrete placed into forms.

Cones

Cones have a circular base and a uniformly curved surface, which tapers to a point called the apex.

Pyramids

Pyramids are solid shapes with a base consisting of a straight-sided figure (square) and triangular sides, which terminate at the apex.

Surface development and area of solid shapes

When the surface of a solid object requires measuring, or a true shape is required to create a template, the simplest way to provide an accurate detail is to develop the surface. This requires the true shape of all sides to be laid out flat as a continuous surface.

Drawing the detail in a two-dimensional view is not the only way to determine the amount of material required to form the surface shape. For example, calculation of the surface area would be a more effective method. The idea of laying the shape of each surface out flat still applies, but it's the calculation of each area added together that provides the information required. The calculation of these surfaces is required when a roof tiler needs to order roof tiles for a conical or pyramidal style roof surface, or a formworker needs to work out how much form ply is required for a prism-like member, or a tank maker needs to work out how much Colorbond® custom orb is required to create a water storage tank.

EXAMPLE 6.17

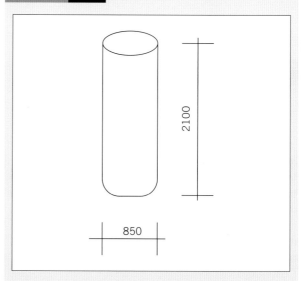

Calculate the surface area of a cylinder with a height of 2.100 m and a base radius of 850 mm:

Area of cylinder $= (\pi r^2 \times 2) + (2\pi r \times \text{height})$
$$= [(3.142 \times 0.723) \times 2]$$
$$+ [2 \times (3.142 \times 0.850)] \times 2.100$$
$$= 4.540 + 11.217$$
$$= 15.757 \, m^2$$

Ratio

Ratio is a way of expressing the relationship or a comparison between two or more quantities, measurements, or proportions of materials. Ratio is normally written with the weight, length, size or amount of the first material followed by a colon (:) and then the weight, length, size or amount of the second material. Below are some examples of how ratios might be used in construction:

■ *Scale ratio.* A site plan takes up a lot of room on the ground so by using a scale a smaller representation of the site plan can be drawn so it can fit onto a piece of paper. Usually, the site dimensions are reduced 500 times. Therefore, the scale ratio of 1:500 is used, or 1:200 depending on the size of the paper. When measuring off a paper site plan, use a scale rule which is labelled that will tell you the actual size on the ground.

■ *Materials ratio.* An example of a materials ratio is how we mix concrete. Concrete consists of coarse aggregate, fine aggregate and cement powder. We decide on the ratios of those three materials based on the type of concrete that we need. We could have a ratio of four parts coarse aggregate to two parts fine aggregate to one part cement powder to give us a ratio that is written as 4:2:1. As long as we know which materials go in which order when we write ratios such as this, we can use ratio as a way to increase or decrease the amount of the finished product.

Ratio is another way of proportioning or comparing different things in the construction industry and it's important that you have an understanding of ratio because it is used from the regulatory design stage of building right through to the final stages of comparing such things as business profit ratios.

Percentages/fractions

A percentage is a common way of expressing a fraction (e.g. ¼, ½, ¾) as a part of the whole (100). Percentages are written as a number followed by a percentage symbol; for example, 25%, 50%, 75%.

Percentages are used commonly in building applications, such as to describe the blend of bricks required to obtain a certain appearance or style, e.g. 60% Cadman cream, 30% Rustic red and 10% Manganese. Whatever the percentage, it must always be a proportion of the whole (100%).

The simplest way to determine a percentage is to write down the fraction you want converted, then multiply it by 100 over 1, as follows.

EXAMPLE 6.18

$\frac{1}{2} \times \frac{100}{1}$ (now multiply the numerators together, then the denominators)

$= (1 \times 100)$ over $(2 \times 1) = \frac{100}{2}$

(now divide 100 by 2) $= 50$

Therefore, $\frac{1}{2}$ is equal to 50%.

EXAMPLE 6.19

Convert $\frac{3}{8}$ to a percentage of 100 :

$= \frac{3}{8} \times \frac{100}{1} = \frac{300}{8} = 37.5\%.$

EXAMPLE 6.20

Convert $\frac{5}{16}$ to a percentage of 100 :

$= \frac{5}{16} \times \frac{100}{1} = \frac{500}{16} = 31.25\%.$

This process may also be used in reverse to convert a percentage to a fraction, as follows.

EXAMPLE 6.21

$45\% = \frac{45}{100}$

(This fraction should always be broken down to its simplest form, i.e. divide the top and bottom figures by a number common to both; e.g. 5.)

$= \frac{9}{20}$

Percentages may also be converted to decimals by determining their value in relation to 1, as follows.

Note: Always round off decimals to three decimal places.

EXAMPLE 6.22

$45\% = 45 \div 100$

(move the decimal point, which automatically would be after 45, i.e. 45.0, back to the left by the number of zeros in 100; i.e. 2.)

$= 0.45$

LEARNING TASK 6.3

FENCING AND FLOOR PLANNING

Task 1

Using a set of plans, determine the total length of temporary fencing to go around the entire building site. Record your results in a way that others can understand.

Task 2

Using the same set of plans as you used in Task 1, determine the surface area for the entire floor/slab. Record your results in a way that others can understand.

Task 3

This task requires a computer with internet access.

- Go to https://www.dlsweb.rmit.edu.au/Toolbox/Volume/start_here.htm
- Read through the various topics listed on the left-hand side.
- Complete the exercises in the 'Volume calculations' tab.

On this website you will find a number of different volume exercises for you to do. If you have difficulty you may need the assistance of your teacher or supervisor.

 Beware when measuring around electricity with a metal tape measure.

 COMPLETE WORKSHEET 3

2.2 Convert measurements in metres to millimetres and measurements in millimetres to metres

When converting measurements from one scale of measurement to another it is important to find the factor that relates to both of those measurements. With metres and millimetres this can simply be found by adding up all the marks along the 1-m rule and to get to 1000. The term 'millimetre' consists of two words: *milli* and *metre*. The definition of milli is 1000, or thousandth. The factor to be used for converting between millimetres and metres is 1000.

- If you are converting measurements from metres to millimetres, multiply the metres by 1000. For example, 1 metre multiplied by 1000 equals 1000 millimetres.
- If you are converting measurements from millimetres to metres, divide the millimetres by 1000.

For example, 1000 millimetres divided by 1000 equals 1 metre.

It's helpful at this stage to explain the main situations where you might need to convert measurements between millimetres and metres. In construction, we usually first see measurements in written documentation regarding a building job. We might then need to convert measurements for ordering quantities, estimating, quoting or contract work. We need to be very careful and clear with our recording of measurements as we go through the process of calculating and converting any measurements. An error in writing things down could become very costly later in the job.

When beginning to do measurement conversion it's always helpful to write your measurements to three decimal places. For example, 1 metre should be written as 1.000. The problem with not being consistent when you write down measurements is that you, or someone else who might read your written measurements, could make an error. Generally, with building work we do our calculations based on metres not millimetres. If we do our calculations in millimetres, we end up with huge, long numbers of millimetres in our answers. Generally, get in the habit of entering calculations into your calculator as metres, and don't forget the decimal point. Table 6.2 shows some examples of converting from metres to millimetres and millimetres to metres. Practise doing the conversions on your calculator to check that you have mastered this skill. After a while you should be able to do this in your head without using a calculator.

TABLE 6.2 Conversion examples

100s of mm in m	10s of mm in m	Individual mm in m
1000 mm = 1.0 m	90 mm = 0.09 m	9 mm = 0.009 m
900 mm = 0.9 m	80 mm = 0.08 m	8 mm = 0.008 m
800 mm = 0.8 m	70 mm = 0.07 m	7 mm = 0.007 m
700 mm = 0.7 m	60 mm = 0.06 m	6 mm = 0.006 m
600 mm = 0.6 m	50 mm = 0.05 m	5 mm = 0.005 m
500 mm = 0.5 m	45 mm = 0.045 m	4 mm = 0.004 m
400 mm = 0.4 m	40 mm = 0.04 m	3 mm = 0.003 m
300 mm = 0.3 m	35 mm = 0.035 m	2 mm = 0.002 m
200 mm = 0.2 m	30 mm = 0.03 m	1 mm = 0.001 m
100 mm = 0.1 m	25 mm = 0.025 m	0.5 mm = 0.0005 m

 COMPLETE WORKSHEET 4

2.3 Check calculations for accuracy and record calculation workings and results

To check calculations for accuracy and record workings and results it's important to understand why you might want to record your calculations and your results. One of the main reasons we write down our calculations or use the history thread in your calculator is to check for errors and to be able to make minor adjustments to calculations without having to do the whole calculation all over again. In some cases, with quantities of materials it may involve quite a bit of work to come up with the initial quantities and we don't have to do that again if just one small factor changes; for example, the cost of the material. If we must go back and do all the quantities again just because the costs changed and because we didn't record our calculations in the first place that is very time consuming. So, make sure you neatly, carefully and logically record your calculations for the various things you're doing on the construction site.

To accurately check calculations, it's important to have mastered the skill of estimating whether your calculations look correct and what the answer might be before you even go to the calculator. For example, if you were to calculate 1.9 times 2.0 then your answer should be somewhere near what the result would be for 2 times 2 which equals 4. If your calculator shows something a lot different to 4, then you may have made a mistake when entering your measurements. This is a very simple example, but the principle will apply during the rest of your life regarding measurements and calculations. The following are some steps you can use to check your calculations for accuracy:

- Double-check your calculations and review them to ensure that you have entered the numbers correctly and applied the correct mathematical operations. Enter your calculations into a scientific calculator and see if the answer comes up the same. Use technology to your advantage to check your calculations by using photo calculation apps that you may find on your app store that can be used to check your calculations. A good principle is: the more costly the mistake if you do something wrong in your calculations then the more time and effort you should put into making sure that everything you calculate is accurate.
- Employ estimating or rounding for your answer. As mentioned earlier, this is a quick way to check whether your calculation and your answer looks reasonable or not. Consider whether the answer you have looks possible for the calculations you did. You may do the same calculations with simple numbers that you know the answer to, to check that your method is correct. This particularly applies to more complex calculations such as ratios and trigonometry, geometry and formula substitution. We don't go into these more complex mathematical calculations in this chapter, but the principle remains that you need to develop some method of estimating to see whether your answer looks reasonable or not.

- Record the calculation workings step-by-step, expanding your calculation on a new line as you slowly unfold the calculation to come up with a result. Use spreadsheets to your advantage and record your full calculation. An advantage of the spreadsheet is that you can 'fill down' your calculations to be able to do very repetitive calculations very quickly.
- Take the time to review and proofread your calculations and record and file your results. You may reuse these calculations at some later stage for another job by just changing some small factors within your workings to help you with the next job.

By following these simple steps and talking to a competent person when you first start doing calculations and recording the results you will be able to fine-tune your methods to complete accurate and reliable calculations for your work in the building industry.

Quantities is a term used in the building industry for the calculation of materials required for a particular task, and may also include the cost of supply of the materials or a total figure that represents the total cost of supply and fixing of the materials.

Typical examples of calculations of quantities and costs are as follows.

EXAMPLE 6.23

Using the diagram in **Figure 6.20** and the following information, calculate the quantities required and the cost of materials to form up and pour concrete for the slab-on-ground.

Quantities:

screed boards 200 × 32 rough sawn Oregon concrete slab to be 200 mm thick (excludes waste allowance)

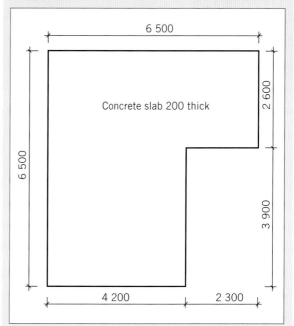

Source: © Cengage

FIGURE 6.20 Irregular-shaped concrete slab

Material costs:
200 × 32 Oregon @ $8.65 per metre
concrete @ $132.00 per cubic metre

1 Calculate the lineal metres of screed board by totalling the perimeter measurements of the slab.

Note: When ordering timber, as described previously, the lengths will be in increments of 0.300 m; therefore, the length of each side of the floor slab would be increased to the next increment.

Perimeter or screed board
$$= S_1 + S_2 + S_3 + S_4 + S_5 + S_6$$
$$= 6.600 + 2.700 + 2.400 + 3.900 + 4.200$$
$$+ 6.600$$
$$= 26.400 \text{ m}.$$

2 Calculate the cost of the screed board.

Cost $= 26.400 \times \$8.65$
$\quad = \$228.36$ for 26.400 m of 200 × 32 Oregon.

3 Calculate the volume of concrete by separating the floor into two sections.

Volume of concrete slab $= L \times W \times D$

Section (a) : $6.500 \times 2.600 \times 0.200 = 3.380 \text{ m}^3$

Section (b) : $4.200 \times 3.900 \times 0.200 = 3.276 \text{ m}^3$

Add sections together:
$3.380 + 3.276 = 6.656 \text{ m}^3$.

4 Calculate the cost of the concrete.

Cost $= 6.656 \times \$132.00$
$\quad = \$878.59$ of concrete
$\quad$ (with no allowance for waste).

EXAMPLE 6.24

Calculate the cost to lay the bricks for a wall 4.500 long by 2.100 high (Figure 6.21). The wall is one thickness (110 mm) and there are 50 bricks per square metre. The cost to supply and lay the bricks is $1345.00 per 1000 bricks.

1 Calculate the area of wall surface.

Area of brick wall $= L \times H$
$\quad = 4500 \times 2.100$
$\quad = 9.450 \text{ m}^2$

2 Calculate the number of bricks.

Number $= 9.450 \times 50$ bricks/m^2
$\quad = 472.5$ (473 bricks are required with no allowance for waste).

3 Calculate the cost. Divide the number of bricks by 1000 as the total number of bricks is less than 1000.

Cost $= \$1345.00 \times 0.473$.
$\quad = \$636.19$ to supply material and labour to lay 473 bricks.

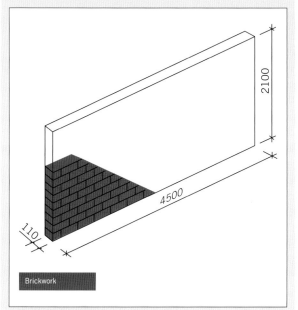

FIGURE 6.21 Single skin of brickwork

COMPLETE WORKSHEET 5

Source: © Cengage

SUMMARY

In Chapter 6 you have learnt how to plan and prepare to measure, obtain measurements, perform calculations and then estimate approximate quantities. You have also learnt that:

- it is important to plan before starting to measure and to understand the four basic questions that need to be considered prior to starting.
- Workers need to have a good knowledge of basic measuring tools, including the scale rule, the four-fold rule, retractable tape measure, long tape measure, the trundle wheel and the laser distance-measuring device.

- Workers must be able to identify the correct method of measuring (accuracy is important) and how to use the metric scale (International System of Units – SI units).
- Workers need to know how to perform calculations of linear measurements, area measurements (including regular and irregular shaped objects), and volume measurements.
- Workers should also be able to estimate approximate quantities and costs, and estimate percentages and fractions.

REFERENCES AND FURTHER READING

Acknowledgement

Reproduction of the following Resource List references from DET, TAFE NSW C&T Division (Karl Dunkel – Program Manager – Housing and Furniture) and the Product Advisory Committee, is acknowledged and appreciated.

Texts

Australian Institute of Building (1985), *Code of estimating practice for building work*, AIB, Canberra

Australian Institute of Quantity Surveyors and the Master Builders' Association (1987), *Australian standard method of measurement of building works*, Australian Institute of Quantity Surveyors and the Master Builders' Association, Canberra

Marsden, P.K. (1998), *Basic building measurement*, University of New South Wales Press, Sydney

Milton, H.J. (1992), *Australian building and construction definitions*, Standards Australia, Sydney

Sierra, J.E.E. (1998), The A–Z guide to builders' estimating, Australian Institute of Quantity Surveyors, Canberra

Smith, J. (2000), *Building cost planning in action*, Deakin University Press, Melbourne

Web-based resources
Resource tools and VET links

http://www.cpsisc.com.au, Construction and Property Services Industry Skills Council

http://training.gov.au, Skills training

GET IT RIGHT

In the photo below, the workers are doing their measuring and marking out task with some incorrect practices. Identify these incorrect methods and provide reasoning for your answer.

Source: Richard Moran

_____

_____

_____

_____

_____

_____

WORKSHEET 1

Student name: _____

Enrolment year: _____

Class code: _____

Competency name/Number: _____

Task: Read from the start of the chapter up to and including the section *Select most appropriate equipment and method for obtaining the measurement*, then complete the following questions.

1 How are linear measurements usually expressed in the building industry? (Circle the correct answers.)

   a millimetres

   b centimetres

   c metres

   d kilometres

2 Which of the following units of linear measurement are *not* used in the building industry? (Circle the correct answer(s))

   a millimetres

   b centimetres

   c metres

   d kilometres

3 Complete the table below.

Name of measuring device	Usage
Scale rule	
	Used to measure and mark out dimensions less than one metre
Retractable tape	
Long tape measure	
	Used to obtain quick estimates of distance up to several kilometres
Laser distance-measuring device/range finder	

4 Name the measuring device that is not an accurate measuring tool.

   _____

5 How many litres of water equal 1 m³? (Circle the correct answer.)

   a 100 litres

   b 500 litres

   c 1000 litres

   d 2000 litres

6   What is the unit of measurement for the mass of an object? (Circle the correct answer.)

a m^2

b m^3

c   N (Newtons)

d   kg (kilograms)

7   What is the unit of measurement for the volume of an object? (Circle the correct answer.)

a m^2

b m^3

c   N (Newtons)

d   kg (kilograms)

8   What is the unit of measurement for the surface area of an object? (Circle the correct answer.)

a m^2

b m^3

c   N (Newtons)

d   kg (kilograms)

9   In Australia our measurements are all: (Circle correct answer)

a   metric

b   centimetre

c   Imperial

 **WORKSHEET 2**

Student name: _____

Enrolment year: _____

Class code: _____

Competency name/Number: _____

Task: Read through the sections beginning at *Use a ruler or tape to obtain linear measurements accurate to 1 mm* then complete the following questions.

1   What is the most common length tape measure used to measure wall frame studs accurate to 1 millimetre: (Circle correct answer)

   a   30 metre tape

   b   5 metre tape

   c   1 metre tape

2   Why might the end of a tape measure be a bit loose? (Circle correct answer)

   a   It is probably worn

   b   To allow for measuring up to something and over something

   c   It doesn't matter how loose it is. It will still be okay to use for measuring.

3   How should you hold the tape when measuring? (Circle correct answer)

   a   It doesn't matter where the tape starts from as long as it is close to the end of the material you are measuring

   b   You must measure in a straight line from the start of the measurement to the end of the measurement

   c   As long as the tape is hooked over the end of the material you are measuring at the start point it doesn't matter whether the tape rides up over a scrap of material or not

4   It's more accurate to use a tape measure for measuring longer lengths than it is to use a rule end to end to measure along length.

   TRUE   FALSE

 **WORKSHEET 3**

To be completed by teachers

Student competent ☐

Student not yet competent ☐

Student name: _____

Enrolment year: _____

Class code: _____

Competency name/Number: _____

Task: Read through the section *Take basic measurements and calculate quantities of materials in a construction environment, using basic formulas for each of: weight, area, volume, perimeter, circumference, ratio and percentage* then complete the following questions.

1   Worked example:

How many fence posts, 2.7 m apart, would be needed to build a fence 20.0 m long across the front garden of a house, and what would be the total cost if each post costs $6.50?

$$\text{No. of fence posts} = \left(\frac{\text{length of fence}}{\text{post spacing}}\right) + 1 \; (\textit{always include the formula first}$$

$$= \left(\frac{20.0}{2.7}\right) + 1$$

$$= 7.4, \text{say } 8 + 1 = \mathbf{9}$$

$$\textit{Cost} \quad = 9 \times \$6.50$$

$$= \mathbf{\$58.50}$$

2   If there is a 75 × 50 hardwood top and bottom rail in this fence to connect the posts, how many 5.4 m long lengths of timber rail are needed to build the fence?

No. of lengths = _____

3   What will be the cost of the fence rail material if 75 × 50 hardwood is $2.85 per metre?

Cost = _____

4   If the timber palings on the fence are 1500 mm long × 100 mm wide, how many palings will be required to cover the fence?

No. of palings = _____

5   **Surface area**

The flat surface of a skillion roof is 24.0 m long with a rafter length of 12.0 m. If a 12.0 m length of corrugated iron roof sheeting covers a width of 762 mm, how many sheets are required to cover the roof?

No. of sheets = _____

6   If the corrugated iron costs $1.75 per square metre, how much would it cost to cover the roof when each actual sheet width is 820 mm?

Cost = _____

7   In the back garden, how many panels of treated pine lattice 2.4 m × 1.2 m would be required to build a fence 32.0 m long by 1.2 m high to keep the dog off the vegetable patch?

No. of panels = _____

8   If each lattice panel costs $3.85 per square metre, how much will it cost to buy panels for the back garden fence?

Cost = _____

9   A luxury mansion has a semicircular stencilled concrete drive that is 3.0 m wide. The inside radius is 17.0 m and the outside radius is 20.0 m. How many metres of edge board would be required to form the drive?

No. of metres = _____

10   **Volume**

Calculate the volume of soil to be removed from a trench for a sewer pipe. The trench is 25.0 m long by 800 mm deep by 450 mm wide.

Volume of soil = _____

11   Calculate the volume of sand a drainer will require to lay a bed 150 mm deep for the length of the trench.

Volume of sand = _____

12   If a cubic metre of filling sand costs $28.00, how much will the sand cost for the bed under the sewer pipe?

Cost of sand = _____

13   Calculate the total volume of timber in the following order:

100 × 50 sawn hardwood – 7/5.4, 12/4.8, 9/3.6 = 127.8 m

200 × 75 sawn hardwood – 20/2.4 = 48 m

300 × 38 sawn hardwood – 9/3.0, 4/2.7, 2/1.8 = 41.4 m

Volume of timber = _____

14   How much will the timber cost if unseasoned hardwood is $240.00 per cubic metre?

Cost of timber = _____

15   **Brickwork**

Calculate the number of bricks in a single-skin wall 8.5 m long by 1.8 m high, when there is an average of 50 bricks per square metre.

No. of bricks = _____

16   How much will it cost to purchase the bricks if they are $965.00 per 1000?

Cost of bricks = _____

17   Calculate the quantity of bricks in a solid 230-mm thick front fence that is 15.0 m long by 900 mm high, when there is an average of 50 bricks per metre squared per skin of brickwork.

No. of bricks = _____

18 **Concrete**

A house has a portico entrance with two grand round columns supporting the roof. The columns are 600 mm in diameter and 4.95 high. How much concrete would be required to build them?

Volume of concrete = _____

19 How much will the concrete for the portico cost when concrete costs $247.00 / m³ (including GST) delivered to the site?

Cost of concrete = _____

20 The bricklayer is building a fishpond that has a diameter of 3.6 m. Before placing the brickwork, he needs to pour a circular concrete slab with a base 125 mm thick. How much concrete will be required?

Volume of concrete = _____

21 The homeowner has a number of ornamental garden statues to be placed around the garden on six triangular prism-shaped plinths. The height of the plinths is 1.2 m and the triangular-shaped ends have a base of 500 mm and a perpendicular height of 433 mm. How much concrete will be required to make the bases?

Volume of concrete = _____

22 To build piers to support the floor of a house, the builder must excavate 24 pier holes 400 mm × 400 mm × 550 mm deep. How much soil must be removed?

Volume of soil = _____

23 If the builder places concrete 450 mm deep in the pier holes from Question 22, how much concrete will be required?

Volume of concrete = _____

24 What will be the total cost to excavate and then pour the concrete for the pier holes from Question 22 if the cost to excavate each hole is $12.00, the cost of the concrete (refer to Question 23) is $245.00 per cubic metre (including GST) and a labourer is required for 3 hours to pour the concrete at $26.00 per hour?

Cost to excavate and pour concrete = _____

25 Convert the following percentages to fractions.

75% = _____

7% = _____

35% = _____

140% = _____

26 Convert the following fractions to percentages.

3/5 = _____

7/20 = _____

27 Convert the following percentages to decimals.

60% = _____

22.5% = _____

3.5% = _____

28 Calculate the amount of concrete required for a slab-on-ground that is 1.273 m³ if an allowance of 5% is added on for wastage.

_____

_____

_____

29 A carpenter has to lay 3550 m of wall-lining boards onto timber studs. Determine how many metres of lining boards the carpenter needs to order if there is a waste allowance of 10% added on.

_____

_____

_____

30 An excavator has to load soil taken from a trench. The soil has an in-situ volume of 2.112m³ but takes up more volume when bulking allowance of 25% is added. Determine the full amount of volume to be loaded onto a truck.

_____

_____

_____

WORKSHEET 4

Student name: _____

Enrolment year: _____

Class code: _____

Competency name/Number: _____

Task: Read through the sections beginning at *Convert measurements in metres to millimetres and measurements in millimetres to metres* then complete the following questions.

1 Using the conversion examples shown in the table, convert the measurements in the exercise below from millimetres to metres and metres to millimetres respectively.

CONVERSION EXAMPLES		
100s of mm in m	**10s of mm in m**	**Individual mm in m**
1000 mm = 1.0 m	90 mm = 0.09 m	9 mm = 0.009 m
900 mm = 0.9 m	80 mm = 0.08 m	8 mm = 0.008 m
800 mm = 0.8 m	70 mm = 0.07 m	7 mm = 0.007 m
700 mm = 0.7 m	60 mm = 0.06 m	6 mm = 0.006 m
600 mm = 0.6 m	50 mm = 0.05 m	5 mm = 0.005 m
500 mm = 0.5 m	45 mm = 0.045 m	4 mm = 0.004 m
400 mm = 0.4 m	40 mm = 0.04 m	3 mm = 0.003 m
300 mm = 0.3 m	35 mm = 0.035 m	2 mm = 0.002 m
200 mm = 0.2 m	30 mm = 0.03 m	1 mm = 0.001 m
100 mm = 0.1 m	25 mm = 0.025 m	0.5 mm = 0.0005 m

EXERCISE					
Convert from millimetres to metres			**Convert from metres to millimetres**		
1100 mm	=		0.090 m	=	
16 400 mm	=		5.775 m	=	
5580 mm	=		3.900 m	=	
400 mm	=		0.006 m	=	
3225 mm	=		0.600 m	=	
8 mm	=		9 m	=	
7780 mm	=		0.990 m	=	
890 mm	=		7.150 m	=	
90 mm	=		57 m	=	
18 mm	=		0.005 m	=	

WORKSHEET 5

Student name: _____

Enrolment year: _____

Class code: _____

Competency name/Number: _____

Task: Read through the section *Check calculations for accuracy and record calculation workings and results* then complete the following questions.

1   Explain why it is good to write down your measurements and calculations.

_____

_____

_____

2   Why is it important to record your calculations step by step.

_____

_____

_____

3   How can the principle of estimating be used to help you check your calculations for accuracy?

_____

_____

_____

4   Why is it a good idea to record your calculations and file them with your job records?

_____

_____

_____

PERFORMANCE OF SKILLS EVIDENCE

Student name: _____

Enrolment year: _____

Class code: _____

Competency name/Number: _____

Task: Working with your teacher, prepare to demonstrate your skills by completing the following task. To demonstrate competency, a candidate must satisfy all the elements, performance criteria and foundation skills of this unit by:

1   Taking basic measurements and performing basic calculations to determine quantities of materials for construction work using each of the following:

   a   weight

   b   area

   c   volume

   d   perimeter

   e   circumference

   f   ratio

   g   percentage

2   Demonstrating converting measurements in metres to millimetres and measurements in millimetres to metres.

   All work must be performed to the standard required in the workplace.

Perform this simple activity to practise and demonstrate your ability to take measurements and perform calculations:

1   Measure a rectangular box-shaped surface with your measuring tape, recording the length, width and height in metres and then calculate the volume. Write these steps down.

2   Estimate the weight of a small box and then weigh it and record the weight. Note how much different your estimate was from the measured weight.

3   Measure the circumference of a circular object and note this down using area and circumference formulas. Apply that measurement of the circular object to determine the area of the object surface and the circumference around the object and record it.

4   Calculate the perimeter of a simple object such as a desk and write down how you calculated it.

5   Thinking about the slope of a ladder on a wall being 1:4 (1 m horizontal to 4 m high), how far out from the wall should the ladder be if it is 5 metres high? How far out from the wall should the ladder be if the ladder is 3 metres high?

6   Perform simple percentage calculations to demonstrate your ability to find percentages. For example, find 10% of $100. Find 10% of $157. Practise this and then discuss percentages with your teacher to ensure you understand.

7   Demonstrate that you can convert measurements in metres to millimetres and millimetres to metres on various lengths in your work area with your teacher.

7 READ AND INTERPRET PLANS, SPECIFICATIONS AND DRAWINGS FOR CARPENTRY WORK

This chapter covers the following topics from the competency CPCCCA3025 Read and interpret plans, specifications and drawings for carpentry work:

- plan and prepare
- interpret construction plans, drawings and their features
- locate key features on site plan
- determine project requirements and plan project.

This unit of competency specifies the skills and knowledge required to read and interpret plans and specifications for carpentry work in order to plan and sequence the work, meeting all relevant requirements of the National Construction Code (NCC), Australian Standards, work health and safety (WHS), and Commonwealth and state or territory legislation.

1. Plan and prepare

1.1 Locate and access plans, specifications and drawings

A construction project involves a lot of in-depth and detailed planning before any work can begin. In most cases, approvals need to be obtained – therefore documentation needs to be generated detailing the project. This process may involve government authorities, lending authorities and other organisations so the project is created in plans, specifications and drawings well before any work starts on a site. Usually plans, specifications and drawings are prepared by specialist technical businesses in consultation with all the parties involved in the project. Your job, as the constructor of the project, is to locate and access any relevant documentation that involves the project. Below are some examples of how to locate and access plans, specifications and drawings:

- Identify the parties involved in the project and seek out documents that were created for the project, going as close as possible to the original source whether it be a government agency, architectural firm, engineering company or the client. You need to obtain the information so you can start to get a picture of the overall project.
- Contact the various parties involved in the project and obtain and study the documentation, noting any different versions so you can ensure you have the latest edition/addition of the various documents that are relevant to the project.
- Check all the documents and make sure they are the correct ones referred to for the project. For example, there may be a few editions/versions of the plans and only one current edition of the engineering details. Make sure those engineering details refer to the correct version of the plans.
- Collaborate with any professionals to ensure you have the correct documents and that you understand all the details before you begin construction.

In the building industry, drawings are used to convey large amounts of technical information between the designer of a building (or structure) and the builder. This technical information must be able to be conveyed without the risk of any misunderstanding. This can be done successfully only if the technical language of drawings is understood by everyone who is required to use them (see Figure 7.1). The technical language of drawings is a standard language and must be learned, just as written and spoken language must be learned.

The technical language of drawings is expressed through the use of:

- standardised drawing layout
- standardised symbols
- standardised abbreviations of terms.

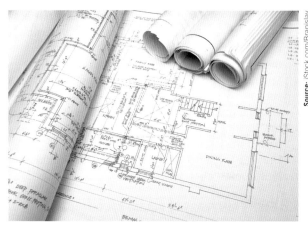

FIGURE 7.1 Plans and specifications

With study, practice and experience, the technical language of drawings and **plans** may be as quickly and easily understood as the written word.

When a drawing of a proposed building or structure has been prepared, many copies are made for the people who will use them.

- **The owners of the proposed building** need a drawing to see that the design is as they imagined it.
- **Structural, hydraulic, electrical and mechanical engineers** need copies so they can design their part of the works correctly.
- **Council health and building surveyors** will require copies of drawings and specifications to make sure that the building conforms to building codes and council regulations.
- **Council town planning officers** will require copies of drawings and specifications to make sure that the building conforms to council planning regulations.
- **Bank or building society** officers will require copies of the drawing before giving approval for finance for construction.
- **A builder** will require copies of the drawing in order to cost the building and to prepare a quotation and then to construct the building (refer to Figure 7.2).

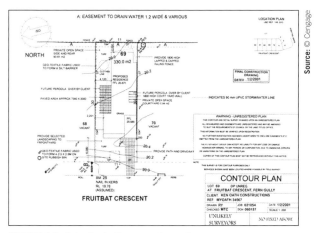

FIGURE 7.2 A typical site or contour plan

- **Contractors and subcontractors** such as concreters, bricklayers, electricians, carpenters, tilers and painters receive copies from the builder so they can quote their prices to carry out work during construction.
- **Suppliers** of prefabricated building components such as wall frames, roof trusses, windows, doors, air conditioning and heating will need copies of the drawing in order to calculate prices for their part of the job.

Drawings are important documents and should be carefully looked after.

1.2 Verify currency of plans, specifications and drawings

The plans, specifications and drawings may be in print or electronic form. You need to search these documents thoroughly to make sure you have the most current versions for the project. Indications of this may be in the file name or in more detailed information in the electronic file. Obtain confirmation from the client, and any other relevant parties, that the plans, specifications and drawings you have are identified as the most current and correct versions. There are normally dates indicating when documents were created or modified, and this can be a good indicator of the currency of the document.

Title block

The title block on a set of plans for a house may contain a minimum of information, such as the name of the owners, the lot number and street number, street name, suburb and scales used in the drawing (Figure 7.3).

The title block for a commercial project may contain:
- the name of the client or company for whom the project is to be constructed
- the lot number and address of the project
- scale or scales used on the plans; for example, 1:100
- numbers on the drawing sheets, if more than one is used; for example, sheet 1 of 3
- the name of the person or drafting service that has prepared the drawings

- a filing system in the form of numbers or letters (for use by the person who prepared the drawings)
- warnings (in some cases) against scaling from drawings to prevent incorrect measurement when using scales rather than using figured dimensions.

From plans and drawings the work schedules can be determined. Schedules establish the overall work activities that will be conducted throughout the construction and are provided to a range of trades who will work onsite. Coordination of trades in the construction timeline allows efficient use of time and space. Activities could include tasks such as excavation, foundation pouring, framing, electrical wiring and plumbing installations. A detailed list of the work activities can be provided to the trades and a supervisor or project manager often sequences the activities to avoid bottlenecks and ensure smooth progress.

GREEN TIP

Before you proceed too far with planning your project, it's important to make sure that all your environment and waste management plans for the job have been designed according to the most current requirements. Check with sources such as the NCC and your local council, state or territory environment regulators.

COMPLETE WORKSHEET 1

2. Interpret construction plans, drawings and their features

2.1 Determine key features of plans, specifications and drawings

As stated previously, a site or block plan is essential for determining the location of the building on the

Proposed dwelling for: Mr & Mrs B. Good, Lot 12 Aldgate St Prospect NSW 2149		
Scale: 1:100	Fly-by-night Builders, Lic. No: 4054	Sheet 1 of 3
Drawn by: R. Joven Date: Jan 2020	• Do not scale off drawings • All dimensions shown are theoretical only and are subject to site measure during construction.	Job No: RJ 06/01/57

FIGURE 7.3 Typical title block showing required information

building block. There are other essential pieces of information that are of equal importance. These include:

- the name of the street or road on which the site is located
- the side of the street or road where the block is located. Is it on the North, East, South or West side?
- allotment number (lot number) – this may be marked on the kerb or even a peg on the site
- boundary dimensions of the block – there should be registered surveyor pegs marking the corners of the block. These should be the horizontal measurements found on the site plan
- number of trees on the plan – this should match the number of trees on the block.

All of these items are important as they help the builder to identify and then build the structure on the correct block of land (Figure 7.4). If the builder gets it wrong it could be very costly to make it right.

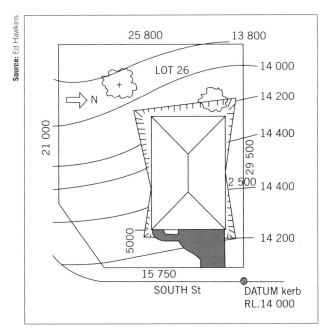

FIGURE 7.4 Site plan

Another important aspect of the site plan is that there may be information that identifies items that are referred to as 'services' (water, gas, sewer, electricity and telecommunication lines). These services may be buried under the ground on the site and could be dangerous and costly if they are accidentally dug up.

Construction methods and ideas for building to be carried out are first set down as drawings and, together with a specification, become the instructions for the different tradespeople to follow without the need for further referral to the client or architect.

Completed drawings and specifications must be approved by the local council before construction work can begin. Drawings must be in a standard format for accurate interpretation by builders and tradespeople from any area or background.

Drawings can be divided into two groups:

1 pictorial representation
2 working drawings.

Pictorial representation

These drawings or illustrations are used by architects and designers to determine, with the client, the final design and appearance of the project. A perspective view (Figure 7.5), or a **pictorial representation** (Figure 7.6), is often used to assist in visualising the building or structure. In these drawings or illustrations the faces or sides of the object appear to taper away or recede to a vanishing point. Parallel lines seem to converge, in the same way as railway tracks appear to converge at a single point in the distance.

Perspective drawings are the closest to what the eye would see, and are often used by architects or builders to present a project for sale.

Isometric projection (Figure 7.7) is also a pictorial view, with lines drawn parallel to the axis at 30° because of the ease in using 60–30° set squares. Perspective view and isometric projection are used as working or freehand sketches to make it easy to interpret the house design or building construction techniques.

Oblique projection (Figure 7.8) is similar to isometric drawing, but with the lines drawn at 45° rather than 30°.

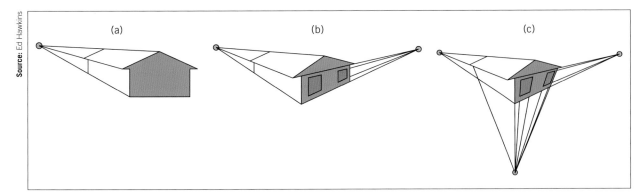

FIGURE 7.5 Perspective views: (a) 1-point, (b) 2-point, (c) 3-point

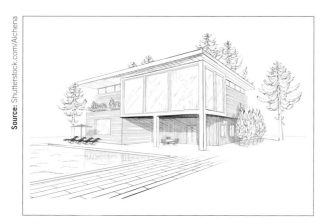

FIGURE 7.6 Pictorial representation

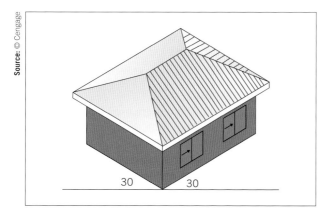

FIGURE 7.7 Isometric projection

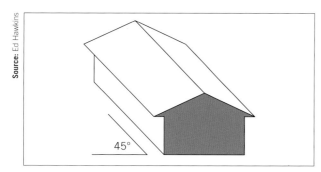

FIGURE 7.8 Oblique projection

Working drawings

Working drawings are produced so that users may:

- gain an overall picture of the layout and shape of the building
- determine setting-out dimensions for the building as a whole
- locate and identify the spaces and parts of the building; for example, rooms, doors, cladding panels, drainage
- pick up references leading to more specific information, particularly about junctions between parts of the building.

The most basic form of working drawing is **orthographic projection** (single angle), which consists of three related views – plan, **elevation** and section – to give a complete understanding of the building.

Working drawings may be divided into the following views:

- plans – site plan, floor plan, contour plan. These are all views looking down from above
- elevations
- sections
- construction details.

Site or block plan

Site or block plans are essential for determining the location of the building on the building block.

Information contained on a site (see **Figure 7.9**) or block plan includes:

- boundary dimensions of the block
- distance from street to boundary
- set-back distance from front boundary to building line
- distance from side boundary to building

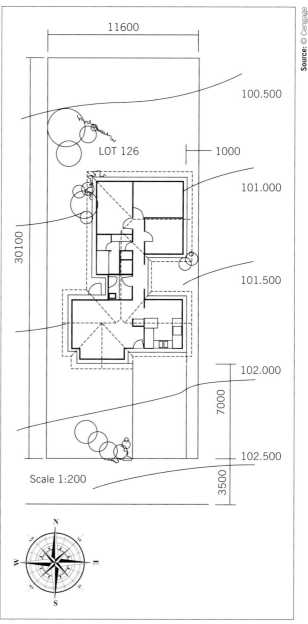

FIGURE 7.9 Typical site plan

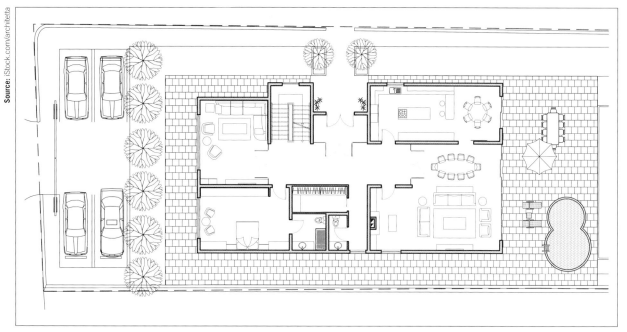

FIGURE 7.10 Typical floor plan

- contour lines and their heights
- position of paths and driveways
- trees
- direction of North
- lot number.

Version control must always be checked to ensure the latest version is being used or changes should only be made to the current version.

Floor plan

The floor plan is a horizontal cut-through section of the building as viewed from above and is the most important of all the related views as it contains most of the information for construction (**Figure 7.10**).

Items shown on a floor plan include:
- overall dimensions to the outside of the walls
- door and window positions and opening sizes
- thickness of external and internal walls
- internal room dimensions
- position of cupboards, stoves, laundry tubs, etc.
- function of each room, such as kitchen, bedroom
- floor surface and type of floor covering
- position and direction from which section lines are taken for sectional elevations.

Elevations

Elevations provide information relating to vertical measurements and external finishes. Each view is generally identified according to the direction it faces – North, East, South and West. This direction is based on the North point shown on the site plan.

Elevations give a projection of the building at right angles and show:
- height of finished floor level (FFL) to finished ceiling level (FCL)

- design of the building
- roof shape and width of roof overhang
- position of doors and windows
- window sill height above floor level
- type or function of windows
- roof covering and slope
- floor height above ground
- finish to external walls.

Note: Standard working drawings in orthographic projection generally require a minimum of two elevations – the front and side of the building – thus enabling correct interpretation of design. Typical information may be indicated on one or both elevations (**Figure 7.11**).

Sections

Section drawings are elevations that cut through the building in the position and direction indicated on the floor plan. The section is a cross-section from the bottom of the footings, through the walls, ceilings and roof structure (**Figure 7.12**). Sections give information such as:
- footing sizes
- wall thickness and construction
- design of sub-floor
- floor construction
- roof construction; for example, trussed
- roof pitch
- section sizes and spacing of structural members.

Construction details

Details are sectional views drawn to a larger scale than sectional elevations, and detail specific requirements that cannot be drawn accurately to scale on sectional elevations (**Figure 7.13**).

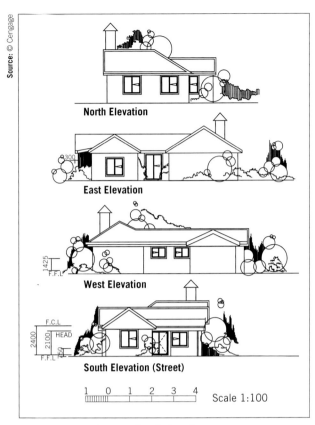

FIGURE 7.11 Typical details of elevations

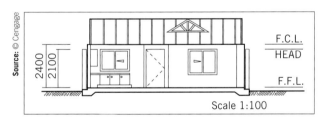

FIGURE 7.12 Typical section taken through kitchen and living/dining areas

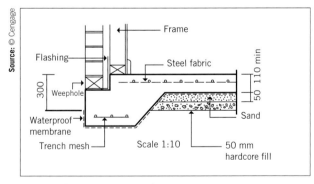

FIGURE 7.13 Slab edge detail

Other information to be shown on working drawings includes:

- scale used for drawings
- dimensions
- title block.

Scale drawings

A **scale drawing** represents a full-sized object reduced to a suitable scale to enable it to be reproduced on drawing sheets. Working drawings state the scale or scales that have been used on the drawing. However, in some cases there is also a warning that all measurements should be taken as read from the drawing, rather than determining lengths using scale rules.

A scale of 1:50 (Figure 7.14) indicates that the drawing is reduced to 50 times smaller than full size.

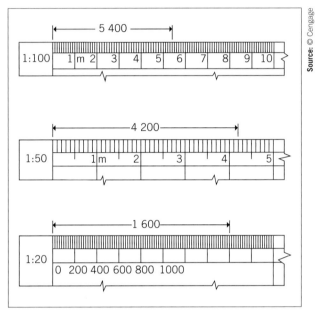

FIGURE 7.14 Common reduction scales

The following list shows standard scales adopted when producing working drawings.

Site plans	1:500	1:200	
Plan views	1:100		
Elevations	1:100		
Sections	1:100		
Construction details	1:20	1:10	1:5

Calculating scales

When a scale rule with the appropriate scale cannot be found, or an unusual scale needs to be used, it may be necessary to make your own. This may be done simply by using a calculator and a rule with millimetres or a scale of 1:1, which is full-size.

Dimensions

Dimension lines on drawings enable scales to be used to determine lengths that are not shown. The forms of dimension lines vary (Figure 7.15), but all are shown as a line parallel to the drawing. Lines at right angles to the main line indicate the position at which the dimension is taken.

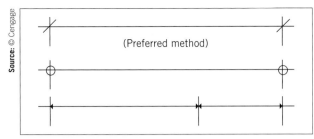

FIGURE 7.15 Examples of dimension lines

(Preferred method)

TABLE 7.1 Drawing sheet sizes

Sheet type	Length (mm)	Height (mm)
A0	1189	841
A1	841	594
A2	594	420
A3	420	297
A4	297	210

EXAMPLE 1

Scale required = 1:100
Measurement to be scaled = 5.400 m.
Step 1 Change the measurement from metres to millimetres, i.e. 5.400 m, change to 5400 mm.
Step 2 Divide the millimetre measurement by the desired scale, i.e. 5400 ÷ 100 = 54.
Therefore, the 1:100 scaled measurement = 54 mm (full size in millimetres).

EXAMPLE 2

Scale required = 1:50
Measurement to be scaled = 4.200 m.
Step 1 Change the measurement from metres to millimetres, i.e. 4.200 m, change to 4200 mm.
Step 2 Divide the millimetre measurement by the desired scale, i.e. 4200 ÷ 50 = 84.
Therefore the 1:50 scaled measurement = 84 mm (full size in millimetres).

Drawing sheet sizes

Plans or working drawings are produced to scale on drawing sheets, which range in size from A0 (1189 × 841 mm) to A4 (297 × 210 mm), with margin or border lines and title blocks (see Figure 7.3). The margins assist in the folding or filing of the plans, while the title block contains information essential to the project. Refer to Table 7.1 for standard paper sizes available.

A specification is a precise description of all construction and finishing, including workmanship, which is not shown on the drawings (Figure 7.16). This includes composition of concrete in footings, species and grades of timber, brick type, mortar quality, paint colours, size and type of hot water system, and the number and positions of power points. The specifications are arranged according to trades, and cover all information relevant to the particular trade in the sequence of construction. Specifications must be kept with the drawings and are read in conjunction with them.

Inevitable variations will occur during the construction process while still maintaining the overall quality and functionality of the structure. Tolerances can allow for variations in dimensions for various components such as walls, beams, columns and other structural elements. Guidelines for tolerances for gaps between elements such as expansion joints, doors, windows and other openings will be located in the workplace documents.

Supplementary specifications also known as supplementary conditions or special provisions, are additional documents that accompany the main construction plans and provide further clarification, requirements and instructions for specific aspects of the project. These specifications can vary depending on the project's complexity and requirements. These can include instructions on variations to materials and equipment, testing and inspection, special construction techniques, environment and sustainability requirements, and WHS regulations and controls.

LEARNING TASK 7.1 READING PLAN DIMENSIONS

1 Obtain a set of plans from the job you are currently working on or use a set from a previous job (ask your boss for a copy).
2 Check to see if the following dimensions are drawn to scale. Write down your findings in the space provided.

Using the floor plan: Measure and check to see that the *total length of the building is correct. If it is correct, what is the measurement in metres?*	
Using the floor plan: Measure and check to see that the *total width of the building is correct. If it is correct, what is the measurement in metres?*	
Using the site plan: Measure and check to see that the *longest side of the block is correct. If it is correct, what is the measurement in metres?*	

Source: © Cengage

Items	
clothes lines	see drawings …
letterboxes	1 per dwelling
shower screens	see drawings
bathroom cabinets	minimum 1 per bathroom
garage doors	see drawings

05 CARPENTRY & JOINERY

5.1 GENERALLY
Timber inspections and bandings: Refer PRELIMINARIES—Materials and workmanship
NSW Timber Framing Manual: The current edition of the NSW Timber Framing Manual
may be used in lieu of AS 1684.

To AS 1684/NSW TFM

5.2 MATERIALS
Timber stress grades: The timbers used must comply with the Timber Marketing Act, and
be graded to the appropriate SAA grading specification.

To AS 1720. AS/NZS 1748.
Refer table in ASDC.

Timber Species: Do not use tropical rainforest timbers. Refer to SCHEDULES— Schedule
of Timber Species and Durability Rating.

Preservative Treatment:
—All Lyctus susceptible sapwood in local rainforest timbers.
—All Lyctus susceptible sapwood in hardwoods other than milled exceeding 20% of
  the perimeter of the piece.
—All Lyctus susceptible sapwood in milled hardwood products.
—Radiata pine used externally and for bearers and joists.

To AS 1604.
Obtain Superintendent's
approval for treatment
details.

Panel and sheet products:
Plywood and Blackboard: Interior use, type D: exterior use Type A. Use particle-board
grades designated by the manufacturer to have moisture resistance appropriate to the
conditions of use. Melamine surfaced particle-board shall be finished with melamine,
surface bonded to all faces.

To AS/NZS 1859.
Refer SAPPC.

External cladding: hardboard planks, fibre cement flat sheets, fibre cement planks.

Refer SAPPC.

Laminated plastic sheet: Fix to background with contact adhesive.

To AS/NZS 2924, AS 2131.

5.3 WORKMANSHIP
General: Perform the operations and provide the accessories necessary for the completion
of woodwork items. Ease and adjust moving parts, lubricate hardware, and leave the
completed work in a sound, clean, working condition.

Joinery:
—mortice and tenon joints in doors, frames, sashes and other parts.
—mitre joints in mouldings, skirting, etc., but scribe internal angles
—dress joinery stock and mouldings, hand finish exposed surfaces and remove arrises to
  provide smooth surface for painting.
—all moulded runs of 800 mm and less must be in single lengths.

Sizes and tolerances:
—Maximum possible tolerance for dressing to be 3 mm per face.

To AS 1684.

Reference Specification for Detached Dwellings—Public Housing
March 1994

Page 8

FIGURE 7.16 Extract from a standard Department of Housing specification

Specifications are normally written up under the headings of the various trades or services that are involved in the construction of a project. A specification contains the wording of things that can't be drawn easily. For example, it is easier to write down the timber type and termite treatment requirements for the timber wall frame than it is to draw a picture of the H2 treated machine graded pine that is used for wall framing. This information would be written in a specification document.

A toolbox talk or site induction are used to provide subcontractors with detailed information about work activities, plans, drawings and specifications, materials, tolerances and quality control processes. Specific information related to access, and waste and recycling facilities need to be provided from information contained in the workplace documents and schedules.

2.2 Interpret legend symbols and abbreviations

Symbols and abbreviations appear on plans, elevations, sections and details. Common symbols are shown in Figures 7.17–7.26. They represent standardised details that are common knowledge in the building industry. Symbols enable the transfer of relevant technical information between the designer and the builder, and they reduce the risk of any misunderstanding of the designer's intention.

To further reduce the risk of misunderstanding, many plans have a 'key' or 'legend' included on the first sheet of the set of drawings. This key or legend lists or itemises the common symbols and abbreviations used in the drawings.

Symbols used in drawings for restoration of, or extension to, existing structures are coloured to distinguish the new work from the existing structure.

Water and other liquids Small scale	Cobalt blue
Packed or piled materials; e.g. rubble, ballast Small scale	Indigo
Earth Large scale	Sepia
Sand Large scale	Yellow ochre
Rock Large scale	Vandyke brown
Graded hardcore Large and small scale	
Fill Large and small scale	Raw umber
Fill Large scale	As above
Fill Small scale	As above

FIGURE 7.17 Symbols for sections in-ground

Symbols on plans indicate details such as:

- direction of opening and operation of doors and windows
- wall or floor construction, such as timber, brick, reinforced concrete
- foundation soil composition: earth or filled material.

Abbreviations are used on drawings to reduce the written content, thereby minimising the congestion of information necessary to convey the correct interpretation of the drawings. The abbreviated form should be used only where confusion or misinterpretation are not likely to occur.

Table 7.2 shows a list of common abbreviations that are commonly used.

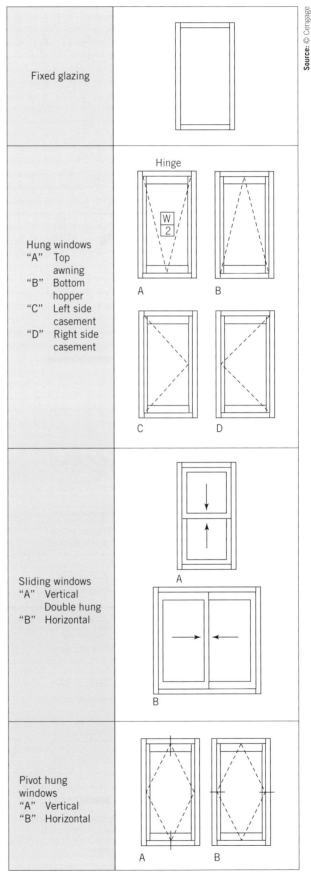

Fixed glazing	
Hung windows "A" Top awning "B" Bottom hopper "C" Left side casement "D" Right side casement	Hinge A B C D
Sliding windows "A" Vertical Double hung "B" Horizontal	A B
Pivot hung windows "A" Vertical "B" Horizontal	A B

FIGURE 7.18 Symbolic representations of windows for elevations

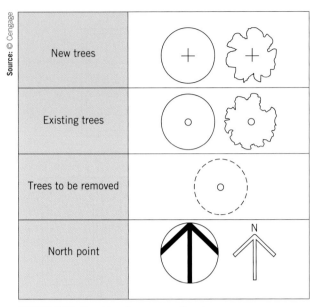

FIGURE 7.19 Graphics for use on site plans

New trees	
Existing trees	
Trees to be removed	
North point	N

Timber stud	"Shaded grey" Chrome yellow
Brickwork single skin	Vermilion
Brickwork cavity wall	Vermilion
Brick veneer	Chrome yellow and vermilion
Concrete block single skin	Prussian green
Concrete block cavity wall	Prussian green
Concrete	Hooker's green deep
Stone	Vandyke brown
Existing wall (alterations & additions) Alternatives a, b *Note:* Colour new work	Heavy black for prominence Light outline New work Ghosted

FIGURE 7.20 Symbolic representations for floor plans and details

A more detailed list of standard abbreviations can be obtained from Australian Standards:

- AS 1100.101–1992 Technical drawing – General principles
- AS 1100.301–2008 Technical drawing – Architectural drawing.

2.3 Check plans, specifications and drawings dimensions against workplace site for accuracy

Before any construction work begins it is very important to check the plans, specifications and drawings dimensions against your work site for accuracy. It is quite common to have errors in drawings and so it is important that you follow some simple processes to check all your documents before you start the hard work of assembling a construction project. The following are some steps to take once you have access to all the documents and access to the work site:

- Obtain all the relevant documents, the plans, specifications, and drawings that are the most current information for the job. You need to have these in a format that is easy to use onsite so you can check everything against the actual place where you will be doing the construction. This format may be electronic with a large-screen device, or printed documents. You may even glue the documents onto a large site board that can be kept onsite for all persons involved in the construction. This is a good way to get an overall view of the project because many pages in a project cannot be read solely by themselves and need to be referred to other pages to clarify finer construction points. Ensure you are very familiar with the documents and make sure you have checked everything, added any ambiguities you clarified and any errors corrected before you start work. For example, get a calculator and add up all the measurements on one side of the floor plan for the project and then subtract all the measurements down the other side. Thoroughly check all information before you proceed. A lot of this work can be done in the office before you even go to the site.
- When you get to the work site make sure that you have inspected the whole area and that you have checked it to get a picture of where the final project will be. Visualise the finished project, considering the height, the width, the depth, and if there is a 3D view of the completed project line it up with what is already on the site and see if it makes sense.
- Using measuring tools appropriate to the size of the project, take measurements to confirm that the actual onsite measurements coordinate with the documents that you are familiar with. Check angles, heights and lengths, taking note of any discrepancies or variations from your documents. If needed, contact the document's authors or the specialists for the task and seek clarification regarding

Fitting	Symbol	Fitting	Symbol
KITCHEN FIXTURES & FITTINGS All coloured French ultramarine		Water closet	
Stove	S	Bidet	B B
Wall oven Hot plates	O	LAUNDRY FITTINGS All coloured French ultramarine	
Refrigerator	R	Sink/Tub	
Dishwasher	DW	Washing machine	WM
Sinks (with drainers)		Clothes dryer	CD
		MISCELLANEOUS FITTINGS All coloured French ultramarine	
SANITARY FITTINGS All coloured French ultramarine		Cleaner's sink	CS
Bath		Slop sink	SS
Basin		Hot water unit Alternatives	HW HW
Shower recess		Rainwater tank	RWT 2000 litre
Urinal		Fire hose rack/reel Recessed	FHR
Urinal stall		Fire hose rack/reel Free standing	FHR

FIGURE 7.21 Symbolic representations of fixtures and fittings

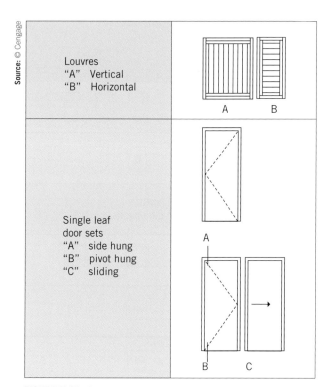

Louvres "A" Vertical "B" Horizontal	A B
Single leaf door sets "A" side hung "B" pivot hung "C" sliding	A B C

FIGURE 7.22 Symbolic representation of doors for elevation

Plywood large scale	As above
Alternative plywood large scale	As above
Dressed timber large scale	Burnt sienna
Sawn timber large scale	Chrome yellow
Composite boards large scale	Discretionary
Particle-board large scale	Indian yellow

FIGURE 7.24 Symbolic representation for sections (b)

Lightweight concrete	Hooker's deep green
Structural concrete	As above
Structural concrete alternative	As above
Terrazzo (on RC slab) Large scale	Emerald green

FIGURE 7.23 Symbolic representation for sections (a)

Glazed partitioning	
Window in single skin wall	W 4
Window in cavity wall (Internal skin)	
Window in cavity wall (External skin)	

FIGURE 7.25 Symbolic representations for floor plans and horizontal sections

Single swing door Arcs continued to indicate doors to fasten	D 4	Single swing door	
Sliding doors "A" Sliding into a pocket "B" Sliding exposed on wall face	A B	Archway	arch over
Opening extending from floor to full height of wall		All other openings not included under headings of doors, windows & archways	

FIGURE 7.26 Symbolic representations for floor plans and horizontal sections

TABLE 7.2 Common abbreviations

aggregate	aggr	minimum	MIN
angle	L	north	N (NTH)
approximate	APPROX	not to scale	NTS
at	@	out of	O/O, ex
average	AV, AVG	over-all	OA
bench mark	BM	pad footing	F
bottom	BOT	prefabricate	PREFAB
brick	BK	quadrant moulding	quad
brick veneer	BV	quantity	QTY
brickwork	BWK	rainwater pipe	RWP
building	BLDG	reduced level	RL
ceiling joist	CJ	reinforced concrete	RC
ceiling level	CL	retaining wall	RW
centre line	LC or C	right hand	RH
concrete	CONC	sewer	SEW
concrete reinforced	RC	shower	SHR
countersink	CSK	sliding door	SD
damp-proof course	DPC	temporary bench mark	TBM
detail	DET	tongue and groove	T&G
diagonal	DIAG	typical	Typ
diameter	DIA	underside	U/S
dimension	DIM	universal beam	UB
distance	DISt	universal column	UC
equal to	EQ	vent pipe	VP
existing	EXST	vertical	VERT
expansion	EXP	volume	VOL
external	EXT	wall oven/oven	W/O, O
finished floor level	FFL	waste pipe	WP
fixed glazing	FG	water closet	WC, W
floor level	FL	window (number)	W
floor waste	FW		
ground floor	GF		
ground level	GL		
ground floor level	GFL		
hardwood	HWD		
height	HT or HGT		
horizontal	HORIZ		
hot water supply	HWS		
include	INCL		
joist	J		
kitchen sink	KS		
left hand	LH		
level	LEV		
longitudinal	LONG		
main switchboard	MSB		

the differences/variations. Check any variations and make an assessment on the impact that it might cause on the final project and try and come up with some solutions yourself before you consult the experts. Make sure that all possible solutions comply with regulations. Then consult all relevant stakeholders providing suggestions or clarification to ensure agreement between the physical site and the documents that relate to that site.

■ Have the documents updated and any variations and/or amendments noted so that all documents refer to the most current up-to-date versions. Ensure all communication is detailed and kept in case of dispute later.

Following simple processes like this and working with competent people who have experience will ensure that your documentation can reflect as accurately as possible the real site that you will be working on. This can aid in solving potential problems that may occur down the track that could be costly and delay the project.

2.4 Check plans and drawing dimensions against specifications for accuracy and inconsistencies

Checking the specification document(s) against the plans and drawing documents should be done as early as possible in the project. As mentioned earlier, a specification is a written description of the works that can't be drawn easily.

Usually the client for a project will relate most closely to plans and drawings as they can often visualise themselves living within the project once completed. It's quite hard for a lot of people to picture themselves in a project that is just described in written text in a document. Therefore, the client will normally have the most input into the plans and drawings and it's the builder's job to write the specification(s) based on the regulations, the material availability, and other decisions that the client has made regarding items in the project. Ensuring the drawn description of the project (the plans and drawings) matches the written description of the project (the specifications) can be aided by following these steps:

- Gather the plans, drawings and specifications and ensure that you are only working from the most current versions.
- Carefully review the specifications making sure you understand the regulations for the job, the requirements, measurements, tolerances, and any other relevant details that are required by codes or standards. Note anything you are not sure about or any inconsistencies.
- Study the plans and drawings and make sure they meet any relevant requirements, codes or standards. Pay close attention to any measurements, notes, symbols, or any other information ensuring there are no conflicts between the specifications and the plans and drawings. Make sure you've compared dimensions across all the jobs looking for discrepancies or inconsistencies.
- Note any errors or variations in any of the documents and determine whether they are insignificant and can be solved without outside consultation, or whether they may impact upon the project and may need further communication before the job begins.
- Communicate with any relevant stakeholders as needed and record all communication clarifying any in accuracies or inconsistencies. Resolve any inconsistencies yourself, if you can, or come up with suggestions for the specialists or client if appropriate.
- After recording any changes, conduct a final review for the quality and verify to ensure that everything is accurate and there no known inaccuracies or inconsistencies in the job documentation.

Checking plans and drawings dimensions against the specifications for accuracy and inconsistencies might seem like a time-consuming exercise when you have a big project to get started. However, time spent clarifying things before the job starts can save a lot of time and expense later and the job can run much smoother with fewer problems.

 COMPLETE WORKSHEET 2

3. Locate key features on site plan

3.1 Orient the plans, specifications and drawings with the site

All construction on a site occurs at a specific location which has an orientation in relation to North. The site plan for a project has a symbol or note that indicates the direction of North for the site. When you take this site plan document to the site and orient it with where North is on the site you will start to build a picture of how and where the project will be located. You may need a compass, a GPS or just a location of the nearest street on the site plan so you can turn the site plan around to orient it with North. By familiarising yourself with the site, and any notes on the site plan, you should be able to locate existing structures, landmarks or survey markers that can help you to align the document with the site. You will need to take your documents and visit the actual site to observe all of this.

Preliminary work can be completed using satellite images from either the internet, or many local government authorities have very up-to-date satellite imagery of the local government area. The advantage of this imagery is that it shows a lot of local government authority information about the site. For example, if you search 'Wagga online mapping' in your favourite search engine you can find out a vast amount of information about every site within the local government area. Satellite imagery, land boundaries, development application plans, estimated cost of construction, topographic flood planning overlays, fences, sheds, existing buildings, all these can be shown on local government authority websites such as this.

Once you have the documents and you are on the site, you can establish the most accurate baseline for the boundary for that site. Confirm with measuring tools or surveying equipment if needed that all the details on the site plan match the actual site. It's a good idea to seek professional assistance at this stage. A peg-out survey by a registered surveyor is invaluable for locating the job on the exact property. Make sure you have any documentation that the surveyor provided so you can confirm that the pegs you are looking at are the actual pegs described on the documentation the surveyor provided (and that

LEARNING TASK 7.2 PLAN INTERPRETATION

Using the same set of plans you used in Learning Task 7.1 find the following information on the set of drawings and write your answers in the space provided.

What is the name shown in the **title block** to identify the **owner or client**?	
What is the **scale** of the site plan?	
What direction does the front of the building face? That is, does it face North or East, etc.	
What is the **name of the street** that the building faces?	
Does the site plan show a **bench mark** (BM) or **datum**? If it does, what is the **height**?	
On the site plan are there any **trees**? If there are, how **many trees** are on the site?	
On the floor plan is there a finished floor level (FFL)? If there is, **what is the height?**	

the pegs, for example, have not been moved by some kids over the weekend). It's a good idea to document your efforts at this stage and keep records of how you are confirming that you have oriented the plans, specifications and drawings with the actual site.

3.2 Locate site services, main features, contours and datum from the site plan

To locate site services, main features, contours and a datum from a site plan you need the most up-to-date site plan for the project. Make sure you understand all the details and have approval to go onto the physical site to verify details on the site plan as they appear on the actual site, or as they are planned to appear. A good site plan should list all existing or planned services, features, contours and items; however, some of these items may only be indicated by symbols on the plan. You need to be familiar with the legend for those symbols so you can identify these items. Some of the services that you may find on a site plan include:

- water lines
- sewer lines
- electrical connecions
- guest supply
- communications
- storm water drainage.

Once you have located these services on the site plan check to see if they exist, or where they may be situated, on the construction site. Label them if they're already there so work can proceed around these services or in conjunction with them. If the services are not installed, then make sure your plan includes leaving room to install these services later. Don't store a lot of materials, for example, in a space where a supply trench will need to be dug.

A site plan should contain existing features, proposed future features, and anything else significant

that may impact the construction. Features such as buildings, roads, fences, trees should all be accounted for in the overall construction plan and knowing where they are on the site is essential. Orient the site plan on the site and examine both the plan and the site for the main features so you can ensure they have been incorporated into the plan for your construction.

Contours should be indicated on a site plan so that all personnel involved in a construction can understand the slope of the land both at the beginning and when the construction is completed. Contour lines are curved lines drawn on a site plan that join points of the same height. They have been drawn on the site plan by the draftsperson based on site-levelling readings taken before the site plan was drawn. The contours are drawn to scale across the site so that construction personnel can understand the slope of the land. On each contour line a height will be indicated as a number above the site's datum or benchmark. The datum or benchmark for the site has been chosen as a location from which to base other heights on that site and a location that is out of the way of machinery and material storage areas so it can be referred to regularly and easily with levelling instruments. When you look on a site plan for contour lines you should note the numbers on the lines because as the numbers increase that means those contour lines on the site are getting higher and higher than the previous lower numbered contour lines. The closer together the lines are the steeper the site. The height difference between the contour lines will normally be nice even round numbers like 1 metre spacing or half metre spacings.

The datum point on a site plan will often be indicated only by a symbol. That symbol may look like a circle divided into four quadrants with two opposite quadrants coloured black. The datum point may be a random height that has been chosen as an even number from which all other heights on the site can be related to. Or the datum point may be a height that relates to a state or territory survey height which could be located on a nearby kerb, gutter or large public structure. In this case, the datum point and the state or territory survey height will be in relation to the Australian height datum system across Australia.

A small construction site may not need a height that relates all the way back to the Australian height datum and a local temporary benchmark or datum may be sufficient. However, if the project involves some risk, such as possible future flood risk, then the finish floor level of the project may need to be a specific height above Australian height datum for insurance and flood safety reasons. Most towns and local government authorities know what the possible flood heights for different flood risks are, for example 1 in 100-year floods, and these heights are published on the councils' websites and may require your finished floor level for your project to be above these heights. The person who draws the site plan for your project will know this and will have indicated the finished floor level. You can confirm by relating back to the datum point that is shown on the site plan.

It is essential that you use site visits to verify your measurements, dimensions, and any features or services you find. Make sure you document your findings by taking photographs and notes of your site visits and your site plan, and have any discrepancies or variations confirmed and fixed before construction proceeds. If you need to consult professionals, then it's worth paying to get the confirmation that you need for your construction job to begin. Some of the professionals you may need to consult are surveyors, engineers, architects, competent people, local government authority town planners and so on.

COMPLETE WORKSHEET 3

4. Determine project requirements and plan project

4.1 Review drawings, plans and specifications to determine construction details and dimensions for project

To help you understand domestic construction practice and the graphical presentation of details, the following components of a building have been identified in segments or structures which, when grouped together, become the complete project.

Footings

Footings are the lowest part of a building, designed to distribute the load of the building evenly over the foundation (Figure 7.27). There are several types, which are designed according to the load of the building to be

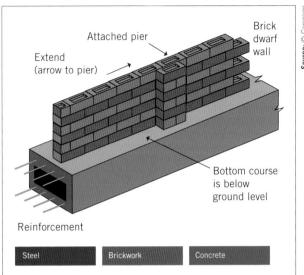

FIGURE 7.27 Typical detail of a reinforced concrete strip footing

supported and the ability of the foundation material to support that load. Common types of footing in building are as follows.

Strip footings

This is a continuous reinforced strip of concrete around the outside of a building to support the external walls. The width and depth of concrete and the amount of reinforcement needed are calculated by a structural engineer, based on the design of the building and the known soil classification, as stated in AS 2870 Residential slabs and footings. The bottom of a strip footing is approximately 600 mm below ground level to prevent erosion of the foundation material. Timber frame, brick veneer and cavity brick buildings are constructed on strip footings.

Slab-on-ground

This combines the floor and the footing into one reinforced, monolithic concrete unit. This method of construction is used on level ground or sloping sites that have been cut to a level surface. The edge beam can be increased to support greater loads or to pass through top reactive soils to more stable foundation material (Figure 7.28).

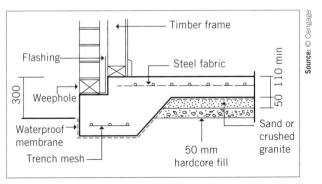

FIGURE 7.28 Slab edge detail in stable soil

An alternative is the 'waffle-pod' system, consisting of a series of beams running at right angles to one another with a thin working slab cast over them to form a monolithic slab with a grid support system (Figure 7.29).

Services such as plumbing and electrical pipes are placed in the slab before the concrete is poured, or are built into the walls during construction. Slab penetrations must be adequately protected to prevent the entry of termites; for example, with the use of termite prevention products such as Granitgard™, Termimesh or Kordon®. Slab-on-ground has become a popular alternative in floor construction of timber, brick veneer and cavity brick buildings (Figure 7.30).

Blob/pad footings

These are square, rectangular or round footings placed under piers or posts and may contain reinforcement, depending on the load to be carried (Figure 7.31). Their size should be calculated so that the same pressure is applied to the foundation material as to strip footings when both are used on the same building. Blob footings are commonly used for supporting brick piers and bearers and joists in timber floor construction. Alternatives to this type of support include steel adjustable piers, pre-cast concrete piers and treated timber stumps (generally used in states other than New South Wales).

Flooring systems

A flooring system is the floor surface and the method of floor framing used as support to the ground floor. Common systems in use are as follows.

Suspended timber floor

One system consists of flooring boards in narrow strips laid on a timber or steel framing of bearers and joists and supported by brick walls or piers. The flooring is placed between the walls and is cramped and nailed in position when the building is advanced enough for the

FIGURE 7.29 Typical waffle-pod slab system

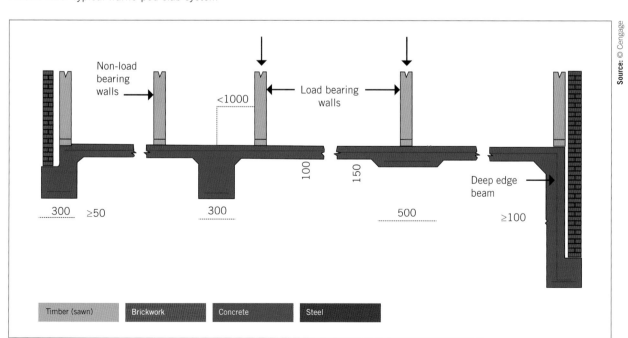

FIGURE 7.30 Slab-on-ground for masonry, veneer and clad frames

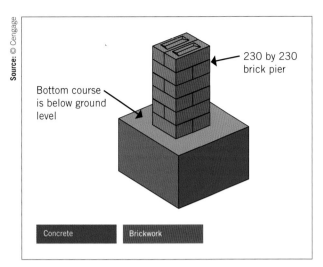

230 by 230 brick pier

Bottom course is below ground level

Concrete   Brickwork

FIGURE 7.31 Typical blob/pad footing with a minimum depth of 200 mm

flooring not to be affected by the weather, i.e. generally when the roof and wall cladding has been installed (see Figure 7.32).

Another system consists of sheet flooring of plywood or particle board laid on a timber or steel framing of bearers and joists before the walls are erected; this system has the advantage of providing a platform to work on. Both of these methods require a minimum 400 mm clearance above the ground to the underside of bearers, which provides ventilation of the area beneath the floor framing to prevent decay in the timber framing. The timber bearers and joists used to support the flooring are placed at specified centres apart and are of a sectional size according to AS 1684.2 Residential timber-framed construction – Non-cyclonic areas.

Slab-on-ground

A reinforced concrete floor is placed directly onto the ground. One advantage of this method is the reduced building height. See Figures 7.28–7.30, and previous information on footings for details.

Suspended slab floor is a reinforced concrete floor suspended above the ground and supported on brick walls (Figure 7.33). The amount and type of reinforcement in a suspended concrete floor will be greater than that used in a slab-on-ground as a suspended slab must carry the floor loads between supports.

Wall structures

This is the composition of the external and internal walls. The walls may be constructed from one or several materials. Common wall structures are as follows.

Timber frame

External walls are constructed of timber framing, with a cladding on the outside of timber or fibre cement weatherboards or sheets, and a lining on the inside of plasterboard. Internal walls are constructed of timber framing and are lined on both sides with plasterboard (Figure 7.34).

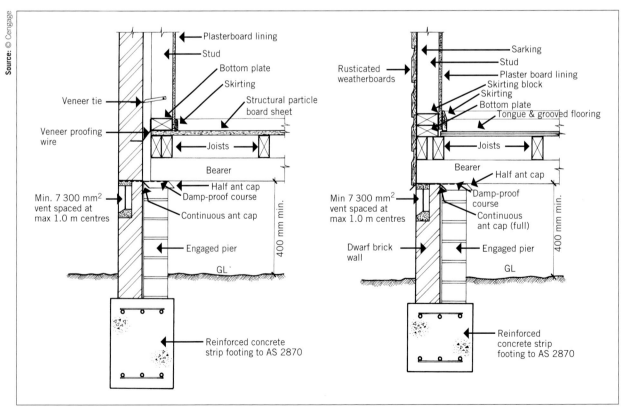

FIGURE 7.32 Vertical section through external walls of brick veneer and timber-frame construction

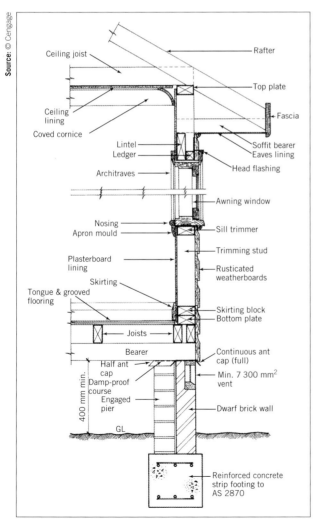

FIGURE 7.33 Suspended concrete floor with balcony projection at first floor level

Brick veneer

This is for external walls only, consisting of a timber frame lined internally with plasterboard and an external skin or veneer of brick (**Figure 7.35**). The brick skin and the timber frame are separated by a gap (cavity) of around 40 mm (25 mm minimum – 60 mm maximum) to allow for ventilation and to prevent contact between the timber and the bricks. The timber wall is load-bearing, carrying the roof and ceiling, while the outer skin of brick is used for weathering, security and visual effect. The brick skin is tied to the timber frame by veneer ties built into the brickwork and nailed to the frame. A barrier of wire mesh (vermin wire) is placed at the base of the timber frame and built into the brick wall to prevent rodents from entering the frame through the cavity. Internal walls are constructed in the same way as for a timber-framed construction.

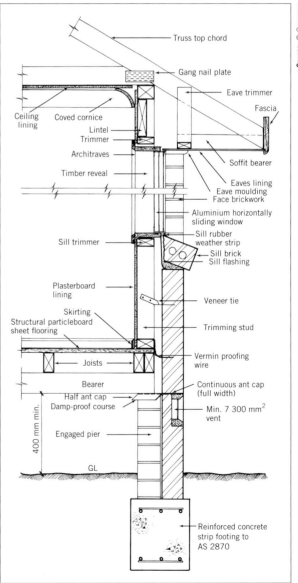

FIGURE 7.35 Vertical section through an external brick veneer wall and horizontally sliding aluminium window

FIGURE 7.34 Vertical section through an external timber frame wall and timber awning window

Full brick

External walls are constructed of two skins of brick, separated by a gap (cavity) of 50 mm and held by galvanised wire cavity ties with a drip groove to prevent moisture travelling from the outer skin of brick to the inner skin when the outer skin becomes wet (Figure 7.36). The inner skin of brick is the load-bearing wall, and the outer skin is for weathering, security and visual effect. Cavity brick construction has the advantage of good thermal and sound insulation. Internal walls are constructed of one brick thickness, which can either be left as it is or have a 13 mm thick cement render applied over it and be painted or decorated.

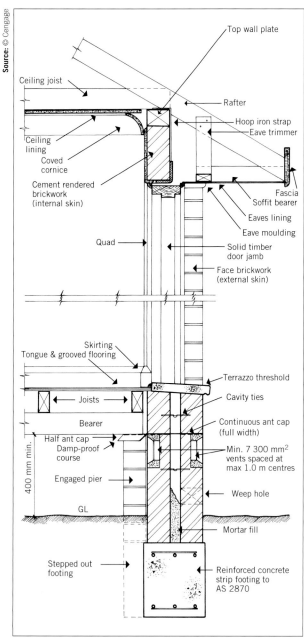

FIGURE 7.36 Vertical section through an external cavity brick wall and timber door jamb

Alternative construction methods

Current building trends allow for conservation of timber resources, resulting in the use of metal as an alternative to timber in the manufacture of ground floor framing, wall framing and truss fabrication. Each system is prefabricated in panels or sections off-site, with final assembly carried out onsite. There is also a wide range of alternative lightweight cladding options from fibre-cement sheeting systems to panels of styrene and autoclaved concrete systems.

Roof structures

Roof structure is the term given to the roof framing, eaves and roof covering. Conventional and trussed roof construction allows for freedom of design, with the floor plan of the building determining the final shape of the roof. Common shapes in either conventional or trussed roof buildings are:

- gable roof
- hip roof
- hip and valley roof
- gambrel or Dutch gable
- jerkin head.

Common types of roof structure are as follows.

Conventional roof

The timber roof framing is cut out and assembled (pitched) on site. Structural framing members used in conventional roofing are rafters, ridge boards, hips, purlins, struts and collar ties (see Figures 7.37 and 7.38). Ceiling framing consisting of ceiling joists, hanging beams, strutting beams and trimmers combine with and tie the roof framing to the walls. Sectional size and spacing of structural members are determined by the Timber Framing Code. The external finish of conventional roofing is provided by fascia boards, barge boards, gable cladding for gable roofs, and by eave or gable linings. Roof coverings, such as cement or clay tiles and corrugated metal sheeting, will influence the design and construction technique of conventional roofing.

Trussed roof

Roof trusses are fabricated off-site, then transported to the site and lifted into position. This enables fast construction, the use of less material and the added advantage of internal design flexibility, as trussed roofs are supported on the external walls only. The structural design of a truss enables the top chord, bottom chord, struts and ties to support and distribute the roof and ceiling loads to the external walls. Trusses are designed according to the span between the external walls, the ceiling and roof loads, and the roof pitch. Roof coverings and external finish are the same as in conventional roofing. Trusses are fabricated from either timber or steel.

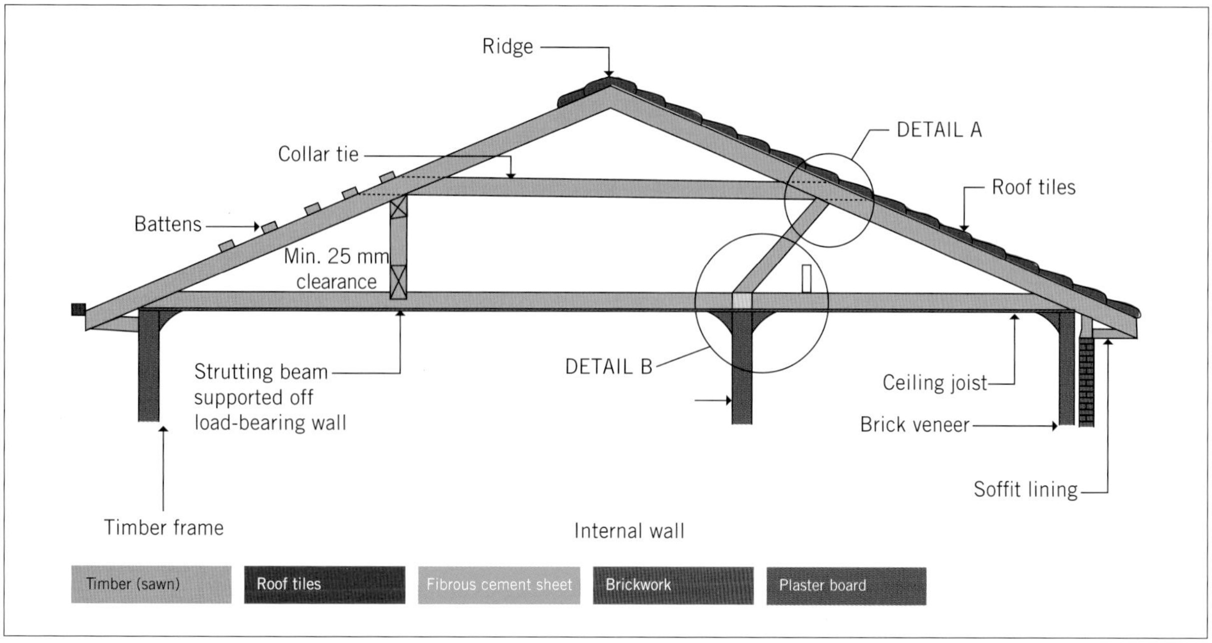

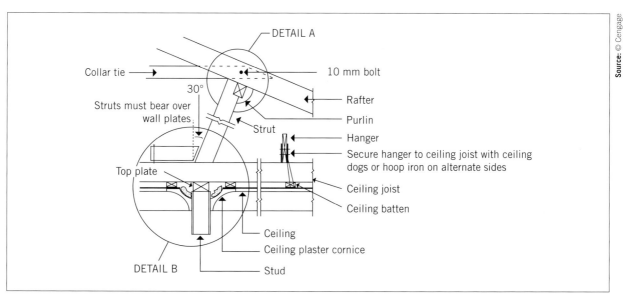

FIGURE 7.37 Vertical section through a conventional roof

Source: © Cengage

FIGURE 7.38 Details A and B from Figure 7.37

4.2 Determine location, dimensions and tolerances for ancillary works

Ancillary works for a project are works that are not on the main construction project but support the building of the main construction project. For example, a temporary site shed, a temporary power pole, temporary storm water diversion and erosion control structures and temporary fencing for the protection of the site and other similar items are considered ancillary works.

To determine the location, dimensions and tolerances for this ancillary work you need to have the project documentation on hand, particularly the site plan, and ensure that you understand the overall project. Make sure you have visited the site to locate any features or structures that need to be incorporated into the plan. Check your site plan and your progress schedule for the job and review which ancillary works need to be placed where on the site to support the main project. By having a clear understanding of the overall project and the support works that need to be in place to help that project proceed, you can identify the location of ancillary works and the timing for that work so the whole project can run smoothly.

You may need to conduct some site surveys and make some extra notes in your planning using measuring tools and marking equipment for where the ancillary work is going to be located on the site.

Always ensure that the main project takes precedence, and all support works are intended to keep the main project flowing smoothly. You should know the desired location, dimensions and tolerances of the various supporting ancillary works that will help you complete the main project.

You need to consider any regulatory standards, such as electrical supply, water supply or sewerage connections, when you plan these ancillary works. Make sure you know what the requirements are for location dimensions and tolerances including drainage pipe slopes. Consult with regulatory authorities, other trades, technical advisers, supply authorities, neighbours, and any other stakeholders as you are doing your planning. Document and record all the outcomes of your research and planning and regularly inspect for quality and compliance as the ancillary work starts, progresses, and is removed once the job is finished. Ensure all approvals are planned well ahead of when they are required, and sign-offs are completed. Remember that some works on a construction site require high levels of accuracy, compliance with project requirements and the location, dimensions and quality of the work needs to be correct the first time it is completed.

A site plan, as drawn by the project planner, may not note any locations or details of ancillary works. The site plan may only show the completed project. If the ancillary works are relatively minor and not structural then you may not need detailed site plans; however, if you do need detailed ancillary works plans then ensure these are planned for and obtained well before the project starts.

4.3 Identify environmental controls and locations

Identifying environmental controls and locations involves obtaining any environmental guidelines from your local government authority regarding the project that you are planning. There are a lot of resources available from the various state and national environment regulators on how to ensure your construction site does no harm to the environment. Once you familiarise yourself with your local government authority's requirements, regulations and standards you can plan for your particular construction site; for example, for controls in water management, noise management, waste management, sediment control and air quality. The local government authority will normally be the main regulatory body that controls environmental site issues, so you need to work within their guidelines.

GREEN TIP

Every local government authority in Australia has information on their environmental requirements for building works and you must be familiar with these when planning a building and the construction of the building.

Make sure you understand any environmental-related controls and locations that are noted on the project documentation. Highlight these and remember to incorporate them in your overall site planning. As you examine the physical site make sure the documentation matches what you observe on the site and that you have a thorough understanding of how you can meet the regulatory guidelines outlined in your documentation for the physical site and the available materials, tools and equipment. The documentation may highlight areas where controls are required; for example, sediment control on your site plan will show where disturbed soil is going to be retained on your site should there be any sudden downpours causing erosion issues. The broad rule is that you are not allowed to let any soil from your site escape into council storm-water drains or onto the roads as this can cause blockages of drains in sudden downpours and rain events and the flooding of other people's property. This is regulated frequently, and heavy fines can occur for breaching your local council's requirements.

As you study the site plan and consider the documentation for the job, remember to consider and observe the surroundings on the site, the slope of the land, any nearby water bodies and any sensitive habitats that may or may not have been noticed when the documentation was prepared. Identify any potential hazards that may be encountered. Be on the lookout for dust pollution, noise pollution and especially asbestos and silica hazards. Consult with experts if needed to determine best practice and measures to mitigate any potential environmental impacts. Obtain the regulatory codes of practice or guidelines to ensure you're thoroughly familiar with your responsibilities to cause no harm to the environment.

 With a set of site plans you should thoroughly examine the building location to make sure you take note of any hazards or items that may cause injury that you need to plan for.

If not documented already, make sure you document all of the environmental protection research you are doing so that you have evidence that you have done your best to protect the environment as your job proceeds. Measures such as pollution prevention systems, noise barriers, erosion control measures and waste management plans are essential on worksites. Make sure you document that you have communicated your job requirements to all stakeholders, especially contractors, subcontractors, visitors and suppliers for the job. Use signage and observe people's behaviour so you can make it as easy as possible for them to do the right thing. For example, placing the applicable bins near where the waste will be generated or recycling bins near the lunchrooms.

If you need any regulatory approvals for environmentally sensitive work, such as asbestos removal, then ensure you obtain these approvals following the procedures by the regulatory authority well before you need to do the work. These approvals may require some detailed planning and take time to be approved.

 If you encounter asbestos, or any other potentially hazardous material on a site, make sure you get expert advice on how to deal with it and protect people from the hazard.

4.4 Determine specifications for materials, standards of work, finishes and tolerances

Plan and document reading is a skill that requires practice. Accurate and quick interpretation of plans and specifications requires a sound understanding of basic drawing symbols and abbreviations. Knowing where to look for the information required is a definite advantage.

You can look on the floor plan for the external size of the building, and on the site plan for the position of the building on the building block. Heights of walls and windows are found on elevations, while specific construction practice can be found in sections or details. Technical information such as material size and spacing not found on the plans will be described in the specifications.

Ancillary works can be located on drawings or indicated in the Environmental Plan. Ancillary works on construction sites refer to secondary or supporting activities and structures that are necessary for the successful completion of the main construction project. These works are not the primary focus of the construction but are essential for various reasons, such as ensuring safety, functionality, compliance with regulations and overall project efficiency.

These works can include land clearing, grading or soil preparation to prepare for construction, erecting temporary structures such as a site office, temporary facilities such as toilets, electricity and water supply, safety barricades and fencing, and access roads and pathways.

Environmental works may be required to prevent erosion and run off. Waste management skips and areas for storage and recycling should also be installed before construction. Information about the ancillary works are be provided in the construction documentation and schedules.

Most states and territories have a guide to standards and tolerances that is very useful for the builder and the client to see if the standard of finish and the tolerances of that finish are suitable for the intended use. For example, a microscope will show up minor blemishes in paint work on a wall; however, that is not how you normally view a wall and therefore a microscope should not be used as a tool to examine the quality of the wall paint in most cases. These guides to standards and tolerances are freely available and can be an invaluable tool to help you know what standard the work you are doing needs to be at.

Table 7.3 gives examples of some typical questions and answers that are required when reading a plan before and during construction. Refer to Figure 7.39 to confirm that the answers and locations are correct.

TABLE 7.3 Sample question sheet

	Statement	Answer	Plan location
1	The meter box is found on which detail?	South Elevation	South Elevation
2	What is the section size of the brushbox flooring to the porch?	Ex 75 mm × 25 mm	Floor Plan
3	What are the window frames made of?	Aluminium	North Elevation
4	What are the internal dimensions of the bedroom?	4830 mm × 3190 mm	Floor Plan
5	What is the thickness of the internal walls?	110 mm	Floor Plan
6	What is the kitchen floor covered with?	Vinyl tiles	Floor Plan
7	What is the sill height of window 5 from the floor?	1200 mm	North Elevation
8	Window 1 is located on the east elevation in the ....?	Bathroom	Floor Plan
9	What is the overall size of the external walls?	280 mm	Floor Plan
10	What is the section size of the galvanised steel rainwater pipes?	75 mm × 50 mm	Floor Plan
11	What is the head height of window 4 from the floor?	2143 mm	North Elevation
12	What is the width of door 4?	970 mm	Floor Plan
13	What is the width of window 2?	2650 mm	Floor Plan
14	What is the width of window 1?	1210 mm	Floor Plan
15	What is the section size of the rafters over the carport?	100 mm × 38 mm	South Elevation

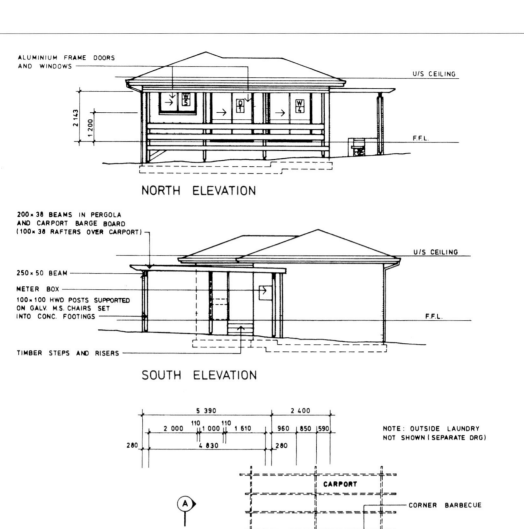

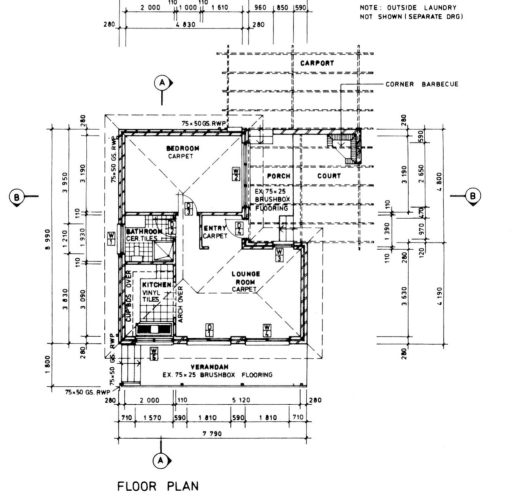

FIGURE 7.39 Cavity brick cottage

LEARNING TASK 7.3 **SPECIFICATION READING**

1   This task requires a computer and access to the internet. Visit your favourite search engine using the keywords > NT> standard specification for small building works 2018 (you can limit your search results to only give you responses from a Northern Territory government website by typing site: nt.gov.au)
    View this document online or download the file.
2   Using these specifications, locate the information below and write your answers in the space provided.

What is the number of the Australian Standard for timber framing and flooring? (See Part 6 TIMBER – Section 6.1 General)	AS
On what pages of the specification can the section on 'CLADDING AND LINING' be found?	
What method of fixing will be used for timber frames? (See Part 11 CLADDING AND LINING – Section 11.1.2 General – Fixing – general)	

4.5 Determine material requirements and processes to be followed

Understanding the material requirements and processes to be followed for a job requires access to all the job documentation, ensuring you have also obtained all the documents that relate to the various materials you will use to construct the project. All manufacturers of materials in Australia are required to produce documentation that tells you important information about their materials and products. For example, if you buy a toaster there's a booklet in the toaster box that tells you safety information about how to use the toaster, detailed specifications of the toaster, what to do if it doesn't work or what are the common faults, and usually there's some warranty information. Building products and materials have the same information produced; however, it is often not supplied with the product but is readily available from the manufacturer's website or from the supplier.

You should refer to documentation for all products and materials used on a site until you are so familiar with the documents that you don't need to read them any more and you just review them from time to time. The following are some common documents that you need to be familiar with:

■   Safety data sheets (SDSs). Materials have SDSs to inform you of the safety information for the product, any PPE that might be required, use and disposal information and storage information, etc.
■   Instruction manual/installation guides. Products and materials have instruction manuals on how to fit and install that product or material to the manufacturer's specifications. For example, installing an aluminium window with a timber reveal is often done by firing fix out nails into the timber reveal and wall frame to hold the window in place. However, upon closer inspection of the

instruction manual, for most aluminium windows the supplier will require different fixings and in most cases the fix out nails might only hold the window in place until there's a major storm or extreme weather event. As a professional builder, it's much better for you to obtain instruction manuals and be familiar with them for all the products and materials that you use on a construction site. It's very important that you use the correct fixings for materials and the correct spacing for those fixings.
■   Consult relevant standards and codes of practice. Many materials and processes are mentioned generically within standards and codes of practice and a safe method of installation and use for that material is documented for your safety and for satisfactory product performance.

The schedule for your project will list the materials that are going to be used in the project and the timing for the use of these materials will be based upon your work plan. You need to check the schedule to ensure the materials meet the requirements for your work process. For example, you may need to consider:

■   Material factors such as strength, durability, compatibility, environmental impact and cost. Gather information from suppliers, standards and the manufacturers so you can make an informed decision.
■   Talking to your local supplier to confirm material availability and ensure you can get them onsite when needed. You may need to consider alternatives if there's a possibility of any delay.
■   Ensuring the material quality, sizes and measurements meet the requirements for the project and will enable you to meet the various specifications, dimensions and tolerances required for the job. Make sure you have the correct materials handling and working equipment available.

- The compatibility of the various materials so you are not compromising aspects of the project. For example, some metals cannot be placed above other metals because the rain runoff from the higher metal will cause corrosion on the lower metal. You may need to get expert advice or conduct some research on this. Ensure the treatments and finishes or any modifications are all compatible across all your various materials for the whole project.
- The efficiency of the various materials you would like to use, evaluating the time–cost factor between different products. For example, if you have a lot of sheet wall bracing to fix, it may be cheaper in the long run to purchase a special nail gun that will shoot the flathead nails into the sheets to secure them, as opposed to nailing them by hand and not purchasing an expensive nail gun.
- Making sure you document the material requirements referring to any quantity lists you might have available from the job costing stage. Ensure that your detailed lists of materials works in with the schedules and specifications and you have information on when the materials are required to keep the job flowing smoothly. Monitor and adjust your progress and plan as needed to keep the project on track. A Gantt chart can be invaluable for this type of planning.

 COMPLETE WORKSHEET 4

SUMMARY

In Chapter 7 you have learnt how to read and interpret plans, specifications and drawings for carpentry work. You have learnt to:

- plan and prepare
- interpret construction plans, drawings and their features
- locate key features on site plan
- determine project requirements and plan project.

This unit of competency specified the skills and knowledge required to read and interpret plans and specifications for carpentry work in order to plan and sequence the work, meeting all relevant requirements of the National Construction Code (NCC), Australian Standards, work health and safety (WHS), and Commonwealth and state or territory legislation.

REFERENCES AND FURTHER READING

Texts

Department of Infrastructure, Northern Territory of Australia (2012), *2012 standard specification for small building works*

Training Sector Services Department of Training and Workforce Development, WA **http://www.dtwd.wa.gov.au/** and search for the resource Read and interpret plans and specifications

Major, S.P. (1995), *Architectural woodwork: details for construction*, Van Nostrand Reinhold, New York

Noll, T. (1997), *The encyclopedia of joints and jointmaking*, RD Press, Sydney

Styles, K. (1986), *Working drawings handbook* (3rd edn), Architectural Press, UK

Web-based resources

Search for Buildsum videos for the plan reading series

Regulations/codes/laws

http://www.epa.nsw.gov.au, NSW Environment Protection Authority

http://www.standards.org.au, Standards Australia

Resource tools and VET links

http://training.gov.au, Skills training

https://hschub.nsw.edu.au; search for construction resources for Read and interpret plans.

Industry organisation sites

http://www.aiqs.com.au, Australian Institute of Quantity Surveyors

https://hia.com.au, Housing Industry Association

https://www.mbansw.asn.au, Master Builders Association

 Relevant Australian Standards

AS 1100.101–1992 Technical drawing – General principles
AS 1100.301–2008 Technical drawing – Architectural drawing
AS 1684.2 Residential timber-framed construction – Non-cyclonic areas

GET IT RIGHT

In the photo below, the workers are reading the plans.

Identify how this could be done more efficiently and provide reasoning for your answer.

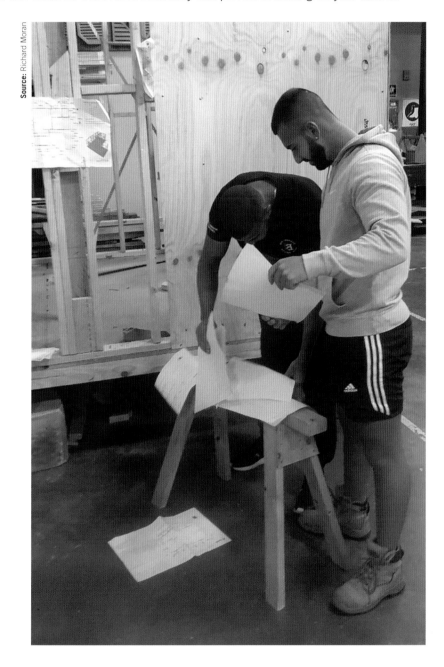

Source: Richard Moran

_____

_____

_____

_____

_____

WORKSHEET 1

Student name: _____

Enrolment year: _____

Class code: _____

Competency name/Number: _____

Task: Read the sections beginning at *Plan and prepare* then complete the following questions.

1   What are drawings used for in the building industry?

_____

_____

2   The technical language of drawings is expressed through the use of three standardised items.
    List these three (3) items below.

1   _____

2   _____

3   _____

3   List the key users of drawings in the building industry.

1   _____

2   _____

3   _____

4   _____

5   _____

6   _____

7   _____

8   _____

4   Name one item that should be found in the title block on a set of plans for a house.

_____

WORKSHEET 2

Student name: _____

Enrolment year: _____

Class code: _____

Competency name/Number: _____

Task: Read through the sections beginning at *Interpret construction plans, drawings and their features*, then complete the following questions.

1   What are the two (2) groups that drawings can be divided into?

1   _____

2   _____

2   Why do architects and designers use pictorial drawings or illustrations?

_____

_____

3   Isometric projection drawings are used as working drawings. (Circle the correct answer.)

TRUE   FALSE

4   Which of the following pictures or views is an isometric projection? (Circle the correct answer.)

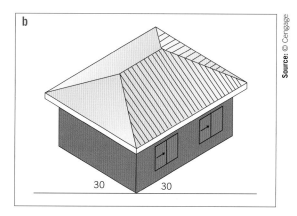

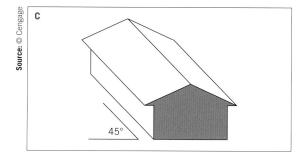

5 Which of the following pictures or views is an oblique projection? (Circle the correct answer.)

a

Source: © Cengage

b

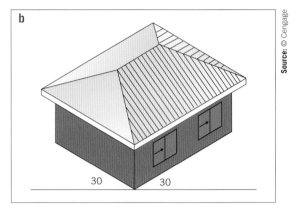

Source: © Cengage

c

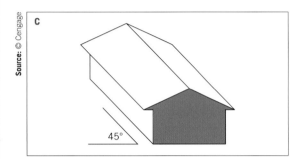

Source: © Cengage

6 Circle the correct answer to the following statement:

The most basic form of working drawing is:

a orthographic projection

b isometric projection

c oblique projection

d perspective projection

7 Circle the items that you should find on a site plan.

a internal room dimensions

b contour lines and their heights

c window sill height above floor level

d roof pitch

e direction of North

8 Circle the items that you should find on a floor plan.

a internal room dimensions

b contour lines and their heights

c window sill height above floor level

d function of each room; e.g. kitchen, bedroom

e floor surface and type of floor covering

9   Circle the items that you should find on an elevation.

a   type or function of windows

b   contour lines and their heights

c   window sill height above floor level

d   function of each room; e.g. kitchen, bedroom

e   height of finished floor level (FFL) to finished ceiling level (FCL)

10   Circle the items that you should find on a section.

a   type or function of windows

b   footing sizes

c   floor construction

d   roof construction; e.g. trussed

e   roof pitch

11   What scale are working drawings of plan views drawn to? (Circle the correct answer.)

a   1:5

b   1:10

c   1:20

d   1:100

e   1:200

f   1:500

12   What scale are working drawings of construction details drawn to? (Circle the correct answer(s).)

a   1:5

b   1:10

c   1:20

d   1:100

e   1:200

f   1:500

13   At a scale of 1:50, how many full-scale millimetres would there be for 4.200 metres?

_____

14   An A3 drawing sheet is exactly half the size of an A2 drawing sheet. (Circle the correct answer.)

TRUE   FALSE

 **WORKSHEET 3**

Student name: _____

Enrolment year: _____

Class code: _____

Competency name/Number: _____

To be completed by teachers

Student competent ☐

Student not yet competent ☐

Task: Use the following floor plan and refer to the section *Locate key features on site plan* to complete the following questions.

1   The name of the view shown on the floor plan is the _____.

2   In the kitchen area, the abbreviation 'w/o' identifies the position of the

_____.

3   The abbreviation 'DP' identifies the position of the _____.

4   How many DPs are shown on the plan? _____

5   The abbreviation 'HWS' stands for _____.

6   The HWS is positioned outside the _____ (room).

7   The internal dimensions of the rumpus room are _____ × _____.

8   The external walls of the house are covered outside (clad) with what material?

_____

9   The abbreviation 'o/a' refers to the _____ size of the entry door frame.

10   In the space below, draw the symbol for the kitchen sink.

11   The width of the window to the dining room area is _____.

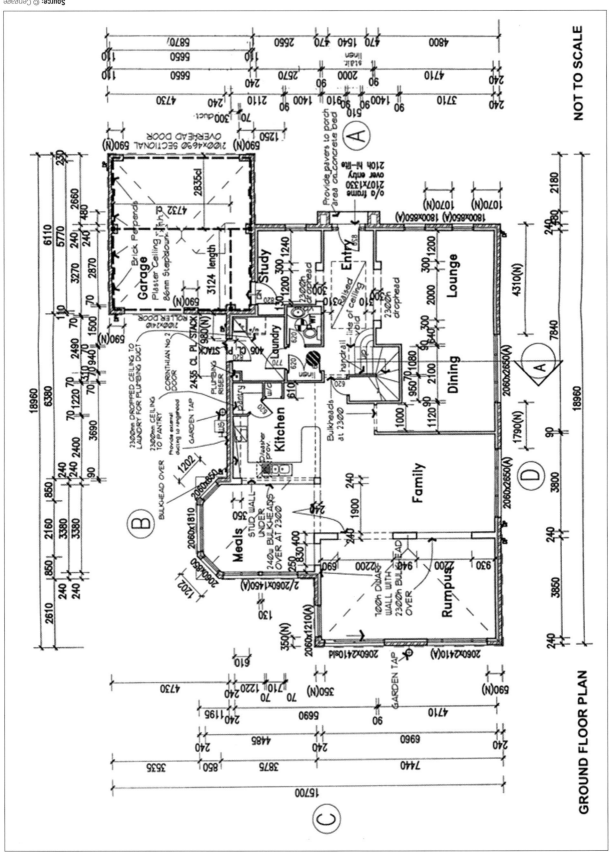

GROUND FLOOR PLAN

NOT TO SCALE

 **WORKSHEET 4**

Student name: _____

Enrolment year: _____

Class code: _____

Competency name/Number: _____

Task: Use the following site plan and refer to the section *Determine project requirements and plan project* to complete the following statements.

1   The boundary formed by survey pegs A and B is the _____ facing side of the block (state orientation) and is _____ m long.

2   The name of the street or road that the site is found on is _____ and the site is found on the _____ side of the street or road.

3   The abbreviation RL stands for _____.

4   The council building line is set _____ m in from the street alignment.

5   The total area of the building block is _____ m².

6   The total floor area of the building is _____ m².

7   There are _____ (number) trees originally found on this block.

8   The total number of trees to be removed is _____.

9   The length of the west side boundary is _____ m.

10   The approximate fall along the west side boundary is _____ m.

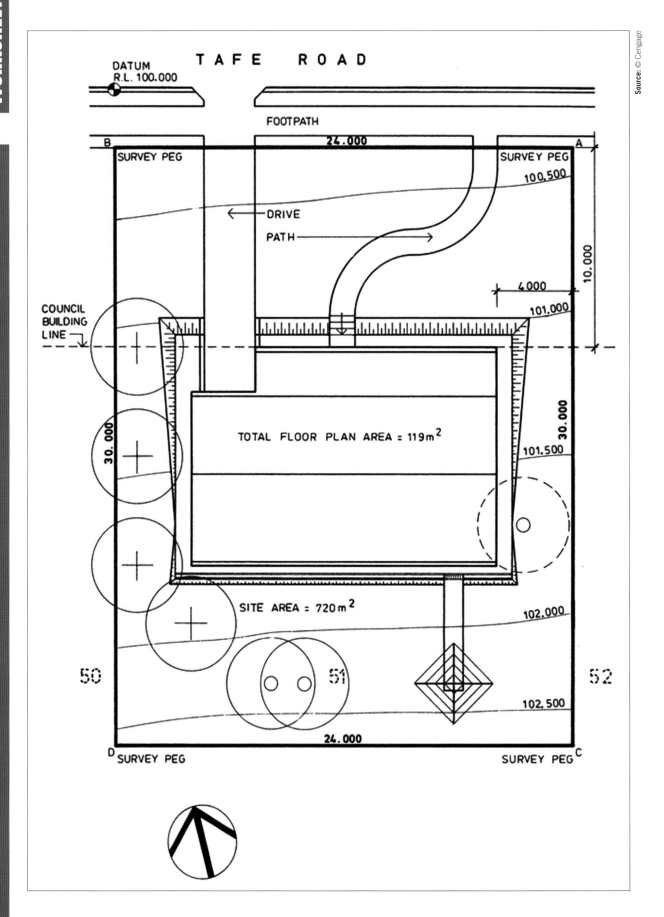

TAFE ROAD

DATUM
R.L. 100.000

FOOTPATH

24.000

B SURVEY PEG

SURVEY PEG A

100.500

DRIVE

PATH

4 000

10.000

101.000

COUNCIL
BUILDING
LINE

TOTAL FLOOR PLAN AREA = 119m^2

101.500

30.000

30.000

SITE AREA = 720m^2

102.000

50

51

52

102.500

24.000

D SURVEY PEG

SURVEY PEG C

PERFORMANCE OF SKILLS EVIDENCE

To be completed by teachers

Student competent ☐

Student not yet
competent ☐

Student name: _____

Enrolment year: _____

Class code: _____

Competency name/Number: _____

Task: Working with your teacher, prepare to demonstrate your skills by completing the following task. To demonstrate competency, a candidate must satisfy all the elements, performance criteria and foundation skills of this unit by reading and interpreting plans, specifications and drawings for two, minimum 30 m², carpentry projects. Each project must have a minimum of seven materials. A candidate must prepare a work plan for each project that should identify the dimensions, material requirements and processes to be followed.

Instructions:

1   Gather the equipment you will need to perform this task. As a minimum you may need:

The plans, specifications and drawings for two carpentry projects, each with a minimum size of 30 square metres.

Paper and pen. You may choose to use a spreadsheet program or word processing program to demonstrate your ability to read and interpret plans and prepare this work plan.

You may choose to access some Australian Standards or technical documents such as AS1100.101, AS1100.301 and AS1684. These documents form the basis for residential house drawing.

2   Review the plans and identify key elements such as dimensions (check these for accuracy and note any inconsistencies), materials (choose seven materials noted in the plans or specifications for this activity and write them down) and note the main processes involved in the construction.

3   Prepare a work plan for the project, particularly making note of the following:
   a   Project description
   b   Brief overview of the size of the project
   c   List the seven materials you have chosen and provide an estimate of the quantities, sizes and specification of those materials. Make a brief note of the source of supply for those materials. Reference the plan specifications or documents for these materials.
   d   Note down the step-by-step processes broadly required for the construction of this project, referencing the plans and specifications.
   e   Note down any potential hazards and special safety precautions that may need to be taken during the project.
   f   Note down how you will maintain quality control, briefly estimate approximate timing for steps of the project, and how you would communicate to any stakeholders during the project.

This task is about how to read and interpret plans, specifications and any other documents related to a building project. You don't need too much detail here, as this is more about providing an overview.

PART 3

USING MATERIALS AND TOOLS

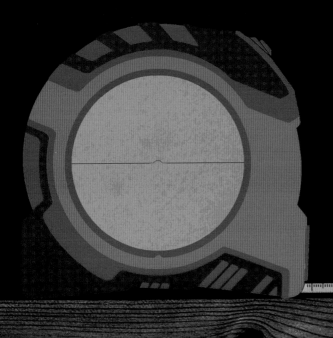

8 HANDLE CARPENTRY MATERIALS

This chapter covers the following topics from the competency CPCCCA2011 Handle carpentry materials:

- plan and prepare
- manually handle, sort, stack and store materials
- prepare for mechanical handling of materials
- check and store tools and equipment.

This unit of competency specifies the skills and knowledge required to safely manually handle and store carpentry materials, meeting all relevant requirements of work health and safety (WHS) and Commonwealth and state or territory legislation.

The unit includes preparing material for mechanical handling and applying environmental management principles associated with carpentry materials.

The unit is suitable for those with basic skills and knowledge undertaking routine work tasks under the direction of more experienced workers.

1. Plan and prepare

For centuries, builders have used traditional materials like timber, brick, iron and stone for homes, and government and public buildings. Today, however, the materials being used in the building industry are changing. Older traditional-type materials are now being combined with new technology or even newly developed materials to form prefabricated or pre-assembled products, which are also given new uses. Building codes and standards are being modified to accept these new materials and combinations of materials and to allow for improved cost-effective construction.

There is an ever-widening choice when it comes to selecting the best or most appropriate material for the job (see **Figure 8.1**).

Source: Shutterstock.com/Rafal Olechowski

FIGURE 8.1 Contemporary residential building, using a variety of materials

Factors affecting the selection of appropriate materials for use in building include their physical characteristics and economic considerations:

- Physical characteristics
  - the density or mass of the material (i.e. how light or heavy it is)
  - strength and ability to carry loads and resistance to being stretched or compressed
  - electrical resistance or conductivity
  - durability (i.e. how well it stands up to the elements)
  - fire resistance
  - insulation properties
  - moisture resistance
  - general appearance.
- Economic considerations
  - manufacture or production costs
  - retail costs
  - whether or not the material has a sustainable source
  - what effect the use of the material has on the environment.

Without knowing the physical characteristics or the economic consideration of the materials to be handled, stacked and stored, it could be easy to incorrectly store a material in the wrong environment or location. An example is bags of cement left stacked outside in the rain. By being aware of these characteristics the material can be correctly handled, stacked and stored when it arrives onsite, thus avoiding it being damaged onsite and causing costs to the builder as they will need to replace or repair the damaged items before work can continue.

GREEN TIP

The safety data sheet (SDS) for materials or products contains important information about how the material, and its use affects the environment.

1.1 Read and interpret work instructions and planned sequence of work

When handling carpentry materials, it is important to have studied any work instructions relevant to the job and to plan the sequence of work. The work instructions may have been noted on safe work method statement (SWMS), a schedule for the project, codes of practice for the task being undertaken, installation instructions for the materials you are working with provided by the manufacturer, or any other documents that may be available for the materials you are working with and the jobs you are going to be doing. It's important that you carefully read through any instructions and job specifications taking note of safety procedures for working with material storage requirements, specific handling requirements, and any tools that may be required. Be careful to make sure that you have all the documentation that relates to the various materials you work with and make sure that the fixings and fittings and fasteners for those materials have all been planned so that they all work in together to form a good solid job. It's no good using nails that will be rusted by the material that you are fastening because they were not specified for that fixing application, and they are going to react to other materials. If need be, seek advice from supervisors or colleagues or other competent people to make sure the work you plan for and the work you do is of a good quality.

Ensure also that you have considered any potential hazards and the sequence of your work is going to fit in with the other trades in the plan for the work you are doing. To plan the sequence of work, it's important to break the project down into smaller tasks and steps so you can logically proceed from start to finish. Work out which steps have dependencies; that is, tasks that need to be completed before another task can be started. For example, the drywall sheeting for a bathroom or wet area must be completely fixed and set smoothly before waterproofing can be applied. You can't get this out of sequence otherwise the waterproofing may fail. You also must allow for drying or curing times between your different tasks and this will become obvious as you study the installation instructions and the material fixing requirements for the different tasks you are doing. Consider the availability of tools and equipment and resources that you might need at each step. If you need scaffolding, this must be arranged well before the day you want it on the job. If need be, create a visual plan or schedule (a Gantt chart) so that all participants can understand and be organised and can visualise the sequence of the work.

By reading and interpreting the work instructions and planning the sequence of work you can make sure that you have a clear understanding of the task you are going to do, and this can help to minimise errors, improve safety, and ensure a productive workplace.

1.2 Plan all work to comply with laws and regulations, work health and safety (WHS) and environmental requirements, manufacturers' specifications and workplace requirements

Timber

Timber is an important building and construction material, which actively competes with companion materials such as steel and masonry products. It is referred to as *wood* or *lumber* while it is in the form of a growing tree and as *timber* once it is felled and milled. It consists of several chemical components which are given by approximate mass, such as: *cellulose* (45%) and *hemicellulose* (22%), which together make up the fibre; *lignin* (30%), which acts as a binding agent or glue for the fibres; and *extractives* (3%), which are minerals, gums and resins.

The main trunk of the tree is cone-shaped, which means that when the tree grows and produces new wood in the form of growth rings, it grows horizontally in girth or increases in diameter, rather than growing vertically. The increase in height occurs as a result of a new layer being added and the activity of special cells at the extreme tips of branches. A guide to the approximate age of the tree can be determined by counting the growth or annual rings found in the cross-section of the tree (Figure 8.2).

Identification of timber

While timber is usually classified as either hardwood or softwood, this has little to do with its hardness or softness.

Hardwood trees

These are broad-leafed trees (Angiosperms) that flower and have either fruit or nuts for seed containment and generally have a slower growth cycle. The timber

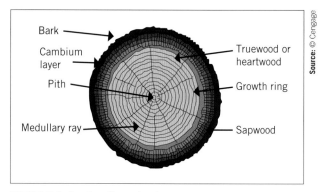

FIGURE 8.2 Section through a log

structure is made up of a series of hollow vertical pipes known as vessels, which are more easily identified when magnified. This can be checked by slicing the end grain of the timber with a very sharp knife or blade to reveal the pores or vessels more clearly (Figure 8.3).

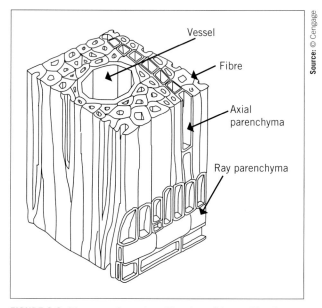

FIGURE 8.3 Diagram of a cube of hardwood (magnification × 250). The pits in the cell walls have been omitted.

Some common types of hardwood in use are tallowwood (Figure 8.4), spotted gum, brush box, meranti, river red gum, Tasmanian oak, ironbark, turpentine, messmate, stringy bark, meubau and, by contrast, balsa wood, which has limited use for building.

Softwood trees

Softwood trees are usually conifers/pines (Gymnosperms), and have needle-like leaves and generally have a faster growth cycle. The timber structure is made up of a series of hollow vertical cells known as tracheids (Figure 8.5). These cells allow the water and minerals to pass up the tree to the leaves where the food is produced.

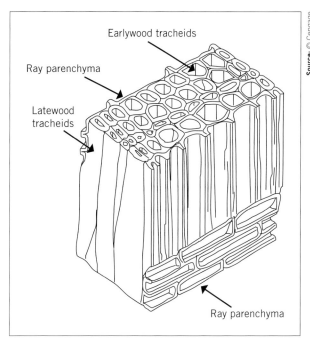

FIGURE 8.5 Diagram of a cube of softwood (magnification × 250). The pits in the cell walls have been omitted.

FIGURE 8.4 Tallowwood tree and timber

Some common types of softwood are western red cedar, Radiata pine (Figure 8.6), Baltic pine, hoop pine, kauri pine, Oregon or Douglas fir and cypress pine.

Uses for timber

Timber and timber products may be used in a variety of situations and in the manufacture of many products. Some of these are extracted oils, fuel, paper, resins, manufactured boards, poles, posts, framing, furniture, boxes, toys, boats, weapons, sporting equipment, claddings and linings.

Structural uses within the building industry include:

- sub-floor framing
- wall framing
- ceiling framing

FIGURE 8.6 Radiata pine trees and timber

- roof framing
- window and door frames
- formwork for concrete
- roof truss construction
- finishing uses:
  - architraves
  - skirtings
  - scotia and quad mouldings
  - floor timber
  - kitchen and built-in cupboards
  - stairs and panelling
- Other uses:
  - fencing
  - pergolas and decking
  - retaining walls and screens
  - solid timber door construction.

Durability of timber

The durability of a timber relates to its ability to withstand the destructive action of moisture and the elements, of abrasion, and its resistance to insect attack. When varieties of timber are exposed to hazards, such as insects, their durability will vary. The Hazard Level (H) is given a number, with H1 for low hazards and H6 for high hazards.

Some timbers, such as cypress pine and ironbark, are very durable and are grouped together with a *Class 1 durability rating*. Other poor-durability timbers such as Radiata pine and Oregon (otherwise known as Douglas fir) have a *Class 4 durability rating* (see Table 8.1). These less durable timbers may require preservative treatment to make them more durable. These treatments may include CCA (copper chromium and arsenate) – a pressure treatment that turns the

timber a distinctive green colour; or LOSP (light organic solvent-borne preservative) – this preservative does not discolour the timber but may leave a waxy residue. Preservatives are available from your local hardware store (Figure 8.7).

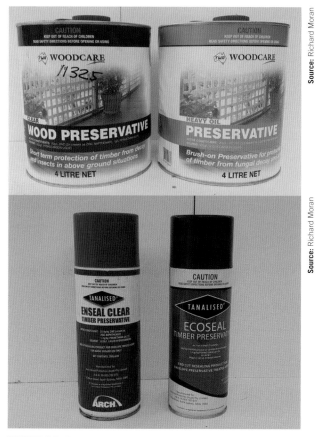

FIGURE 8.7 Rot stop wood preservative and Enseal

TABLE 8.1 Natural durability classification of heartwood of some commonly used timbers

Class 1	Class 2	Class 3	Class 4	Hazard level of timber (treated timber)
White cypress pine	New England blackbutt	Brush box	Brownbarrel	H1: inside, above ground, fully protected
Ironbark	Blackbutt	Mixed open forest	Caribbean pine	H2: inside, above ground, protected from elements, no termite protection
Tallowwood	Kwila (Merbau)	Hardwoods from Nth NSW or Sth Qld	Douglas fir (Oregon)	H3: outside, above ground
Turpentine	Spotted gum	Rose/Flooded gum	Radiata pine	H4: in ground contact
Grey gum	Western red cedar	Sydney blue gum	Slash pine	H5: contact with ground or fresh water
	River red gum	Silvertopped stringybark	Mountain ash (Tasmanian oak)	H6: salt water contact
	Stringybark (yellow and white)		Alpine (Vic.) ash	
			Unidentified hardwood or softwood	

Untreated Class 3 and 4 timbers should not be used for weather-exposed structural members; i.e. posts, joists and bearers of decks or unprotected beams protruding from the house.

The heartwood or innermost section of a tree has inherent natural durability, whereas the sapwood or outermost section has poor durability qualities and is especially susceptible to insect attack.

Strength of timber

Timber is strongest along the straight length of its grain. It will be weakened where short, wavy- or cross-graining occurs and also where knots, caused by branches of the tree intersecting the trunk, occur. Solid timber is generally not used for load-bearing spans greater than 4.8 m, as it deflects or bends excessively and becomes too expensive. Steel may be used as an alternative, or, if timber is preferred, engineered timber products such as laminated veneer lumber (LVLs) or Glulam beams may be used, allowing for greater spans to be bridged than if using natural timber.

Fire rating

All timber will eventually burn, but some timbers, such as teak, karri and jarrah, have a degree of resistance to fire. When timber is heated the temperature rises steadily to about 100 °C, where it remains until the considerable amount of water vapour the timber contains is evaporated. The temperature rise then continues without much further effect until, at about 200 °C, flammable gases begin to form; by 250°–300 °C ignition will happen if sufficient oxygen is present. At 350°–450 °C, depending on the density of the timber, spontaneous ignition occurs.

While small pieces of timber burn through quickly, a large piece of timber will develop a layer of char, which tends to insulate its interior. Large timbers (e.g. 250 × 250 mm) with equal flat faces may stand up to fire better than steel in the same conditions. Timber may be further protected by treating with mineral salts, which will stop the timber burning after the source of ignition is removed, or the application of intumescent paint, which forms a gas on the surface when heated to remove the presence of oxygen.

Timber cutting

Timber may be cross-cut (across the grain) or ripped (along the grain) by using a hand or power saw (**Figures 8.8** and **8.9**). Blunting of these tools may be

FIGURE 8.9 Safe use of a power saw to rip timber

caused by the high silica content in the timber of most hardwoods, which acts like an abrasive. Hardened or tungsten teeth on the blades will help overcome this problem.

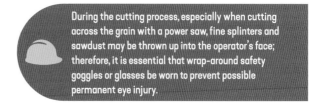

During the cutting process, especially when cutting across the grain with a power saw, fine splinters and sawdust may be thrown up into the operator's face; therefore, it is essential that wrap-around safety goggles or glasses be worn to prevent possible permanent eye injury.

Timber handling

Care needs to be exercised when handling timber, especially in its rough-sawn (RS) state, as it will still have furred edges. This may cause the painful entry of splinters, resulting in possible skin infection or serious damage. Also, some species of timber, mainly unseasoned eucalypt hardwoods, will stain the skin a bluish colour during handling due to the moisture and extractives contained in it. Normal washing of the hands followed by rubbing with a cut lemon should remove most of the stain.

Both of these problems can be overcome by wearing leather gloves (see **Figure 8.10**), at least during manual handling procedures. Most long lengths of timber can be carried safely and comfortably by placing them on

FIGURE 8.8 Cross-cutting timber with a hand saw

FIGURE 8.10 Wear leather gloves to prevent splinters

the shoulder and wrapping one arm over the top. To prevent physical injury, always use your legs to do the lifting and keep your back straight (see **Figures 8.11** and **8.12**).

Another way to handle large quantities of timber is by mechanical means; through the use of cranes (usually a truck-mounted crane referred to as a hiab) or a forklift truck (see **Figure 8.13**).

Source: Ed Hawkins

FIGURE 8.13 Hiab lifting a sling of timber

Timber stacking

Timber in its green or unseasoned form is very unstable, and this causes problems with its shape during the drying-out or seasoning process (**Figure 8.14**). It is therefore necessary to stack timber lengths with evenly sized and spaced pieces of timber, called 'gluts', between each layer to allow for even air circulation. Timber should also be stacked with the longest lengths at the bottom and the shorter lengths on the top.

Timber storing

To store correctly, timber should be stacked neatly off the ground and protected from the elements with a waterproof covering if stored outside (**Figure 8.15**). If the timber is ordered as a 'sling', then it should be left strapped, kept dry and stored on gluts off the ground until it is ready to be used. This applies to any timber product, such as wall frames and roof trusses. Remember, moisture over long periods of time will seriously degrade timber.

Timber environmental issues

The environmental issues that need to be considered for timber are disposal and personal hygiene considerations for people handling treated pine.

Source: Shutterstock.com/studioloco

FIGURE 8.11 Lift correctly to prevent back injury

 When disposing of treated timber, it *must not* be burnt, particularly CCA (copper chromium and arsenate) impregnated timbers. When treated timber is burned, the chemicals are released into the atmosphere as gases and fumes.

Source: Shutterstock.com/wavebreakmedia

FIGURE 8.12 Carrying timber safely and comfortably

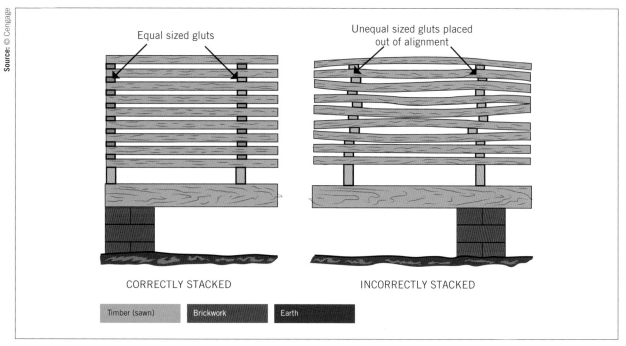

Equal sized gluts

Unequal sized gluts placed out of alignment

CORRECTLY STACKED

INCORRECTLY STACKED

Timber (sawn) Brickwork Earth

FIGURE 8.14 Bowing and bending caused by poorly aligned gluts

FIGURE 8.15 A sling of framing timber stored out of the weather

In enclosed environments the gases from treated timber can have serious health effects if inhaled. There are also concerns relating to the leaching of these chemicals into the groundwater if they are placed into unlined landfill.

Untreated and unpainted timber should be separated from general waste on the construction site. Waste timber should be collected and recycled rather than being used as landfill.

Because of the chemicals used in treated pine, people handling these timbers *must* wash their hands prior to eating.

GREEN TIP

Familiarise yourself with the environmental hazards of treated timber and make sure you use these treated products according to their guidelines.

Engineered timber products

Engineered timber products have many forms and a wide variety of uses. The most commonly used products are:

- particleboard
- fibreboard
- hardboard
- plywood
- laminated veneer lumber (LVL)
- cross-laminated timber (CLT)
- Glulam beams
- 'I' beams
- wood plastic composites.

These products are either slices of timber or timber particles combined with a binder (resin or adhesive) and formed into sheets or sections that usually have a higher strength-to-weight ratio than the timber it came from. In the case of LVL, CLT, Glulam beams or 'I' beams, these products can be ordered in any lengths, only being limited by the method of transporting the material to the site (e.g. semi-trailer or shorter table-top truck).

Manufacture

Particleboard, fibreboard and hardboard sheets are formed using a range of different-sized wood fibres, mainly Radiata pine forest thinnings and sawmill residues (for particleboard and fibreboard) and mainly eucalypt chips for hardboard, which are bonded together using a variety of resins, such as urea-formaldehyde, melamine-formaldehyde and phenol-formaldehyde. The glued wood particles are placed in spreaders. These spread the particles onto mats on a transfer device such as a belt or metal caul plate.

The mats are then transferred to the hot press, where they are compressed under high pressure to the thickness required. The high temperature of the press cures the resin, forming a solid board product, and the rough panels are trimmed and cut to size as required. Boards are usually sanded prior to sale or prior to prefinishing with the various surface and edge treatments available.

Some common trade names for these products include:

- particleboard – pyneboard, pineboard, chipboard
- fibreboard – MDF, Fibron
- hardboard – Masonite, Burnieboard, Weathertex.

Plywood, a glue-laminated wood product, is an engineered panel constructed of thin sheets or veneers that are rotary-peeled from the log. The veneers range in thickness from 0.8 mm to 3.2 mm. The number of veneers used varies according to the thickness required and the intended use of the sheet. As a guide, the minimum number of veneers is three and the maximum 11, hence the names 3 ply, 5 ply and so on. The number of veneers is always an odd figure to allow both faces of a sheet to have grain running in the same direction (Figure 8.16). Each veneer has its grain running at 90° to the previous one, and these veneers are hot-pressed using phenolic resins of varying types to suit an internal or external use of the ply sheet. The resulting composite material is stronger than solid timber, due to the long grain of each veneer being laminated at 90° to the other.

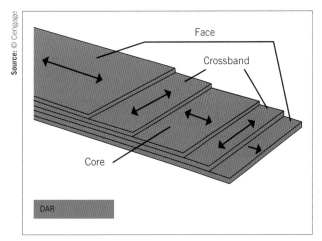

FIGURE 8.16 Typical plywood composition

Plywood products are also available in 'I' section profiles. They are used in both domestic and commercial construction for beams and various other common framing members (Figure 8.17).

Uses

Manufactured timber products can be used in all areas of building construction and fit-out, such as structural particleboard flooring (wet and dry areas), particleboard shelving with edge strips, fibreboard

FIGURE 8.17 'hyJOIST®'

cupboard doors and tops, fibreboard cabinets, fibreboard enamelled furniture, hardboard strip cladding, hardboard wall-frame wind bracing, plywood flooring, plywood decorative wall lining, plywood wall-frame wind bracing and plywood 'Plasply' formwork, to mention a few.

Durability

As the basic ingredient of these materials is timber, the durability rating is similar. The durability may be enhanced by additives to the resins, which can be sprayed onto the particles. Fire retardant, insecticide and fungicide chemicals may be added in small quantities. Paraffin wax also can be added in small quantities, either as an emulsion or sprayed in the molten state, to provide water resistance and to control swelling caused by temporary wetting. Some plywoods, especially those used for structural formwork, can be purchased with the surface pre-sealed with materials such as phenolic film, as used on Plasply. Alternatively, for a very high-quality surface, the raw plywood can be faced with GRP (fibreglass). This surface protection increases the durability level and therefore saves cost over time.

Fire ratings

Many factors need to be taken into consideration when determining the **fire rating** of a material. For example, the timber jarrah has a below-average or poor ability to resist ignition when exposed to flame; it has a fair resistance to spreading the flame over itself; it gives off a medium or fair level of heat when burning, which lowers the hazard to materials around it; and it produces little smoke when burning, which lowers the risk of smoke damage to other materials around it. When all these factors are averaged out, jarrah would be classified as having a fairly good fire rating, even though it burns readily.

Therefore, for ease of identification and rating of materials, a rating of good, fair or poor is given to individual materials based on their ability to resist ignition, spread of flame, heat produced and smoke developed. Some timber materials are rated in **Table 8.2**.

TABLE 8.2 Ignition performance of tested materials

Timber	Resistance to ignition	Resistance to spread of flame	Heat generated	Amount of smoke created
Jarrah	poor	fair	fair	low
Particleboard and fibreboard	poor	poor	poor	low
Plywood	poor	poor	poor	low

Engineered timber products – cutting

All manufactured timber products can be cut using conventional hand and power saws or large, accurately cutting panel saws. It is recommended that a tungsten carbide-tipped saw blade be used to allow for the dulling effects of the resin within the product.

Due to the fibrous nature of these products, small particles and splinters are thrown out when using power saws; therefore, appropriate eye protection must be worn to prevent possible personal injury.

When cutting thin sheet materials with a hand saw, the saw should be used at a fairly flat angle and the off-cut side of the sheet slightly lifted with one hand to prevent the saw jamming in the kerf. This method also helps to prevent excessive splitting or chipping of the underside of the sheet and keeping the saw fairly flat allows you to 'steer' the saw in the cut (Figure 8.18).

Engineered timber products – handling

Care needs to be exercised when handling and moving sheet material as the edges or the face may be damaged. In many instances, sacrificial sheets are placed on the outside of the sheet packs to protect the face of the first and last sheets. Some of the sheet material may be light in weight but will bend and break easily, other sheets are thick and heavy, requiring two people to carry them. If the sheets are in a pack, it may be easier and safer to mechanically move the pack using a forklift or pallet truck (Figure 8.19). Single sheets must be handled manually using correct lifting techniques and using hooked hand-held carrying devices to make it easier and safer for the handler (Figure 8.20). This prevents fingers from being squashed when the sheet is set down edge first.

Engineered timber products – storing

Manufactured timber product sheets should always be laid flat and stacked off the ground on packing timbers, called gluts, or on pallets that allow the sheets to remain dry with easy access underneath for mechanical lifting purposes. Where there are high levels of moisture, sheets should be covered to avoid damage caused by moisture absorption. The stack can be mechanically lifted, using a forklift or pallet truck, to load it onto a delivery vehicle or to move it to another covered position.

Source: Richard Moran

FIGURE 8.18 Correct method for manual cutting of sheet material

Source: Shutterstock.com/Dmitry Kalinovsky

FIGURE 8.19 Mechanical handling of sheets stacked on a pallet

FIGURE 8.20 Manual handling of sheet material

On the job site always choose an area where the material will not be damaged by passing machines, vehicles or people.

Engineered timber products – environmental issues

Manufactured timber products are environmentally efficient, in that they are made from materials that were previously waste, but are now recycled into useful products. This is an environmentally good practice. There is, however, a darker side to the story that relates to how these materials are currently disposed of. These materials are being used for landfill or are being burnt. Therefore the chemical binders and formaldehydes used in producing the products are either released into the atmosphere or into the ground. The question now being asked is: Is it possible that this waste may also be recycled into second- and third-generation recycled material?

There is another consideration, too – when these products are cut using a circular saw or electric planer, dust particles and chemical fumes are released. Over time these may affect the health of the carpenter.

Bricks, blocks, masonry and mortars

Bricks are one of the oldest and most durable forms of building material, dating back to early Egyptian and Middle Eastern civilisations of between 5000 and 4000 BCE. These bricks were handmade from clay and grass or straw, then pressed into bottomless moulds and left to dry in the sun before use. When protected from the weather, these 'adobe'-type bricks would last for many hundreds of years. During the time of the Roman Empire, to produce a more durable product, the bricks were burnt in primitive wood-fired kilns. This high heat enabled the clay particles to fuse together or become vitrified.

Brick making has improved over the centuries from the adobe-type to the handmade sandstocks produced by convicts after 1788 in Australia, up to the present day, where a wide variety of brick types, shapes, colours and textures are available with a high level of durability. The broad classifications of bricks are:

- adobe – made of natural sun-dried clays or earth with a binder
- kiln-burned – composed of clays or shales to which other materials may have been added and fired to hardness (Figure 8.21)
- sand-lime – mixtures of sand and lime hardened under steam pressure and heat
- concrete – solid or cored units composed of Portland cement and aggregate
- autoclaved aerated concrete – made from sand, lime and cement with added aluminium paste to create a gas to aerate the mix. It is cured in an autoclave. Manufacture commenced in 1990.

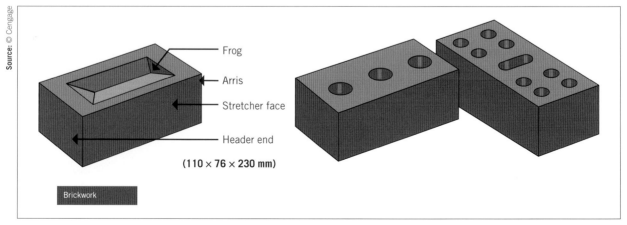

Frog

Arris

Stretcher face

Header end

(110 × 76 × 230 mm)

Brickwork

FIGURE 8.21 Typical forms of modern kiln-fired metric bricks

Bricks can also be identified by their finish, which is determined by the making process:

- clinkers – overburned bricks, very hard and distorted shape
- callows – underburned, light in colour and very absorbent
- sandstocks – handmade by convicts and early settlers (still produced using machine methods)
- commons – regular shape but can have flaws or uneven colouring
- face – regular shape, colour and texture.

Blocks

Blocks, or concrete masonry, are made from a mixture of coarse and fine aggregates with cement and water, manufactured into units of rectangular prismatic shape, hollowed and intended for use in bonded masonry construction.

The first recorded block manufacture was by a builder named Ranger in England in 1832. These blocks were made from sea gravel, broken flints, masons' chippings and other exotic inert materials which were then mixed with powdered lime and boiling water. The mixture was rammed into a mould and would set rapidly due to the hydration effect of the boiling water. The blocks were set aside to cure for about two weeks before being ready for use.

Much experimentation was carried out with block moulds during the 1800s until Jesse Besser of Michigan, USA, began making block machines in 1906. This process of manufacture is still carried out, hence the common title given to *Besser* blocks (**Figure 8.22**).

FIGURE 8.22 Concrete blocks stacked ready for transporting

Stone masonry

Natural stone has been used in walls and structures since the time of the Egyptian and Greek civilisations of about 3000 BCE. Granite, sandstone, slate, limestone and marble are the favoured types of natural stone used for masonry construction. The style is classified according to the shape and surface finish of the stone; for example, rubble, ashlar and cut stone or dimension stone. Within each of the classifications, variations can be used for interest or to bring out the characteristics of

a particular type of stone. Surface finishes range from split face, which is a rough natural face of stone used for masonry walls with small, uneven, narrow stones, to polished face, which is a mirror-like, glossy finish produced by computer-assisted grinding techniques. These can all be achieved using hard, durable, natural stone (see **Figure 8.23**).

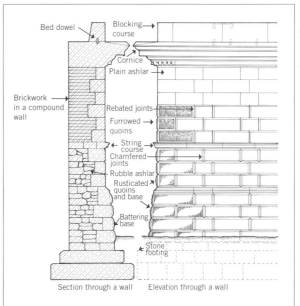

FIGURE 8.23 Typical stone wall configuration

As well as natural stone, there is also reconstituted stone, which is crushed fragments or stone powder that are bonded together with polyester resins or mortar. This material can be shaped into blocks, pavers, brick or even sheets. The example shown in **Figure 8.24** is not real stone but sheets of crushed stone with resin made to look like real stone set in mortar. These are sometimes referred to as *faux stone* wall panels. The advantages of this system is cost (much cheaper than real stone) and lighter weight.

Mortar

Mortars are mixed and used to bond brickwork, blockwork and all types of masonry to form a homogeneous unit capable of carrying great loads. Mortars are made by mixing cement, hydrated lime or lime putty, and clean, sharp sand, normally proportioned by volume, with clean, fresh water. Mortars are made using the following proportions:

- cement:sand – 1 part cement to 3 parts sand. A very strong mix that usually needs a plasticiser to make it workable. Used for load-bearing work.
- lime:sand – 1 part hydrated lime or lime putty to 3 parts sand. Usually a fairly weak mix, depending on the lime used. Hydraulic lime putty tends to give the most strength. This mix may be used on its own for internal walls as it tends to weather badly when used externally, or it may be used as a base mix for composition (compo) mortar.

Source: Alamy Stock Photo/aaphotograph

FIGURE 8.24 Faux stone wall construction

- cement:lime:sand (compo mortar) – 1 part cement to 1 part hydrated lime or lime putty to 6 parts sand. A good-strength general-purpose mortar that is very workable.
- cement:sand plus plasticiser – 1 part cement to 5 or 6 parts sand plus a plasticiser, such as Bycol, added to the water. A general-purpose mix commonly used for both internal and external work, especially on brick veneer construction. It usually contains Type GP cement with a 'larry-mix' or 'brickies' sand', which is normally a mixture of fine and coarse sand particles, plus an aerating admixture such as the detergent-like Bycol – also called liquid ball bearings, due to its action in the mix.

Many variations are possible with these mixes, and the climate and type of work to be performed will determine the mix. All of these mixes can be coloured using powdered oxides to allow variations in joint colour or to match brickwork. Red and black are the most common, but yellow, green and brown are also available.

Jointing
Jointing is the term given to the horizontal (bed) and vertical (perpend) connections between bricks, blocks and masonry. These joints, normally 10 mm thick, can be finished in many ways, depending on the required appearance or weathering properties (Figure 8.25).

Durability
Natural stone is very durable; however, there are environmental considerations to consider, particularly

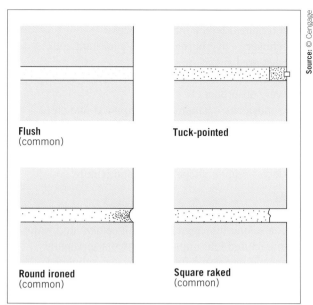

Source: © Cengage

Flush (common)

Tuck-pointed

Round ironed (common)

Square raked (common)

FIGURE 8.25 Commonly used joints for brickwork and blockwork

if using limestone. Some stones, such as limestone, are dissolved and eroded by the action of acid. In cities, where rain catches atmospheric contaminants that change the pH level of the rain water to a slight acid, this 'acid rain' can have an impact on stone structures.

Fire rating
Bricks and masonry material all have a good fire rating. They have been and still are used to line the interiors of chimneys, kilns and most *refractory* units and are able to withstand high temperatures.

Masonry cutting
Brick, concrete and masonry materials can be cut using similar types of tools. Manual tools include lump hammers and bolsters, brick hammers and scutch hammers or chisels (Figure 8.26).

Mechanical tools include the masonry disc angle grinder and the diamond-dust blade brick or block saws for wet cutting (Figure 8.27).

Masonry handling
Due to the weight involved in these products, handling is usually carried out by mechanical means. Generally, the masonry units are transported to the building site on a truck with a hiab (refer to Figure 8.13). Another way that masonry may be moved from the truck is by a truck-mounted forklift. Larger pieces of stone can be moved around using a small front-end loader known as a bobcat (Figure 8.28), or by using a small crane.

The final procedures are usually done by hand. Bricks, blocks and masonry can be moved around with the aid of brick carriers (Figure 8.29). Leather gloves and rubber or leather cots may be used to protect the hands and especially the tips of the fingers, which could end up with 'bird's eyes' (round areas of skin worn away until they bleed).

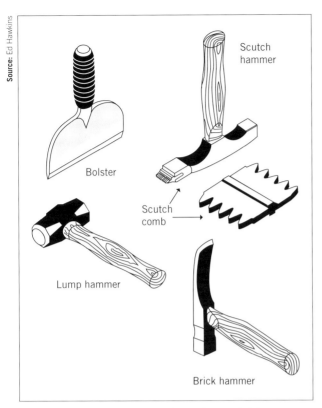

FIGURE 8.26 Common hand tools used to cut bricks and masonry

Scutch hammer

Bolster

Scutch comb

Lump hammer

Brick hammer

FIGURE 8.27 Diamond-dust blade wet-cutting brick saw

FIGURE 8.28 Front-end loader – bobcat

FIGURE 8.29 Typical brick barrow and brick carrier

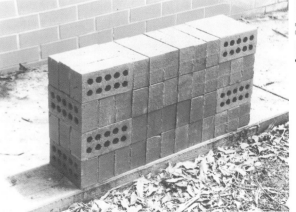

FIGURE 8.30 Preferred method of stacking bricks on the ground

Masonry stacking

Bricks, blocks and masonry should be stacked on timber pallets for delivery and site storage. They can also be stacked onto plywood sheets or directly onto the ground just prior to being laid in a form so that the stack will not collapse (Figure 8.30). Remember, if masonry is stacked onto wet clay, the clay will need to be cleaned off before being laid into the wall as the clay will weaken the bond between brick and mortar.

Masonry storage

As most bricks, blocks and masonry products are porous, they need to be protected from exposure to heavy rain and from lying in water. If these units soak up too much water they will tend to 'float' when laid. This means water in the mortar is not taken up and they tend to float on top of the mortar bed rather than bonding to it. This may result in the bricks or block moving out of alignment. Plastic covers should be placed over stacks outside until they are ready to be laid. This applies also to 'green' or freshly laid brickwork during wet weather. Some covering may be necessary to allow the work to set before excessive moisture is absorbed.

Masonry environmental issues

The problem, or downside, for bricks, is their modern kiln-fired production, and the energy used to manufacture or make them. The upside is that bricks have a long life expectancy and can be reused and recycled easily. In fact, all bricks, blocks, masonry and stone are able to be reused and recycled.

GREEN TIP

When designing buildings, it can be useful to consider the scientific research around use of different materials that is available on government websites. The National Construction Code has mandatory information on house construction that designers must be familiar with.

Concrete components

Concrete is basically artificial or man-made rock, and when it is cast into formed moulds in a plastic or wet state it will set in the shape of the mould with characteristics similar to those of natural solid stone. Concrete has been around since the Roman Empire (first-century BCE to fifth-century CE), when the Romans used a primitive form of it to construct vast structures and works of engineering; first using stones and then bricks to form a shape that was filled with a lightweight concrete made from porous volcanic rock. It has evolved over the centuries, but was not widely used until an improved cement was developed in the early 1800s. Since then, concrete has been used in many shapes and forms and combined with steel reinforcement. Today it is one of the most commonly used building products.

Concrete

On its own, concrete is very strong in compression (when it is being squashed or crushed), but is very weak in tension (when it is being stretched or pulled apart). It consists of four main parts, which when combined create an artificial stone with a high durability level. The components are:

- cement
- fine aggregate – sand in Australia
- coarse aggregate – usually stone or rock fragments
- water.

Cement

This material acts like a glue, bonding all the particles together. The first type of cement was prepared from volcanic ash and lime called *pozzolana*, invented by the Romans, while modern-day cement, developed by Joseph Aspdin in England in 1824 and called Portland cement, is made from limestone and clay crushed and mixed together with water and burned at a very high temperature, where it fuses together to form marble-sized 'clinkers'. It is then ground finely to form the common grey powder known today as Type GP – general-purpose Portland cement (previously known as Type A cement), more commonly available today in 20 kg bags with 50 bags to the tonne. There are two types of general-purpose cement:

- *Type GP* – general-purpose Portland cement
- *Type GB* – general-purpose blended cement.
  There are also four types of special purpose cement:
- *Type HE* – high early strength cement
- *Type LH* – low-heat cement
- *Type SR* – sulfate-resisting cement
- *Type SL* – shrinkage-limited cement.
  All cement must comply with the requirements set out in Australian Standard 3972 Portland and blended cements.

Cement bag storage

Cement may be delivered to the site in bags or in bulk form. It will deteriorate quickly if allowed to absorb water or moisture from the ground or atmosphere, so bags should be stored off the ground, covered with a tarp or plastic sheet and preferably placed in a shed, while bulk cement should be stored in watertight bins or hoppers.

Fine aggregate (sand)

Sand is the name given to fine particles of stone less than 5 mm in size. Natural sands are usually hard and durable because they are the remnants of decomposed rock that have withstood the physical and chemical ravages of time. Sand for concrete use must be a clean, washed, regulated mixture of coarse and finer grains made up of round and sharp-edged particles to create a good workable mix. Some types of sand available for concrete work are:

- pit sand – mined from the sand dunes of beaches, which must be free of salt and clay
- river sand – smooth and coarse particles dredged from rivers and banks
- beach sand – generally not used due to a high content of salt and shells

■ crusher fines – produced as a by-product of rock crushing. Particles are rough and splintery in shape, so only small quantities are used.

Coarse aggregate (blue metal)

This consists of crushed rock such as basalt, granite, diorite, quartzite and the harder types of limestone (Figure 8.31). Special types of coarse aggregate, such as blast furnace slag, expanded shale and clay, may also be used. A good coarse aggregate would be:

■ dense and hard, not brittle
■ durable and chemically inert
■ clean, with no silt, clay or salt
■ rough and of various sizes over 5 mm
■ non-porous to help prevent water penetration of the finished concrete.

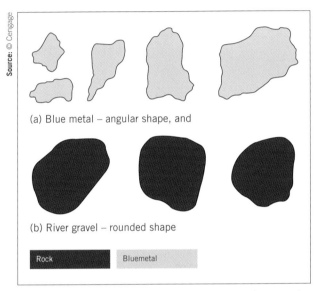

(a) Blue metal – angular shape, and

(b) River gravel – rounded shape

Rock   Bluemetal

FIGURE 8.31 Materials used for coarse aggregate in concrete

The average size of coarse aggregate for general domestic work would be up to 20 mm; for most structural building construction it is up to 75 mm; and for massive structures, like dams, up to 150 mm.

Water

As a guide, only water that is suitable to drink, or potable water, should be used for concrete mixing. The water used should be free of problem-causing substances such as acids, alkalis, oils, sugar and detergents. Water containing decayed vegetable matter should be avoided, as this may interfere with the setting of the cement within the mix. Sea water may be used but is not recommended, as it causes corrosion of steel reinforcement and forms surface efflorescence (a white salt powder on the surface).

Aggregate storage

Aggregate should be stored separately in metal or concrete storage bins and covered with a tarp to keep it dry. This prevents inaccurate weight batching caused by the presence of excessive water, as well as avoiding the material being blown around the site by the wind, or washed away with the rain (Figure 8.32).

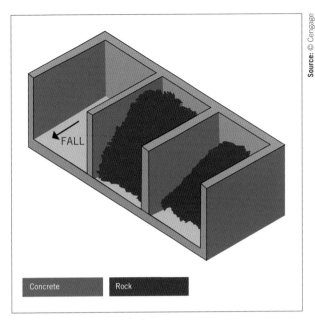

FALL

Concrete   Rock

FIGURE 8.32 Storage of aggregates

Reinforcement

Reinforcement is used within the concrete for the purpose of giving the 'reinforced concrete' strength when tensile and shear forces are applied to it. Reinforcement may be divided into two categories:

■ bars
■ steel fabric sheets.

Bars

Generally reinforcement bars are constructed from hot-rolled 500 MPa or 500 N deformed steel bars; ranging in size from 12 mm to special sizes of 50 mm in diameter (Figure 8.33), and round bars ranging in size from 4 mm to 12 mm in diameter.

There are also now available lightweight fibreglass reinforcement rods ranging in size from 6 mm to 25 mm.

Steel fabric

This is a mesh of steel wires placed at right angles and welded together by an electrical process to form reinforcing sheets. There are two types of steel fabric sheets. The first is trench mesh which is a sheet 6 m long by 200 mm (3 bar) up to 400 mm (5 bar) used for strip footings (see Figure 8.34). The second type of sheet is 6 m long by 2.4 m wide and is used in reinforced concrete slabs, paths and driveways.

Durability

As mentioned above, reinforced concrete as a completed product is extremely durable, and concrete buildings will last many years. The weak link is the reinforcement used in the concrete. If the steel reinforcement is exposed to moisture and air it will corrode or rust which will result, over time, in the failure of the reinforced concrete component. Therefore, it is important to have the reinforcement completely

Source: © Cengage

FIGURE 8.33 Deformed reinforcement bars

FIGURE 8.34 Trench mesh reinforcement

buried in the concrete to an engineer-specified depth referred to as *cover*. In some locations that are highly corrosive (e.g. by the sea) the engineer may specify galvanised or even stainless steel reinforcement. With the advent of the new fibreglass reinforcement, buildings in these highly corrosive environments may soon be using fibreglass reinforcement, because it is not affected by salt and is cheaper than galvanised or stainless steel reinforcement.

Strength

Reinforced concrete is considered to be the strongest building material available today, and this is evidenced by the fact that *all* mega-high rise buildings throughout the world are built using reinforced concrete.

Fire rating

Concrete serves as a good insulator to many materials in a fire. It won't ignite, spread flame, generate heat or create smoke, and is therefore classed as having a good fire rating. Due to its structure, however, it does expand when exposed to extreme heat and may explode and shatter if this occurs.

Steel also has a good fire rating, but at its critical temperature of 550 °C it will soften sufficiently to expand and buckle, and allow a reinforced or pre-stressed concrete member to fail.

Reinforcement handling

On large-scale constructions the quality of the final concrete product must be assured to a set standard. For this reason, all handling of concrete components is carried out by mechanical means so the engineers can be assured that the concrete will meet the required specification. However, on small-scale tasks, such as a slab for a small shed, handling can be performed manually: concrete materials are mixed by hand or in a bricklayer tilting drum mixer. In this case the cement (which comes in 20 kg bags), the sand and the coarse aggregate must be stored onsite prior to mixing the components into concrete.

When handling bags of cement, it must be remembered that these were altered from 40 kg to 20 kg to reduce the incidence of back injuries. Another consideration when handling cement is to remember that wet cement will affect the skin and the eyes and should be washed off as soon as possible. Sand and gravel is brought to a small site in a tip truck or the back of a utility or table-top truck, and may need to be shovelled into place.

Bulk steel reinforcement, due to its mass, is usually mechanically handled by crane or hydraulic lifting arms. Individual bars or sheets can be manually handled, but some care needs to be taken. Due to its mass and length, most steel will need two people to move it (see Figure 8.35). Another difficulty is that steel reinforcement may have a textured surface (caused by deformation

FIGURE 8.35 Sheet reinforcement

patterns created during manufacture) and could be rusty due to being stored outside or being placed on the ground. It also has many sharp edges, especially when it is cut, and may inflict minor or even severe wounds during handling. The surface can even harbour a bacillus called tetanus, which is a microscopic vegetable organism causing muscle spasm and lockjaw. Therefore, it is recommended that leather gloves be worn during both handling and cutting. The fibreglass reinforcement has stiff fibres protruding from the rods, so it is highly recommended that leather gloves are worn when handling these reinforcement rods.

Reinforcement storage

Steel reinforcement should be stored off the ground on timber gluts so that it doesn't become contaminated with vegetable matter, animal matter, oils or mud. It can have a small amount of rust on its surface as this actually aids bonding, but should never have flaky rust, as this will continue to flake and cause **spalding** of the concrete surface, which will lead to 'concrete cancer', as seen in Figure 8.36. (Spalling or spalding concrete refers to concrete that has become pitted, flaked or broken up. This could be caused by poor installation, stress and environmental factors that damage concrete. You can address the spalling problem immediately when the concrete is poured.)

FIGURE 8.36 Concrete cancer

Fibreglass reinforcement should be treated similarly to steel reinforcement, with the additional requirement that it should not be exposed to direct sunlight for extended periods of time.

Reinforcement environmental issues

Concrete is in many ways similar to bricks, masonry and stone in that the ingredients need to be mined. Similarly to brick manufacture, cement production uses vast amounts of electrical energy, but concrete, like brick, has a high life expectancy and will last many years. Concrete can be crushed and recycled as rubble into the next generation of concrete.

LEARNING TASK 8.1 PRACTICAL ACTIVITY FOR A GROUP OF 2–3 STUDENTS – CONCRETE

If possible, obtain either five 20 kg bags of concrete mix, or a small amount of coarse and fine aggregate and then a bag of cement; which you are to hand mix and place into a form made from 90 × 35 timber with the dimensions of 450 mm × 450 mm.

Part 1:
Before mixing the concrete, obtain an SDS for cement and read through this document to be sure that you are correctly handling the material.

Part 2:
Mix the concrete and place into the formwork.

Part 3:
Finish the surface of the concrete as directed by your facilitator.

Part 4:
Correctly clean off all the tools that you have used to mix and place the concrete.

Part 5:
Allow a minimum of 24 hours before carefully stripping off the formwork from the 450 mm × 450 mm square concrete bases.

Paints and sealants

Paint has existed in a simple form as a mixture of earth pigments and animal fat, applied with the fingers, or plant stems crushed at one end to form a fan shape, for many thousands of years. In Australia, these materials and methods have been used for over 50 000 years.

Modern paint has its beginnings around the 1920s. The mixture consists basically of pigment, binder and solvent. The pigment is mined or manufactured in the form of a fine powder that gives paint its colour and opacity (it doesn't allow light to pass through it). The binder is the liquid part of the paint, which may be an oil, such as linseed, or a resin. The binder holds the pigment together and enables the paint to stick to a surface. Solvents are used to dissolve the resins or thin the oils, to make paint workable. Solvents are an extract of petroleum oils, such as mineral turpentine.

Paints for building purposes are available in two forms:

1 oil paints – solvent-thinned. Oil paints contain pigments, binders, solvents, driers and extender pigments. They are designed to soak into the surface, lock into it and provide a key for the next coat. Thinning and washing up is carried out using mineral turpentine, sometimes called turps.

2 latex paints – water-thinned. Latex paints contain pigments, binders, solvents, driers and also extenders. The main differences between oil and latex paints are the limited surface penetration of latex paints and the use of water instead of turps for thinning. Clean-up is also carried out using water.

The range of latex paints available is the same as for oil-based paints, but the latex range also includes acrylic and vinyl paints, such as 100% acrylic paints, vinyl paints, vinyl/acrylic paints and vinyl/latex paints. The variation in this range is due to differences in the binder.

Use and application of paints

Generally, oil-based and water-based paints can be used either internally or externally. Oil-based paints are preferred externally for their ability to soak into materials and seal them from the effects of weathering, but in recent times water-based paints have been used externally for their ease of application, low cost and easy clean-up properties. The range of uses for both types is as follows:

- primer – penetrates and seals the surface and provides adhesion for the next coating
- sealer – designed to stop suction or absorption at the surface and provides a base for the next coat
- undercoat – adds film thickness and contains more pigments for grain/texture filling and provides good sanding ability for a smooth finish
- flat finish – not designed for external use, as it attracts dust and dirt. Mainly used for general ceiling finishes
- satin finish – mainly used internally for areas where regular washing is required, such as in bathrooms, laundries and kitchens. Can be used on walls and ceilings
- gloss finish – can be used internally or externally where a hard, shiny, durable finish is required for appearance and ease of cleaning.

All types of paint can be applied by using a brush, roller, paint pad or sponge, sheepskin or synthetic paint mitten and by spray-gun. The surface, position, cost, use, texture and durability will determine the appropriate application method (Figure 8.37).

Good brushes and rollers are expensive, so the utmost care should be taken to ensure they are kept in good condition (Figure 8.38). They should be cleaned after every use in the appropriate solvent recommended by the manufacturer, and be ready for use when next

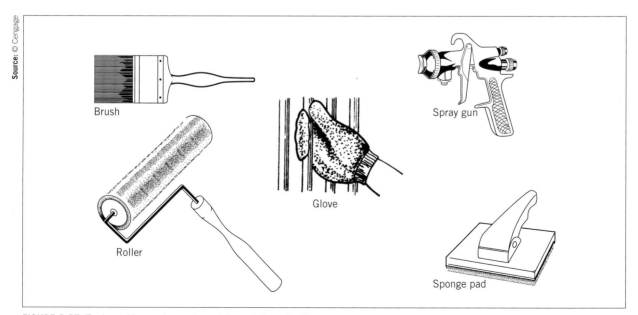

Source: © Cengage

Brush

Roller

Glove

Spray gun

Sponge pad

FIGURE 8.37 Tools and/or equipment used for paint application

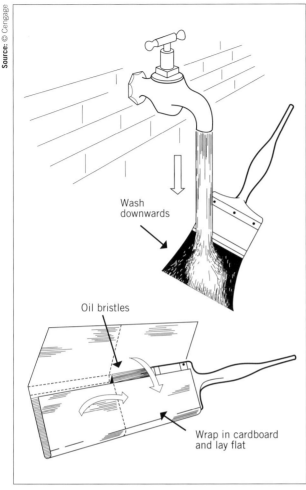

FIGURE 8.38 Cleaning and proper storage of brushes

Wash downwards

Oil bristles

Wrap in cardboard and lay flat

required. After cleaning, they should be stored either flat or hung in such a way that the bristles are not damaged, and in a position where they cannot be contaminated by dust, corrosive liquids or powders.

Fillers

Fillers are smooth paste mixtures used for filling slight surface imperfections. They range from the plaster/cellulose-based (Polyfilla type) to thinned brush or spray fillers. Fillers are used to form a blemish-free surface ready to receive a paint or stain finish.

Fillers can be applied with a spatula or putty knife, a tube and gun as with common caulking or joint/gap compounds, by rubbing into the grain of timber with a cloth, or can be brushed or sprayed on.

Sealants

Sealants are substances that block the passage of liquids or gases through the surface of a material or joints or openings in or between materials. There are a number of different types of sealant, from polyurethane, acrylic and silicone, and each has different qualities including adhesive properties, waterproofing, thermal insulation, acoustic insulation, medium- to high-elasticity properties and as a fire barrier. An example of where a sealant may be used for its elasticity and water-proofing properties is the mastic joint between two panels of brickwork in an articulated masonry wall.

Durability

The durability of paint is dependent on the strength of the coating. There are three important factors that cause the surface coating to break down: these are a combination of UV light, heat and moisture.

Durability of paint is affected by a number of items including the surface preparation, quality of the binders and the quality of the pigments.

Sealants by their nature must be durable, they are required to be exposed and able to cope with movement, heat, cold and moisture changes. If a sealant fails, the joint or surface that it is covering will be left unprotected.

Fire rating

As paints contain at least some oil and solvents they tend to be highly flammable and are, therefore, given a poor fire rating. There are, however, some specialist paints available that are used as fire retardants: that is, the gases given off are in the form of bubbles or a foam when the surface is heated and these act to smother any flame. These are called intumescent paints. Wallpaper has characteristics similar to paper and, therefore, attracts a poor rating.

Some sealants are designed to be fire rated (Figure 8.39) while others are not. It is impossible to make a broad statement, that all sealants are fire resistant or not fire resistant.

FIGURE 8.39 An example of a fire-rated sealant

Paint handling

Paints are supplied in 1-, 2-, 4-, 10- and 20-litre cans or drums. The larger the drum, the heavier it will be. Using a hand trolley to move a number of drums at one time is better than multiple trips, or even trying to lift and move heavy drums. Fillers and sealants usually come in tubes and it is important to clean the skin of any residue after use.

Paint storage

Careless storage of paints and sealants can prove to be expensive and, in some cases, dangerous. Paints containing solvents or volatile thinners should be stored in a cool, dry area, preferably below 15 °C, with tight-fitting lids. Brushes and other materials deteriorate when subjected to damp conditions. To prolong the life and shape of brushes, they should be thoroughly cleaned in the recommended solvent, washed with warm soapy water, dried and wrapped in a cardboard liner and laid flat on a shelf. Rollers and spray guns should also be thoroughly washed and dried before storage.

Steel components

Steel is a combination of iron, carbon and manganese and is easily worked into shape from a molten state. Mild steel may be melted and poured into billets, or short thick sections, that can be cooled, reheated and rolled into shape. As the steel is rolled and worked it gains a finer grain structure owing to the removal of impurities like slag. Steel can be produced with a remarkable range of properties through the addition of only 2 per cent or 3 per cent of other elements. Steel is strong in both tension and compression, which makes it a very versatile material.

Types and uses

See Table 8.3 for an outline of the types of steel and their uses.

Steel may be used in many structural and ornamental areas within a building. It may be welded, bolted, riveted, bent, crimped or clipped together to form a variety of building elements such as reinforcement, door frames, wall frames, roof frames, roof sheeting, brickwork lintels, handrails, fasteners, beams, columns, gutters and downpipes or furniture (see Figure 8.40).

Steel cutting

Steel can be cut using a hacksaw (manual or mechanical), tin snips, bolt cutters (Figure 8.41), hand shears, oxyacetylene torch or a metal angle grinder.

 Care should be taken when cutting with abrasive discs or an oxy-torch, as red-hot pieces of metal may be thrown into the eyes. Safety face shields or goggles should be worn during these operations. Also, razor-sharp edges are formed when cutting sheet and cold-formed materials, so leather gloves should be worn.

Fire rating

Steel is classified as having a good fire rating, but, as mentioned previously, at its critical temperature of 550 °C it will soften and may buckle and collapse. Stainless steel has a low thermal expansion rate and would be a preferred material to use in high heat exposure situations.

Steel handling

Most heavy steel items are lifted using a crane, but manual handling can also occur. Where items are too heavy for one person to lift or hold in position, chain blocks and hoists may be used to gain a mechanical advantage.

Steel storage

Uncoated or unprotected steel items should be stored under cover until ready for use. Some steel products, such as stainless steel door-push plates, have a protective plastic coating, which prevents the surface from being spoiled or marked prior to installation.

TABLE 8.3 Types and uses of steel

Type	Uses
Carbon steel	Usually classified according to the carbon content. Up to a point, the more carbon, the stronger, harder and stiffer the steel is. As the carbon content is increased, the steel becomes less able to be stretched out into wire as it becomes more brittle.
Alloy steel	Alloy steel is carbon steel that has elements such as manganese, nickel, chromium, vanadium and copper added to improve the steel's properties. Alloy steel may have strength, hardness and improved resistance to corrosion.
High-strength, low-alloy steel	Sometimes called weathering steel, there is up to 40% higher strength and resistance to corrosion than carbon steel. Thinner, lighter sections may therefore be used for exposed structural positions.
Stainless steel	There are many different types of stainless steel, but the basic ingredients are carbon steel with around 12% chromium. This provides enough of an inert film to prevent the steel from corroding or staining.
Structural steel	Structural steel is hot-rolled into sections, shapes and plates not less than 3 mm thick. These elements make up the structural frame of a building rather than ornamental elements such as stairs, grates, etc.
Reinforcement steel	Reinforcement steel may form part of some of the previously mentioned types. For example: *A quenched and tempered deformed bar is a low-carbon steel that obtains its strength by a heating and water cooling treatment.* *A micro-alloyed deformed bar is a low-carbon steel that obtains its strength from the addition of small amounts of vanadium during the smelting process.* *Other bars, rods and coils obtain their strength and various properties by modification of the hot-rolling process.*

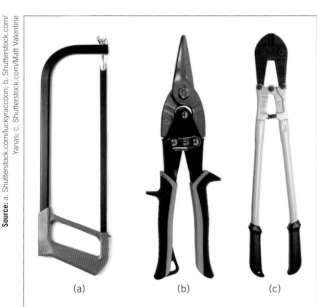

Metal roofing sheets

Reinforcement bar and stirrups

Rolled steel joist (RSJ)

Round

Square

Flat

Equal angle

Cold formed framing

Universal beam (UB)

Unequal angle

Channel

Steel channel lintel

Steel

Brickwork

Source: © Cengage

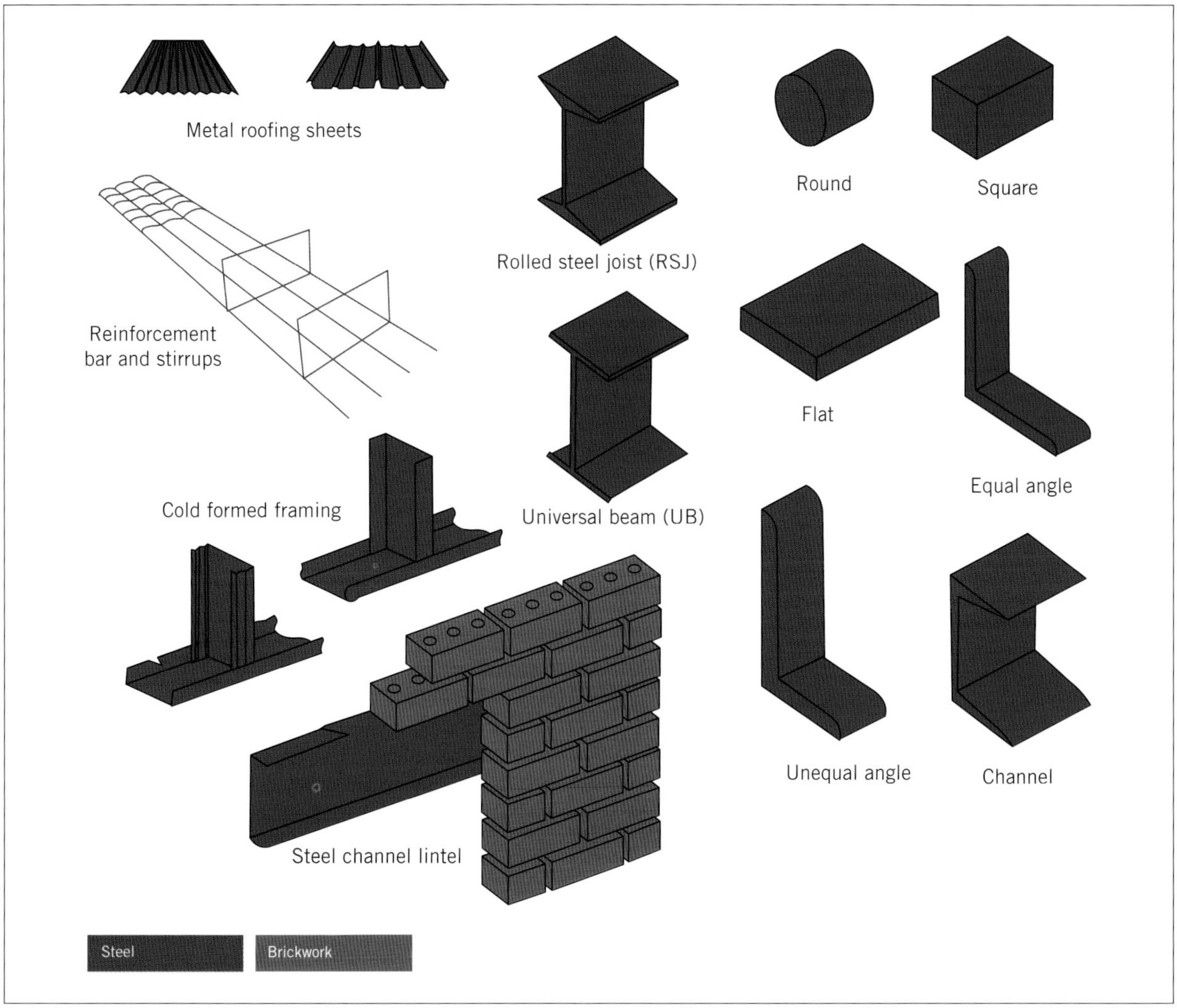

FIGURE 8.40 Some of the many steel products

Source: a. Shutterstock.com/luckyraccoon; b. Shutterstock.com/Yanas; c. Shutterstock.com/Matt Valentine

(a)   (b)   (c)

FIGURE 8.41 Various tools used to cut steel: (a) Hacksaw; (b) Straight-blade tin snips; (c) Bolt cutters

Steel environmental issues

Steel is one product that can be reused and recycled. Scrap steel can be placed into a smelter and reformed into other metal objects.

Insulating materials

The purpose of a building is to provide sheltered protection from the elements and possible intruders. Apart from the structural and security benefits of the building enclosure, the *thermal* and *acoustic* properties of the materials used must be considered when designing the structure. The mass and density of walls, floors and ceilings will aid with the insulation process but will increase the weight and cost of the building. Therefore, lightweight materials with good insulating properties are preferred.

The conduction of heat and sound occurs when energy waves are passed from atom to atom through a material. The lower the density of the material, the lower the conductivity of heat and sound. Air has the lowest density of all the materials present in a structure, and if it can be trapped motionless within a material it will improve the insulation properties of that material. Mineral wool (blown or expanded

TABLE 8.4 Types and uses of insulating materials

Type	Uses
Mineral fibre	Glass, molten rock or slag from a furnace, which is blown with steam or high-pressure air to create mineral fibres that trap air. It can be spun to form batts, blankets or loose material laid between ceiling joists.
Wood fibreboards	Soft, open-bodied boards or pre-prime soft pulp boards such as Caneite, often used for pin boards, provide good thermal and sound absorption properties due to the open weave of the cane fibres. Used mainly for acoustic tiles in suspended ceilings.
Polystyrene	This is a foamed plastic material used for many purposes, such as for hot or cold food product containers, cold-storage compartments, insulation sheets up to 50 mm thick in brick veneer cavities, and sandwich panel (thick polystyrene centre faced with steel or timber sheets).
Blankets or batts	These may be of fibreglass, rock wool, and organic or inorganic materials, and are used to insulate ceilings by being placed between joists or under roof sheeting in the form of a roll or blanket.
Foil	Shiny aluminium may be used on its own from a roll or as a backing to blanket insulation and plasterboard. Although aluminium is normally a good conductor of heat, its shiny surface will reflect back into the air up to 95% of the heat that reaches it. This makes it a good insulator.
Blown or poured-in insulation	Several types of loose insulation can be used such as perlite and vermiculite (lightweight volcanic minerals), glass fibre and cellulose fibre. These are blown or poured between ceiling joists to any desired thickness. The lightweight open-bodied nature provides good thermal and acoustic properties.

rock), organic or wood fibre (cellulose, cork), foamed plastic (polystyrene foam) and air-entrained concrete, as well as many other types of insulation, all work on the principle of trapped air. They are usually treated to improve fire and insect resistance. Double-glazing of doors and windows works on the same principle of trapped air to lower sound transmission from outside the building.

Types and uses

See Table 8.4 for an outline of the types of insulating material and their uses.

There are many other insulating materials available on the market for use in buildings for specific purposes; for example, reflective film can be applied to windows and glass doors to prevent heat conduction through the glass (Figure 8.42). Some insulation materials (e.g. asbestos and urea formaldehyde foam) are no longer used due to their hazardous nature and links with cancer.

Fire rating

Insulation materials vary in their inherent capacity to resist the effects of fire. Materials such as glass fibre, mineral fibres, foamed concrete and cellulose insulation have a good fire rating. Materials such as wood fibre insulating board, polystyrene foam and other plastic foamed products all have a poor fire rating. Materials with a poor rating can be offset by using combinations of materials; for example, by facing polystyrene with metal.

Insulation handling

Many insulating materials do not present handling problems, although great care should be taken when handling any fibre-based material.

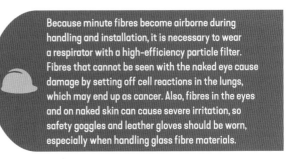

Because minute fibres become airborne during handling and installation, it is necessary to wear a respirator with a high-efficiency particle filter. Fibres that cannot be seen with the naked eye cause damage by setting off cell reactions in the lungs, which may end up as cancer. Also, fibres in the eyes and on naked skin can cause severe irritation, so safety goggles and leather gloves should be worn, especially when handling glass fibre materials.

Insulation storing

In the event that insulation needs to be stored prior to being installed, it is important to keep the insulation material away from moisture and any active work area. By its very nature most insulation material will absorb

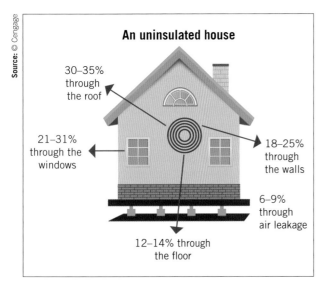

An uninsulated house

30–35% through the roof

21–31% through the windows

18–25% through the walls

6–9% through air leakage

12–14% through the floor

Source: © Cengage

FIGURE 8.42 Typical heat loss of an uninsulated brick veneer cottage in temperate regions of Australia

moisture quite readily, resulting in poor performance of the insulation.

Insulation environmental issues

Insulation is required in all new homes today to reduce the loss and transfer of heat and cold, and to reduce traffic noise. The biggest issue for builders is what to do with the waste. At this time, there is no reuse or recycling technology and so insulation waste generally goes to landfill.

Plasterboard sheeting

Plasterboard was introduced into Australia in the early 1960s and is the modern version of the original plaster sheet known as fibrous plaster, which is plaster of Paris reinforced with a hemp fibre, coconut fibre or horse hair.

Plasterboard is a strong, stable panel product, consisting of a core of gypsum (calcium sulfate dihydrate) sandwiched between two paper surfaces. It is made by feeding wet plaster between the two sheets of paper as they travel along a conveyor, allowing the gypsum plaster to set, then cutting it into set lengths. Panels emerge with the face, back and long edges covered with paper. The long edges are also tapered, enabling joints between sheets to be butted together, reinforced with a perforated paper or cotton tape, filled with a hard base coat (Figure 8.43) and then finished with several softer top or finishing coats prior to sanding smooth. Sheets are available in widths of 1200 mm and 1350 mm and lengths from 2400 mm up to 6000 mm.

Types and uses

Generally, plasterboard sheets are available in thicknesses of 6.5 mm to 25 mm. The types available are outlined in Table 8.5.

Source: iStock.com/Jodi Jacobson

FIGURE 8.43 Setting plasterboard sheets with base coat

TABLE 8.5 Types and uses of plasterboard

Type	Use
Flexiboard	6.5 mm thick sheets specially formulated to construct curved walls, ceiling arches and feature panels with ease.
Standard tapered edge	Available in sheets of 10 mm and 13 mm thickness for use on internal walls and ceilings. A common trade name is Gyprock.
Square edge	Available in sheets of 10 mm and 13 mm thickness for use in office partitions where joints may be covered with aluminium, vinyl or timber mouldings. This enables quick and easy installation where removal or alteration may be necessary.
Extra strength	10 mm thick sheets specially formulated to span greater fixing centres on ceilings, without sagging. Common trade names include Unispan and Spanshield.
Water-resistant	Specially processed plasterboard available in 10 mm thick sheets. The core and lining board facing are treated to withstand the effects of moisture and high humidity. It is used for lining walls and/or ceilings of bathrooms, kitchens, laundries, garages and ceilings of walkways and verandahs. Common trade names include Aquachek and Watershield.
Fire-resistant	Available in 13, 16 and 25 mm thick specially processed, glass fibre-reinforced, mineral core sheets. It may be used for walls, ceilings, partitions, liftwells, stairwells, shafts and ducting where fire may spread. A common trade name is Fyrchek.
Foil-backed	Available in 10 mm and 13 mm thick sheets for walls and ceilings where a vapour barrier or thermal insulation is required.
Special panels	Available in 13 mm thick cut-to-size panels to suit grid system suspended ceilings. They are also available in sculptured panels for decorative wall applications.
Cornices	Available in 55, 75 and 90 mm coved profiles of 10 mm thickness to finish the connection between walls and ceilings.
Glass fibre sheeting	Reinforced gypsum core using glass fibre. This provides a stiff body capable of withstanding impact and offers some fire resistance. It may be used to line the corridors of hospitals where trolleys and mobile beds may constantly collide with walls. A common trade name is Plasterglass.

PART 3

8

HANDLE CARPENTRY MATERIALS **299**

Plasterboard cutting and fixing

Plasterboard sheets may be cut with a hand saw, keyhole saw or jig saw but the usual method for cutting is to score through the surface with a utility knife (**Figure 8.44**), bend the offcut back at 90° and then run the utility knife through the backing paper still attached. An alternative method is to again score the surface with a utility knife, bend the offcut back at 180°, then pull forward quickly to snap it off cleanly. This method however takes practice and experience to master effectively.

Source: a. Shutterstock.com/Gavran333; b. Shutterstock.com/Stocksnapper c. Shutterstock.com/satit_srihin; d. iStock.com/simonkr; e. Shutterstock.com/Anton Starikov. iStock.com/BigJoker; g. Shutterstock.com/amplejs

FIGURE 8.44 Tools used for plasterboard work: (a) Flat steel trowel; (b) Keyhole saw; (c) Broad knife; (d) Hand sander; (e) Corner tool; (f) Small tool; (g) Utility knife

Sheets are butt-jointed with walnuts of stud adhesive spaced 230 mm apart. Wall sheets are nail-fixed, using ring-shank lattice head nails around the perimeter, and temporarily held in the centre by nails driven through scrap blocks of plasterboard.

Ceiling sheets are nailed around the perimeter and double-nailed, not less than 50 mm apart, at 300 mm centres. Screws may also be used, normally when fixing into metal studs or furring channels. All joints are taped and set with external corners reinforced with a perforated metal angle prior to setting. End joints of sheets should be staggered so that a straight joint does not occur in the full height of walls or full width of ceiling.

Fire rating

Generally plasterboard has a fair fire rating. It has a low spread of flame and a low smoke development index due to the non-combustible gypsum core, but it has a high ignition index due to the paper facings. Special fire-resistant plasterboard, such as CSR Fyrchek, has an excellent fire rating and is widely used in fire-prone building situations. It will provide fire-resistance ratings from half an hour to four hours for walls, and up to two hours for ceilings. Double-sheeting will increase the ratings substantially.

Plasterboard handling

When manoeuvring plasterboard, always lift, handle and carry the sheets on edge. All lifting should be done with a straight back, bent knees, and at least, depending on the size of the sheet, in pairs (**Figure 8.45**).

Source: © Cengage

FIGURE 8.45 Always handle plasterboard at the edges to avoid sheet breakage

Plasterboard storage

All materials should be kept dry, preferably by being stored inside the building. If it is necessary to store plasterboard outside, it should be stacked flat, off the ground, properly supported on a level platform and protected from the weather with plastic sheeting or a similar waterproof covering. Care should be taken to avoid sagging and damage to edges, ends and surfaces.

LEARNING TASK 8.2

MATERIAL RESEARCH ASSIGNMENT NUMBER 1

Prepare a folder with a cover sheet containing the following details:

* your name
* your class/group name or code
* name of the assignment ('Material Research Assignment')
* your teacher's name
* the due date.

Part 1

Refer to the section above beginning at 'Engineered timber products', then provide the following details:

1   A brief description of each material
2   A brief description of the manufacturing process

>>

3 At least two uses of the material

4 How the material should be safely handled during use or preparation for use.

The materials to be identified and described are:

- particleboard
- fibreboard
- hardboard
- plywood
- laminated veneer lumber (LVL)
- cross-laminated timber (CLT)
- Glulam beams
- 'I' beams
- wood plastic composites.

Approximately half a page for each of these materials would be acceptable.

Part 2

Include a manufacturer's brochure and/or details obtained from the internet, which relate specifically to each of the materials listed above. These details should be attached to the assignment at the end of the description of each material.

Hand in the assignment to your teacher on the nominated due date.

1.3 Select tools and equipment, check for serviceability and report any faults

Before you select tools and equipment, check for serviceability, and report any faults. You need to determine which tools and equipment are required for the task. You may ask a competent person which tools and equipment are required, or you may refer to installation instructions, codes of practice or other industry information. By referring to these job method documents we can determine which tools and materials we may need to use.

Every tool will have a user manual and all materials should have SDSs and instructions so you can refer to the user manuals for the tools and follow the procedures to ensure those tools are serviceable. If there are faults with the tools, then check with your organisation and find out what processes they have in place for tool servicing and maintenance. For example, store them in an appropriately labelled and tagged box ready to take to the service agent for repair so that those tools are not used until they are repaired and put back into service. All electrical tools need to be tagged according to your state or territory's schedule of time. There should also be a register kept of all electrical tools that are used onsite due to the hazards from electricity. Of course, cordless tools eliminate much of the electricity hazard; however, cordless tools still require mains power to charge the batteries, and this is where the test and tag procedure would occur.

LEARNING TASK 8.3

CLASS/GROUP ACTIVITY: CHECK YOUR TOOL FOR SERVICEABILITY

Take a picture of one of your tools and obtain the user manual from the manufacturer for that tool based on the model number. Provide a short report about your tool showing how it does or does not comply with the manufacturer's documentation.

1.4 Select and use personal protective equipment (PPE) for each part of the task

When performing carpentry tasks, you should have a SWMS that breaks down the task into steps. The SWMS considers any hazards associated with each part of the task and any protections that need to be implemented to minimise or eliminate those hazards. If it is determined that PPE is the most appropriate way of controlling the hazards and reducing them to an acceptable level, then the PPE must be Australian Standards quality and used in accordance with the instructions for that PPE. If you don't have the initial documentation that was supplied with the original purchase of the PPE, you can contact the manufacturer through their website or through the supplier to obtain another copy. In most cases you won't need many copies because once you know how to use PPE you remember each subsequent time you use it, and you may only need to update documentation if the PPE changes or if the task changes. Always remember that PPE is the last line of protection from a hazard when you can't eliminate or reduce or control a hazard using the hierarchy of control.

1.5 Inspect work site, locate services, assess hazards and apply risk controls, including required signage and barricades

At the site establishment stage, inspections of the work site will have been carried out to locate services such as water, gas, electricity, sewer and communications. At the same time, hazards will have been assessed and any hazard controls, which may include signage and barricades, should have been installed. All site inspections should be completed by a competent person and should take into account state or territory codes of practice, site safety management plans and your organisation's procedures for site setup and safety.

An investigation should be done into what services are available on the site so that a thorough understanding of any potential hazards can be planned for. There will be a plan for where everything is going to be stored on the site and where all work is going to occur. There should also be an emergency evacuation location. All barricades should be either physical to stop people from entering the site and becoming injured or, if that is not required, then visual barricades may be satisfactory for the types of hazards that are on the site. All signage should be to the Australian Standard for signs and should inform people about any potential hazards on the site. Signage should be at a height that is easy for people to read, in the most common language for people who are likely to access the site and should be at the entrance to the site or at the location of the hazard that the sign is warning against.

 COMPLETE WORKSHEET 1

LEARNING TASK 8.4

CLASS/GROUP ACTIVITY: SITE SIGNAGE

Take a picture of a construction site that has site signage or barricades and explain why they are located where they are and why they are needed on that site. Use codes of practice or documentation from Work Safety regulators to justify your response.

2. Manually handle, sort, stack and store materials

2.1 Apply safe manual handling techniques to move carpentry materials to specified location

Manual handling

The oldest method of moving an object is simply to pick it up by hand and move it. Among the most common injuries in the construction industry are back-related injuries, including muscle strain, pulled or torn muscles, bulging discs, ruptured discs and dislocated or broken bones. Correct lifting technique may not stop all back injuries but it will reduce their likelihood and frequency.

For one person to correctly lift an object, follow the steps outlined in Figure 8.46.

If there are two people lifting an item, each person should follow the steps outlined in a single-person lift but with the addition that they either carry at the ends of the load or on the same side of the load.

Remember, if you don't need to manually lift something, don't. Drag it, push it or use a machine to do the task for you.

The techniques for safe manual handling of carpentry materials have changed a lot over the years. Materials used to be bagged in maximum of 20 kg weights; however, it was found that 20 kg is too much for some people to carry, and not enough for others.

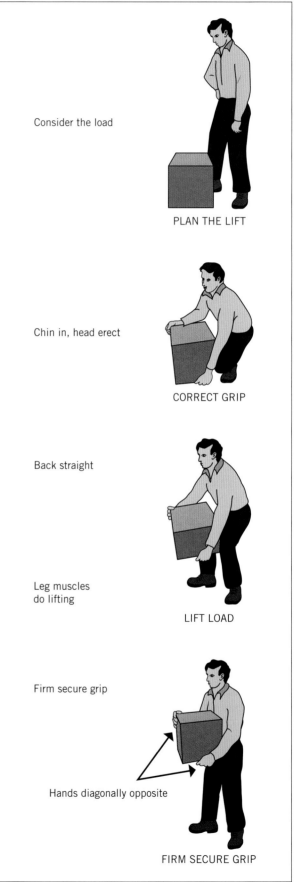

FIGURE 8.46 Correct manual lifting techniques

So the rule now is to have training in manual handling and to size up the load and plan your lift for every individual, so you don't sustain an injury. SafeWork NSW has a good code of practice on hazardous manual tasks where it is emphasised that training is important to avoid musculoskeletal injuries from manual work. It may be that all workers need general training, and then get specific training in tasks that pose greater risk. 'How to lift' training is a common approach to try and capture the hazards for every situation; however, training is now more closely aligned to the nature of the work to be done, what needs to be known about control measures, and how control measures can be implemented to manage the health and safety risks that occur from hazardous manual tasks. Australian codes of practice and compliance documents emphasise that training should cover manual task risk management and sources of risk, specific manual task risks and how to control them, how to perform manual tasks safely by using mechanical aids, tools or equipment, and how to report a problem.

The states and territories have their own laws and documents for manual tasks; however, they all aim to reduce the risks of injury occurring to workers. For example, the NSW Regulation 2017, clause 60 and 61 specify that a person conducting a business or undertaking must manage the risks of health and safety relating to musculoskeletal disorder associated with a hazardous manual task and in determining the control measures to implement, the person must consider all relevant matters that contribute to musculoskeletal disorder including:

- posture, movements, forces in vibration
- the duration and frequency of the hazardous task
- workplace environment conditions
- design and layout of the workplace
- systems of work use
- nature, size and weight of the task.

Manufacturers, designers and suppliers of plant and equipment must design, or supply products that eliminate, or minimise, the need for hazardous tasks and must provide information about how the product eliminates or minimises the need for any hazardous manual task.

The states and territories have codes of practice for manual handling and Safe Work Australia's code of practice is one example. It is helpful to understand the different types of forces that can occur on our bodies when we manually handle objects, and they fall under the following categories:

- forces:
  - repetitive force is doing the same thing repeatedly over a period of time
  - sustained force is a force applied continually over a period of time
  - high force is where increased muscle effort is required

  - sudden force can occur from unexpected or jerky movements when handling an item or load
- movement:
  - repetitive movement is using the same parts of the body to repeat similar movements over a period of time

Posture is ideal when your body is in a neutral position. Postures that are both awkward and sustained are hazardous. Beware of:

  - sustained posture is when the body is kept in the same position for a prolonged period of time
  - awkward posture is when any part of the body is in an uncomfortable or unnatural position
- Vibration:
  - whole body vibration can occur when vibration is transmitted through the whole body
  - hand-arm vibration occurs when vibration is transferred to the body using tools such as jack hammering, chain sawing, etc.

As with all hazards and risks, it is important to plan your work by considering the hierarchy of control measures. Seek to eliminate the risk, substitute the work for something that involves less risk, or isolate the hazard by designing engineering controls. A common strategy for manual handling is to use administrative controls where the task is only performed for short periods of time, and the last method of control PPE; for example, vibration-resistant gloves.

2.2 Sort carpentry materials to suit material type and size, and stack clear of access ways for ease of identification, retrieval, task sequence and task location

To sort carpentry materials according to their size and type, and to stack clear of access ways for ease of identification and retrieval for when and where they are needed requires some planning. All materials delivered to a site should have a delivery docket indicating the quantity and size of the materials, and material suppliers will have available information on how to store their material. All this information is important when you plan how you are going to work with the material onsite.

Consider the schedule for when the various materials are going to be used. All materials need to be easily accessible when needed. Consider the storage requirements and if the materials need protecting from the weather, can be stored directly on the ground or if they should be packed up off the ground. Can they be stacked against a solid object such as a fence or a site shed, or do they need to be stored indoors? Read the storage section of the material information sheets so you can plan where you're going to store the material and make sure the material is ready for use as required.

Generally, we categorise materials by type and size and store similar materials together because their storage requirements will be similar. Timber will need to be stored off the ground and kept straight. Sheet material will normally need protecting from the weather and be stored flat due to the risk of sheets falling on people if they're stored upright or warping. Metal may need to be stored off the ground and kept straight. Glass needs to be kept away from anything that might damage it. Any powder or liquid used on a site such as cement, lime or paints arc normally stored undercover. In summary, make sure you refer to the storage instructions for all the different materials you plan to have on site.

Make sure you have planned for dedicated open air and undercover storage areas which could include shelving and racks. You will normally need a lock-up weatherproof area for material that needs to be kept dry and secure. Ensure your storage areas are out of the way of most of the building work so you don't need to unnecessarily move materials part way through the job. Many construction businesses purchase an old shipping container or have site sheds that they transfer from one site to the next. Their storage spaces may incorporate room for lunch and office accommodation, or lunch facilities may be separate to material storage.

Always stack materials clear of access ways and paths of travel. One of the safety considerations for all sites is to avoid trip hazards and this is done by constantly watching where you walk and where other people walk and planning to ensure that no materials or tools are left in the path of travel. This could be as simple as moving materials or tools underneath saw stools or behind ladders or next to walls while you're doing your work. However, long-term storage of products and materials should be planned so those materials are stored well clear of where work is going to occur and away from pathways. Always stack materials securely so they don't shift and cause accidents. You may need to tie the materials in place.

As you order materials make sure you consider the task sequence and the location and try not to have more materials onsite than what you need for the work coming up. You may consider offsite storage of materials because sometimes purchasing materials when they are in a sale can save you more than the extra cost of storage, and this can be a way to make sure the materials are available when you need them. You should have a plan in your site office that's available to all workers indicating where the materials are stored, and it should be easy to see what materials are needed for upcoming jobs on your schedule.

2.3 Protect carpentry materials against physical and water damage

Protecting carpentry materials against physical and water damage is best done by following the material storage requirements listed on the material documentation. Some materials require greater

protection from the weather and may need to be kept indoors. This could include plasterboard, cement bags, lime, etc. Having knowledge of the expected weather in the area and being prepared is a good way to ensure your materials are not damaged.

Protecting materials from physical damage also involves ensuring they are out of paths of travel, away from machinery movements, and not likely to have things fall on them. This is especially true with brittle or fragile materials such as windows or tiles. You might need to consider purchasing some sheets of plywood and some wrapping to protect some products from damage.

Protecting material from water damage involves either wrapping materials in plastic, storing them inside waterproof containers, or sometimes just coating them with temporary waterproofing agents. In all cases, refer to the materials storage and handling documentation for the specific recommendations for each material.

COMPLETE WORKSHEET 2

LEARNING TASK 8.5

MATERIAL RESEARCH ASSIGNMENT 1

Prepare a folder with a cover sheet containing the following details:
- your name
- your class/group name or code
- name of the assignment ('Material Research Assignment')
- your teacher's name
- the due date.
  Provide some site pictures of materials that
DO COMPLY with the site storage instructions in the manufacturer's user documentation for that material or product and provide the manufacturer's documentation.

LEARNING TASK 8.6

MATERIAL RESEARCH ASSIGNMENT 2

Prepare a folder with a cover sheet containing the following details:
- your name
- your class/group name or code
- name of the assignment ('Material Research Assignment')
- your teacher's name
- the due date.
  Provide some site pictures of materials that **DO NOT COMPLY** with the site storage instructions in the manufacturer's user documentation for that material or product and provide the manufacturer's documentation.

3. Prepare for mechanical handling of materials

3.1 Stack and secure carpentry materials for mechanical handling in accordance with the type of material and equipment to be used

Information regarding stacking and securing carpentry materials has been covered previously in this chapter.

In all cases, you must refer to the storage and handling documents that have been produced for all materials, and any instructions for manual handling equipment. Whether you are manual handling with a forklift, a pallet trolley or a wheelbarrow, instructions for the different machines should be studied and followed. Table 8.6 lists some general guidelines for the main categories of carpentry materials.

TABLE 8.6 General guidelines for stacking and securing carpentry materials

Type of material	Equipment	Stacking and securing requirements
Lengths of timber	Manual stacking, strapping and then move using forklift, machine or pallet trolley	Stack the boards in similar lengths and sizes and strap the boards in as square a shape as you can get. Stack the boards up on blocks so the forklift or pallet trolley can easily lift the whole, strapped bundle. Ensure the pallet, bundle and strapping is properly secured and stable during stacking and moving. The main aim is to keep the lengths of timber as straight as possible.
Sheets of timber-based material	Manually moving one sheet at a time and then using a machine to lift multiple sheets once they are secure	Stack sheets flat on a level surface off the ground. Ensure they line up directly on top of each other and strap together when you have the required number of sheets. Use pallets or blocks to keep the sheets off the ground and to enable mechanical movement of the load. Cover with plastic if not stored indoors.
Sheets of fibre cement sheet material	Manually moving one sheet at a time and then using a machine to lift multiple sheets once they are secure	Stack sheets flat on level surface off the ground. Ensure they line up directly on top of each other and strap together when you have the required number of sheets. Use pallets or blocks to keep the sheets off the ground and to enable mechanical movement of the load. Cover with plastic if not stored indoors.
Sheets of plasterboard material	Manually moving one sheet at a time and then using a machine to lift multiple sheets once they are secure	Stack sheets flat on level surface off the ground. Ensure they line up directly on top of each other and strap together when you have the required number of sheets. Use pallets or blocks to keep the sheets off the ground and to enable mechanical movement of the load. Store indoors.
Bricks and tiles	Move using a wheelbarrow or trolley onsite and manually handling one or two at a time unless using a brick carrier which can carry five or six bricks at a time	Bricks and tiles should be stored on the pallet that they were delivered to site on until needed. Bricks will need to be moved much closer to the location of laying and this is normally only done immediately prior to laying the bricks. They can be set directly on the ground immediately prior to laying; however, they should be kept covered from rain so the brickwork doesn't sag as it is laid. Tiles should be kept covered because they are normally wrapped in cardboard and if it gets wet it will stick to the tiles and be hard to remove.
Metal	Metal includes steel beams, sheet roofing, etc. and is normally moved by hand unless it is too heavy, in which case a machine will be used onsite	Store metal off the ground as straight as possible and depending on the finish it may or may not need covering. Be very conscious of the plastic coating on sheets of zinc aluminium, gutters and fascia and how the sun will deteriorate that sheeting over time and make it very hard to remove. This protective plastic coating is quite soft and easy to remove provided it hasn't been exposed to the sun for too long.
Plastics	Normally moved by hand	Store flat and out of the weather. Stack and secure plastics so they don't roll or fall on anyone. Plastics could be pipe, boxes or other odd shapes and can be quite awkward to store.
Powder or liquid materials	Powder materials are usually in packaging and will be moved one pack at a time unless they're all stacked on a pallet, in which case a machine could move them. Liquids are in containers and can be moved one container at a time unless they are on pallets, in which case a machine would move them.	Store all powder out of the weather because moisture could activate the powder and render it unusable. If it's on a pallet it can be covered with plastic or it may be moved indoors. Liquid materials are in containers and should be fairly weatherproof for relatively short periods of time. If they are on a pallet they can be moved undercover. Containers of liquid can be very heavy and awkward so it's important that you plan your lift and do some research on how to lift and move these products safely to avoid musculoskeletal injuries.

3.2 Unload, move or locate carpentry materials at specified location

As indicated above, materials need to be moved to and from the construction site as well as moved around the site. There are many mechanical aids to assist you, including cranes, forklifts, pallet trolleys, wheelbarrows and brick barrows, just to name a few. Alternatively, material can be moved by hand (manual handling).

Cranes

The most used item to move large components on a job site is a crane, which is used to move packs of wall frames and trusses, individual trusses into place, large structural steel components and packs of floor and roof sheets (Figure 8.47).

FIGURE 8.47 Crane lifting trusses

Forklifts

Forklifts can be divided into two categories, all-terrain and hard-surface trucks. It is important that an all-terrain forklift is used for the delivery of bricks onto all building sites as these machines are designed for this purpose and if used correctly they are less likely to roll over (Figure 8.48). Hard-surface trucks have smaller wheels and are designed to work on level or near-level concrete or pavement; for example, factory units loading material onto a truck.

FIGURE 8.48 All-terrain forklift trucks

Pallet trolleys

Pallet trolleys and platform trolleys (Figures 8.49 and 8.50) are used to move heavy loads around level surfaces safely. The pallet trolley has a hydraulic lifting mechanism that allows it to slide under timber pallets or raised loads. It has a lifting and carrying capacity of 1800 kg. Platform trolleys are fitted with four castors for easy steering and are used to transport items like bags of cement, drums or boxes to a capacity of 500 kg.

FIGURE 8.49 Rubber-tyred pallet trolley with hydraulic lift

FIGURE 8.50 Metal-framed trolley with rubber-tyred wheels

Safety and maintenance of trolleys

- All moving parts and wheels should be oiled regularly.
- Hydraulic lift equipment should be checked for leaks on a regular basis.
- Do not use these trolleys on uneven floors or unpaved ground.

 COMPLETE WORKSHEET 3

4. Check and store tools and equipment

4.1 Check tools and equipment and report any faults

A competent person may be so familiar with their tools and equipment that they can tell quite quickly and easily whether there are any faults or if any maintenance is required. However, the first time you obtain tools and equipment and check them for faults you should use the documentation that came with that tool or piece of equipment to ensure that you are working according to the manufacturer's design. All documentation for tools and equipment is available freely online from the supplier.

To report any faults, you should use your organisation's fault reporting procedures, or if there are no procedures, follow the product's guidelines and ensure the machine is not used while it has faults. This may involve tagging the machine to make sure no one else uses it. The size of your workplace will determine the procedures for reporting faults and ensuring no one uses a machine until the faults are fixed. If it's just you and the boss, you may have a box where you put tools that need maintaining or fixing. If it's a larger organisation you may have a supply of lockout tags to indicate what is wrong and who is reporting that tool for repair.

4.2 Store tools and equipment in accordance with workplace requirements

Storing tools and equipment in accordance with your workplace requirements may involve grouping similar tools together in drawers, shelves or spaces designed for that tool. If you can, use the proper tool storage that came with the tool when it was purchased. Store tools in such a way that they don't become damaged in transit. Protect sharp parts of tools so they don't become blunt and don't injure anyone or damage other tools. Tool storage is a continual process of improvement as tools change and as you obtain more tools. Always clean and inspect tools before you store them and make sure that any defects are repaired or maintenance is done so you are ready to use the tool next time you need it.

Any tool that involves mains electricity will need to be tested and tagged according to your state or territory's requirements and a record of this test and tag will need to be kept with your tools. If you work for a large organisation you may need to label the different locations of the tools so workers who go to retrieve tools do not waste time searching. Ensure the tool storage area is secure from theft or from people injuring themselves by accidentally encountering something sharp. Any hazardous or dangerous tools need to be stored according to safety regulations. An explosive power tool containing gas canisters may need to be stored in a locked storage container. Keep flammable chemical materials and chemicals and fuels stored in appropriate containers and cabinets following the safety guidelines for those products. You can obtain this information from the SDS for the various products.

Keeping an inventory of regular maintenance and periodically inspecting your tools will keep them in good working order. Using tools is how you efficiently perform your work so it's important to keep them in as good working order as you can and always available for use.

 COMPLETE WORKSHEET 4

SUMMARY

In Chapter 8 you have learnt how to safely manually handle and store carpentry materials, meeting all relevant requirements of work health and safety (WHS) and Commonwealth and state or territory legislation.

You have also learnt how to prepare material for mechanical handling and how to apply environmental management principles associated with carpentry materials. We have learnt how to:

- plan and prepare

- manually handle, sort, stack and store materials
- prepare for mechanical handling of materials
- check and store tools and equipment.

Additionally, you have learnt about applying the skills to handle carpentry materials for three different carpentry tasks, including safely handling and sorting and stacking lengths of timber, sheet material, for manual and mechanical handling.

REFERENCES AND FURTHER READING

Acknowledgement

Reproduction of the following Resource List references from DET, TAFE NSW C&T Division (Karl Dunkel, Program Manager, Housing and Furniture) and the Product Advisory Committee is acknowledged and appreciated.

Texts

Bootle, Keith R. (1983), *Wood in Australia*, McGraw-Hill, Sydney

Cement and Concrete Association of Australia (1978), *Basic guide to concrete construction*, Cement and Concrete Association of Australia, Sydney

Construction and Transport Division TAFE NSW (1999), *Building materials and hand tools* (carp 02), Construction and Transport Division TAFE NSW, Sydney

Web-based resources

Resource Tools and VET Links

http://training.gov.au, Skills training

Visit the Australian Glass and Window Association website and search for an industry guide to the installation of residential windows and doors and note the storage and handling recommendations

Visit the Cement Concrete and Aggregates Australia website and search for information about the storage and handling of cement and concrete products

Visit these websites for more detailed information on various products:

- Wood Solutions of Australia
- Timber Development Association of Australia
- Australia Forest Products Association
- Timber Preserves Association of Australia. For example, **https://www.tpaa.com.au/faq**

and search for information on storage and handling of their products. Take note about how to handle preserved products.

Industry organisation sites

ccaa.com.au, Cement Concrete & Aggregates Australia

cement.org.au, Cement Industry Federation

csr.com.au, CSR

australbricks.com.au, Austral Bricks

boral.com.au, Boral

jameshardie.com.au, James Hardie

onesteel.com, OneSteel

smorgonsteel.com.au, Smorgon Steel

safeworkaustralia.gov.au, Hazardous Manual Tasks Code of Practice, October 2018

 Relevant Australian Standards

AS 1478.1 Chemical admixtures for concrete, mortar and grout – Admixtures for concrete
AS 1684.2 Residential timber-framed construction – Non-cyclonic areas
AS 3972 Portland and blended cements

GET IT RIGHT

In the photo below, materials on the job site are stored.

Identify how the materials could be stored better and provide reasoning for your answer.

Source: Richard Moran

_____

_____

_____

_____

_____

_____

 **WORKSHEET 1**

Student name: _____

Enrolment year: _____

Class code: _____

Competency name/Number: _____

Task: Read through the section beginning at *Plan and prepare* then complete the following questions.

1   Name the two (2) factors that affect the selection of the material used in a building and provide an example of each.

    1 _____

    2 _____

2   State the four (4) main chemical components of timber.

    1 _____

    2 _____

    3 _____

    4 _____

3   Circle the correct response to the following statement: 'A tree increases in size by adding growth rings and a new layer at the very tips of branches. This means that the tree grows horizontally, but gives the impression of growing vertically.'

    TRUE   FALSE

4   Name three (3) types of hardwood.

    _____

5   Name three (3) types of softwood.

    _____

6   Timber and timber products may be used for a variety of purposes. List all the uses you can think of:

    _____

    _____

    _____

7   State the name of the chemical used in the process of treating timbers that have a poor durability, to improve their durability rating and which turns the timber a typical green colour:

    _____

8   Complete the following sentence. The durability of a timber relates to _____

    _____

9 Timber is usually classed into groups ranging from good to poor, with Class 1 being the best and Class 4 being the worst. Identify the following timbers by durability class:

Spotted gum   Class _____

Douglas fir   Class _____

Tallowwood   Class _____

Western red cedar   Class _____

Radiata pine   Class _____

Grey gum   Class _____

Brush box   Class _____

Blackbutt   Class _____

White cypress pine   Class _____

Tasmanian oak   Class _____

10 Complete the statement: 'Timber is strongest along the straight length of its _____.'

11 All timbers will eventually burn in a fire, but some timbers have a natural resistance to fire. Name three (3) of these timbers.

1 _____

2 _____

3 _____

12 What may cause blunting of saws and other cutting tools when cutting through timber?

_____

13 Describe how timber should be stored on the job site.

_____

_____

_____

14 List six (6) products that are engineered timber products.

1 _____

2 _____

3 _____

4 _____

5 _____

6 _____

15 List three (3) uses for engineered timber products.

1 _____

2 _____

3 _____

16 What is the easiest and safest way to move a pack of plywood?

_____

_____

_____

17 How should engineered timber products be stored?

_____

_____

_____

18 What are the environmental benefits of using engineered timber products?

_____

_____

19 Complete the following statement:

Adobe bricks are made from _____

_____

20 List the five (5) classes of brick (the first being adobe).

1 adobe _____

2 _____

3 _____

4 _____

5 _____

21 What is mortar used for?

_____

_____

_____

22 What are the common hand tools used to cut masonry materials?

_____

_____

23 What item of PPE should be worn when manually handling masonry?

_____

24 Why is it important to not place bricks or blocks directly onto wet clay prior to laying?

_____

25 Are bricks and blocks able to be reused and recycled? (Circle the correct answer)

YES   NO

26 Complete the following statement:

Concrete is very strong in _____ but is very weak in

_____.

27 Concrete is made by mixing four materials together. Name the four (4) materials used to make concrete.

1 _____

2 _____

3 _____

4 _____

28 The water used for making concrete should be free of problem-causing substances such as:

1 _____

2 _____

3 _____

4 _____

5 _____

29 Name the two (2) categories of concrete reinforcement.

1 _____

2 _____

30 What will happen to steel reinforcement that is exposed to moisture and air?

_____

31 What is the technical term used when reinforcement is completed and buried in the concrete to an engineer-specified depth?

_____

32 Describe how bags of concrete should be stored on the job site.

_____

_____

33 Name the two (2) forms of paints for building purposes.

1 _____

2 _____

34　Each of the two forms of paint use different liquids for thinning and washing up.
Name the two (2) liquids.

1　_____

2　_____

35　What are fillers?

_____

36　Name the three (3) different types of sealants mentioned in the text.

1　_____

2　_____

3　_____

37　Is it possible to have a sealant that is fire rated? (Circle the correct answer)

TRUE　　　　　　FALSE

38　What size containers is paint supplied in? (Circle the sizes that are *not correct*.)

1 kg　　　　2 L　　　　4 kg　　　　6 L　　　　10 L　　　　15 kg　　　　20 L

39　Name three (3) of the six (6) different types of steel.

1　_____

2　_____

3　_____

40　Name the two (2) forms of PPE that should be worn when cutting steel with an angle grinder.

1　_____

2　_____

41　List the six (6) tools that can be used to cut steel.

1　_____

2　_____

3　_____

4　_____

5　_____

6　_____

42　Complete the sentence:

Most heavy steel items should be lifted using a _____.

43 Complete the sentence:

Apart from structural and security benefits of the building enclosure, the

_____ and _____ properties of the materials

used must be considered.

44 How many people should lift a long length of plasterboard (e.g. 6.000 metre length)?

_____

45 Complete the sentences:

Plasterboard will provide fire-resistance ratings from _____ to

_____ hours for walls, and up to _____ hours

for ceilings. Double-sheeting will _____ the ratings substantially.

 **WORKSHEET 2**

Student name: _____

Enrolment year: _____

Class code: _____

Competency name/Number: _____

To be completed by teachers	
Student competent	☐
Student not yet competent	☐

Task: Read through the sections beginning at *Manually handle, sort, stack and store materials* then complete the following questions.

1 What is one of the most common injuries in the construction industry?

_____

2 Should you always lift things? What are some other manual alternatives?

_____

_____

3 When planning and preparing to do manual lifting, what advice or training should you research from the safe work regulator within your state or territory?

_____

4 What are some of the hazards that must be planned for when considering musculoskeletal disorders caused by manual handling?

_____

_____

5 When manual handling, what is one of the first things you need to check when planning how to move something from one location to another?

_____

_____

6 Why might it be important to store some things off the ground?

_____

_____

7 What important document should you use at the planning stage when determining manual handling and storage plans for your site?

_____

_____

8 Is it ever a good idea to wrap materials in plastic when they are stored outdoors?
(Circle the correct answer)

YES   NO

9   Should all materials be coated with waterproofing agents as soon as they arrive onsite?
     (Circle the correct answer)

     YES   NO

10   Is it okay to store unopened bags of cement on a pallet outdoors? (Circle the correct answer)

     YES   NO

11   Complete the sentence:

     Remember if you don't need to manually lift something, _____.

     _____ it, _____ it or use

     a _____ to do the task for you.

12   List the basic steps for lifting something correctly (for one person).

     1   _____

     2   _____

     3   _____

     4   _____

WORKSHEET 3

Student name: _____

Enrolment year: _____

Class code: _____

Competency name/Number: _____

Task: Read through the sections beginning at *Prepare for mechanical handling of materials* then complete the following questions.

1   Describe how lengths of timber should be stored on the job site.

_____

_____

2   List three (3) tips for stacking sheet material onsite.

1   _____

2   _____

3   _____

3   Is it okay to store sheets of plasterboard leaning up against a fence in the open? (Circle the correct answer)

     YES   NO

4   Is it ever okay to store bricks directly on the ground? (Circle the correct answer)

     YES   NO

5   Is it okay to leave the plastic coating on some metals on for long periods of time when storing outdoors? (Circle the correct answer)

     YES   NO

6   Complete the sentence. Forklifts can be divided into two categories:

     _____ and _____ trucks.

7   List three (3) safety and maintenance items for trolleys.

1   _____

2   _____

3   _____

8   List five (5) mechanical aids that may be used to move materials around the building site.

1 _____

2 _____

3 _____

4 _____

5 _____

9   How should plasterboard be stored?

_____

_____

_____

WORKSHEET 4

Student name: _____

Enrolment year: _____

Class code: _____

Competency name/Number: _____

Task: Read through the sections beginning at *Check and store tools and equipment* then complete the following questions.

1 How can you determine whether there any faults or if any maintenance needs to be done on tools and equipment?

_____

_____

2 Where can you get documents for a tool or piece of equipment if you no longer have the original paperwork from when you bought the tool?

_____

3 What should you do if you find that a machine has a fault and shouldn't be used?

_____

_____

4 How do you know what is the correct method of storing tools and equipment?

_____

5 Do all tools need to be tested and tagged?

_____

_____

6 Are there some tools that might need to be kept in a locked storage container, and if so, which ones?

_____

PERFORMANCE OF SKILLS EVIDENCE

Student name: _____

Enrolment year: _____

Class code: _____

Competency name/Number: _____

Task: Working with your teacher, prepare to demonstrate your skills by completing the following task. To demonstrate competency, a candidate must satisfy all the elements, performance criteria and foundation skills of this unit by handling carpentry materials for three different carpentry tasks, including:

1   Safely handling, sorting and stacking:

   i   varying lengths of timber or similarly-proportioned materials onto an Australian standard pallet, a minimum of 0.5 cubic metres, secured ready for mechanical handling

   ii   different sizes and types of sheet material onto an Australian standard pallet, a minimum of 0.1 cubic metres, secured ready for mechanical handling.

2   Preparing the following for mechanical handling:

   i   varying lengths of timber or similarly-proportioned materials

   ii   different sizes and types of sheet material.

9 USE CARPENTRY TOOLS AND EQUIPMENT

This chapter covers the following topics from the competency CPCCCA2002 Use carpentry tools and equipment:

- plan and prepare
- select, check and use tools and equipment
- clean up.

This unit of competency specifies the skills and knowledge required to select and safely use and maintain carpentry tools and equipment, meeting all relevant requirements of the National Construction Code, Australian Standards, work health and safety (WHS), and Commonwealth and state or territory legislation.

The unit includes hand tools, power tools, pneumatic tools and equipment.

1. Plan and prepare

Hand tools have long been used, and are still used, to form functional items from raw materials, with a fairly high degree of accuracy. Hand tools are used for simple tasks and where power tools would be of no advantage. Correctly used hand tools show the ability and skill of the person using them. Hand tools have an important role in setting and marking out the work to be undertaken.

Since the Industrial Revolution (from the late 1700s), machinery or powered tools have gradually taken the lead in making tasks quicker, easier, cheaper and more accurate than ever before. Modern technology has given us tools and machinery that have made our lives and work easier, with the ability to create structures only dreamed of in times gone by.

The modern construction worker makes use of many electrical, pneumatic, fuel cell and battery-operated tools on projects that range from simple residential work to impressive landmark structures seen in most capital cities.

Although power tools are generally considered by many as superior to hand tools, it should not be forgotten that safe use, regular maintenance and an appreciation of function comes from an understanding of using hand tools properly.

1.1 Review work instructions to use tools and equipment

Before starting any work, it is important to obtain the work instructions to check which tools and equipment are best suited to performing that work. All tools and equipment come with user manuals when purchased (and can still be accessed on the manufacturer's website) and these manuals will outline what the tool is used for. This information can be matched up to the job that you want to perform, and you can ask a competent person to assist you with deciding which tool is best suited for a job. The following are some general points that might help you when reviewing work instructions before using tools and equipment:

■ Read the work instructions carefully and take time to work out exactly what needs to be done for the job and then identify the tools and equipment that are best suited for completing the job.

■ Check the tool and make sure it is in good working order using the manufacturer's instructions as a guide.

■ Make sure all safety equipment is in good working order for the tool and that you have the correct personal protective equipment (PPE) required to use that tool. Ensure you understand the steps involved in the work instructions to complete the task and try to anticipate any potential problems or variations you might come across as you do the work.

■ Seek clarification on any of the work instructions or the use of any of the tools and equipment if needed.

Prior to using any tool on the job – whether it be a hand tool, **power tool** or **pneumatic tool** – it is important to prepare and organise the task to be done. There are five things to consider before starting to use the machine or tool:

1 Is the work area safe, clean and tidy, with adequate lighting and ventilation?
2 Has the most appropriate tool been selected to do the task?
3 Is the tool ready for use, with the appropriate cutter or blade?
4 Is the tool safe to use?
5 Is the person using the tool familiar with the tool and able to operate it safely?

1.2 Plan all work to meet all relevant industry compliance requirements

When planning all work to comply with laws and regulations, the National Construction Codes (NCC), Australian Standards, work health and safety and environmental requirements, manufacturer's specifications and workplace requirements, ensure you are aware of those requirements and where to find them as they relate to the various jobs you're about to do. Complying with these requirements helps ensure the work is performed in a safe, efficient and environmentally friendly manner while minimising risk. We do this by identifying any requirements and incorporating them into our work plan and then communicating that to all relevant workers and reviewing the process for the entire job. By doing this we can ensure the work is carried out to a high standard with minimal risk while maintaining efficiency and meeting any compliance requirements.

The laws and regulations that apply to carpentry tools and equipment will be listed on your state or territory's safety regulator website (see Table 1.2 in Chapter 1). Current state and territory WHS Acts and regulations can also be accessed from the Safe Work Australia website via Law and Regulation links on the site https://www.safeworkaustralia.gov.au/law-and-regulation/whs-regulators-and-workers-compensation-authorities-contact-information

The Safe Work Australia website also has the model work health and safety laws as well as regulations and codes. For example, you can view the model *Work Health and Safety Act 2011*, the model Work Health and Safety Regulation 2011, and at last count there were over 20 model codes of practice for various work activities on various worksites.

The NCC concerns construction and building in Australia and sets out the minimum requirements, including for health and safety. The various Australian Standards provide information about how things should be designed and this relates to tools and equipment as well. The manufacturers of tools and equipment must manufacture these things to Australian Standards for safety, reliability and function.

Codes of practice are invaluable for guiding you on how to use tools and equipment safely and there are codes of practice for all the main work tasks carried out on a construction site. For example, there are codes of practice for construction work, demolition work, excavation work, managing electrical risks in the workplace and managing the risks of falls in housing construction. These codes of practice can get very specific and it's important to refer to them before you start your task to make sure you have all the correct information to do your job safely.

Your workplace will have its own requirements for how to perform tasks safely when using tools and equipment and you should be aware of these. Your workplace requirements should always be at least as safe as the industry compliance requirements. For example, if eye protection is specified when using a drop saw then, even if your workplace doesn't want you to use eye protection, you should wear it because the legal compliance requirement is for you to use it as noted in the instructions for that tool.

LEARNING TASK 9.1 GRINDING AND HONING

Using sharp tools for the job is critical to ensure an accurate, neat and professional finish. All tools need some maintenance; however, tools with cutting edges require special attention to ensure efficient use.

Ask your trainer or employer to demonstrate how to correctly grind and hone various tools that have cutting edges.

Now practise the correct techniques yourself to:
- grind and hone wood chisels
- grind and hone plane blades.

1.3 Select and use personal protective equipment (PPE) for each part of the task

The use of PPE can be very effective at controlling risks to health and safety if no other means of protection are practical or reasonable. It should be used as a last resort or as an interim measure while awaiting a more effective measure of protection or as a backup to a higher level of protection. When PPE is being used to minimise a risk to health and safety it needs to be provided by the person conducting business or undertaking (PCBU) for both workers and visitors to a site. The PPE should be selected based on the nature of the work, paying particular attention to any codes of practice for the task and the manufacturer's user manual for the tool.

A worker must use PPE in accordance with the instructions that came with that PPE or any other training by a competent person. A worker must not intentionally misuse or damage PPE. The PPE must be comfortable and must fit properly. The PPE must be chosen to be compatible with any other PPE that might be required; for example, earmuffs, hard hat and eye protection my need to be purchased to all fit together. Some examples of PPE that should be provided to workers includes head protection, foot protection, eye protection, gloves, sun protection, hearing protection and respiratory protection.

1.4 Inspect work site, locate services, assess hazards and apply risk controls, including required signage and barricades

Before you start using tools and equipment on a construction site make sure that you're thoroughly familiar with that site. Find out where the services are, such as electricity, water, gas and communications. Assess any hazards that you may encounter when you're using the tools and equipment on the site and plan your risk controls and then apply them. Your risk controls normally involve signage and barricades, and these should be constructed and positioned according to codes of practice, Australian Standards and a competent person's advice.

While inspecting the site look for potential hazards such as unstable ground, exposed power sources, underground pipes or hazards, trip hazards, environmental hazards such as weather, and anything else that may cause harm while you're using tools and equipment.

When locating services, you may need an expert with technical equipment to locate the services and then mark them on the ground. You may find it helpful to get old plans from previous work that might have been done in the area, or you may look in areas such as the meter box to see if there are any sketches of where services have been positioned on the site.

Assessing hazards on the site requires a competent person with access to hazard information, such as codes of practice, to make a list of potential hazards so they can be planned for. If the site is going to have moving plant working while you're using tools and equipment; for example, concreting, then the code of practice for working on a construction site with moving plant would be a good resource to identify the possible hazards and plan the controls to stop those hazards injuring anyone.

The controls for the hazards that you have identified may include training of workers, using PPE, setting up barricades or signs, setting up administration safeguards; for example, traffic direction and flow rules. In the case of all controls, it's important to ensure that all workers and visitors to the site are informed of the hazards and are provided with all the safety equipment they need to do their task on the site safely.

You need to review and evaluate all the steps you've put in place to control any hazards on the site and be alert for any new hazards as the work on the site progresses.

Safety should always be your top priority on a construction site and regular reviews of your actions and what you have in place is important so you can identify and control anything that may cause injury.

GREEN TIP

Some tools and machinery have the potential to create hazardous dust which can harm the environment. Make sure you are aware of information in a tool's user manual and the SDS so you don't create any hazards.

 COMPLETE WORKSHEET 1

2. Select, check and use tools and equipment

2.1 Select equipment and hand, power and pneumatic tools for the carpentry task, identify their functions and operations, check for serviceability and report any faults

Upon identifying a tool, you will then need to learn what it is designed for, its correct use and any important safety aspects. The following examples outline some common tools that you are likely to use in the construction industry.

Hand tools

Hand tools are used for different tasks and with different materials. For example, a carpenter uses a claw hammer to hit nails into timber, whereas a bricklayer would use a lump hammer and bolster to break bricks in half. To simplify the hand tools available, we will discuss them in groups associated with the material or tasks they perform.

Woodworking tools

The tools listed below are generally used by a carpenter on a construction site.

Squares

There are several different kinds of squares and each kind has its own use but all are equally useful when called upon.

- A combination square is the most versatile type of square. It has a sliding blade and is used for marking 45 and 90 degrees; it can also be used to mark parallel lines from an edge (Figure 9.1).
- A roofing square has a blade 600 mm and a tongue 400 mm, primarily used with the aid of buttons or fence for roofing and stairs (Figure 9.2).
- A quick/pocket square is designed to provide a quick means for development of various cuts on roofing members (Figure 9.3).

FIGURE 9.1 Combination square

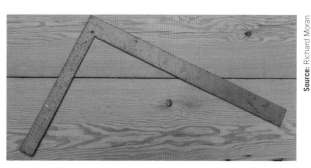

FIGURE 9.2 Roofing square

FIGURE 9.3 Quick/pocket square

- A try square is used for measuring 45 and 90 degree angles (Figure 9.4). It consists of a metal blade fixed at a right angle.

Hammers

There are a variety of shapes and sizes of hammers available to suit specific tasks or operations.

- Solid timber mallets are used in joinery work for hitting or striking chisel handles (Figure 9.5); timber-handled chisels might be damaged if a metal hammer is used on them.
- Claw hammers are used to drive and/or extract nails (Figure 9.6). They have a metal head which is hardened and tempered, and a timber, steel or fibreglass handle that is usually fitted with a leather or rubber grip. It is necessary to wear eye protection when using a claw hammer due to the risk of small pieces of metal flying off nails during the driving process.

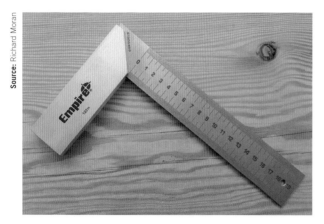

FIGURE 9.4 Try square

FIGURE 9.5 Timber mallet: Mallet head may be made of brush box and the handle of spotted gum

FIGURE 9.6 Claw hammer: head sizes available from 225 g to 910 g

Plasterboard hammers

Plasterboard hammers differ from carpenter's hammers as the head is domed to avoid damaging the plasterboard and may have a hatchet-styled end for rapid trimming of plasterboard (**Figure 9.7**).

- Warrington or cross pein hammers have a small driving face with a wedge-shaped top used to start small nails, like panel pins, held between the fingers (**Figure 9.8**).
- A lump hammer, also known as a club or mash hammer, is used with a bolster to cut bricks and stone (**Figure 9.9**). The handle, which may be made of timber, fibreglass or steel, is forged with the head.

FIGURE 9.7 Plasterboard hammers

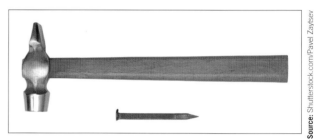

FIGURE 9.8 Small Warrington hammer for light nailing

FIGURE 9.9 Small heavy hammer for masonry work (up to 1.36 kg)

- Sledge hammers have a double-faced steel head with a long handle (**Figure 9.10**). They are used for demolition work or for driving pegs, and may have a timber, steel or fibreglass handle.

FIGURE 9.10 A long handle makes the sledge hammer easier to swing (available in 1.8 kg to 12.7 kg)

Nail punches

Nail punches are used to drive nails below the timber surface, to allow putty to fill the hole and give a smooth finish with no visible nail showing. Available in a variety of sizes and lengths (**Figure 9.11**).

Source: Richard Moran

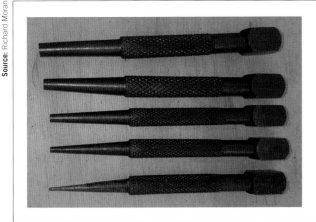
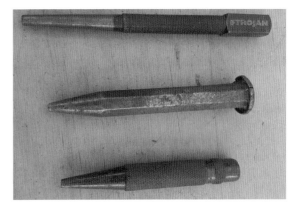

FIGURE 9.11 Nail punches and floor punches

Wood chisels

Wood chisels have a sharpened steel end or blade with a timber or hardened plastic handle designed to be struck with a mallet or hammer (Figure 9.12). Chisels are used to remove sections of timber to produce joints, sockets or holes.

Source: Ed Hawkins

FIGURE 9.12 A selection of chisels that may be used to cut into timber

Check the chisel's condition and ensure there are no nicks or significant damage. If there are deep nicks or chips, consider using a coarse diamond stone or a grinder to remove the damaged metal before proceeding with sharpening.

The angle on the edge should be 25–30 degrees. This is achieved using an oil or diamond stone. Use light pressure and long, smooth strokes across the stone, covering the entire bevel. Adjust your technique if you notice uneven wear.

Once a consistent bevel is achieved on the coarse stone, switch to a fine grit stone. Repeat the same process, maintaining the angle, but using lighter pressure. This step removes the scratches from the coarse stone and refines the edge.

After sharpening and honing, carefully test the chisel's sharpness on a piece of timber.

Hand planes

Hand planes are tools that are either pushed or pulled along the surface of a piece of timber to scrape away or remove timber to a set line or intended surface. There are many planes including jointer or try planes, jack planes, smoother planes and block planes, just to name a few (Figure 9.13).

Source: Ed Hawkins

FIGURE 9.13 (a) Jointer or try plane, (b) jack plane, (c) smoother plane and (d) block plane

Some timber planes may have removable blades, while others might need the entire assembly to be sharpened. If your plane has a removable blade, carefully remove it before proceeding.

Most timber planes have a bevel angle of around 25–30 degrees. You'll want to maintain this angle while sharpening. Try to match the existing angle on the blade. A honing guide can be helpful to maintain a consistent angle while you sharpen. Attach the blade to the honing guide and tighten it securely.

Place the sharpening stones on a flat surface, such as a workbench. If using water stones, soak them in water for the recommended time (usually 5–10 minutes) before using. If using oil stones, apply a few drops of honing oil to the stone's surface.

Hold the plane blade at the desired bevel angle and gently move it back and forth across the stone. Apply even pressure and cover the entire bevel.

Adjust the grit of the stone from course to fine as the blade sharpens. After sharpening on the fine grit stone, test the sharpness of the blade by trying to shave off a thin piece of timber.

Bench grinders

Bench grinders have a variety of purposes including to grind plane blades, chisels and sharpen drill bits, when grinding at the correct angle 25–30 degrees. When grinding, the steel becomes hot and needs to be cooled with water. Once the correct angle has been achieved, the cutting edge needs to be honed at 30–35 degrees on a stone (Figure 9.14).

Stones are used to hone the cutting edge to produce a fine sharp edge. These include oil stones and diamond stones.

FIGURE 9.14 Bench grinder with guide

Oil stones

Oil stones (Figure 9.15) are divided into two main groups:

- Natural stones – these are imported with 'Norton Washita' being the most common. It is white in colour and is very durable.
- Artificial stones – these have two main components:
  (a) abrasive, which does the actual cutting and
  (b) bond, which supports the abrasive grains while they cut. They are made from aluminium oxide and silicon carbide.

FIGURE 9.15 Oil stones

Diamond stones

Diamond stones are manufactured from industrial grade diamonds which are not useful for jewellery. It is bonded to a nickel backing to provide a long-lasting flat surface (Figure 9.16).

FIGURE 9.16 Diamond stones

Hand saws

Hand saws are tools with a spring-steel blade with teeth set along the bottom edge. There is a timber or plastic handle set on one end that allows the user to push and then pull the blade across the timber, producing a clean cut. There is a range of different hand saws available including the rip saw, crosscut saw, panel saw, tenon saw and pad saw (Figure 9.17).

Pinch or wrecking bars

Pinch or wrecking bars have a flat chisel-like end for levering and a straight shank with a hooked end claw for pulling nails (Figure 9.18). They are used as a lever for prising framework apart, lifting heavy objects into place, de-nailing timber and general demolition work. The larger bars are around 800 mm long and the smaller bars, known as *jemmy bars*, are around 300 mm long.

Pincers

Pincers have a pair of circular jaws for holding and removing nails or fasteners, and can have a small claw for levering and extracting tacks on the end of one handle (Figure 9.19). They are very handy for grabbing and holding during fixing operations.

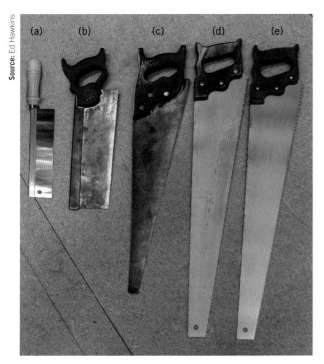

FIGURE 9.17 (a) pad saw, (b) tenon saw, (c) panel saw, (d) crosscut saw and (e) rip saw

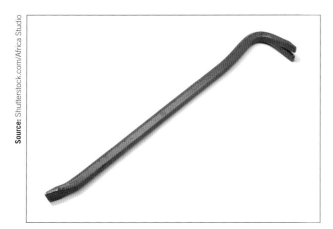

FIGURE 9.18 Round or hexagonal steel shank with specialist ends

FIGURE 9.19 Pincers – a useful tool for extracting fasteners

Tapes and rules

For information regarding tapes and rules refer to Chapter 6.

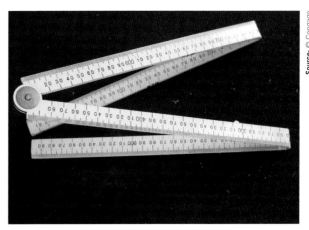

FIGURE 9.20 Rule folds out to a straight 1.0 m length

Marking gauge

This tool is used to impress or mark a line into a piece of timber. It is used by setting the pin the required distance from the block. The pin is then drawn along the timber, marking a line parallel to the selected face or edge of the timber (Figure 9.21).

Utility or safety trimming knife (Stanley knife)

A utility knife is often used for sharpening pencils as well as for accurate marking out; a pencil may leave a ½–1 mm line, whereas a knife blade leaves a much finer cut into the timber. A utility knife blade must be able to be folded or retracted when not in use and the knife should be stored in a secure place when not in use (Figure 9.22).

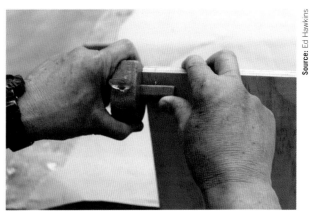

FIGURE 9.21 Marking gauge being used

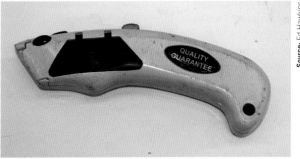

FIGURE 9.22 Utility knife

Note: do not carry this tool off-site on your person as it is considered a weapon in many states and territories.

Planks

Planks may be used with ladders and scaffolding and are made of solid timber (**Figure 9.23**) or boxed aluminium (**Figure 9.24**). Timber planks are a minimum size of 225 × 38 mm (Oregon) and usually up to 3.6 m in length (they may be 32 mm thick in hardwood). Aluminium planks are 225 × 50 mm and are available in standard lengths of 3.0, 4.0, 5.0 and 6.0 m; they can also be custom-made to length.

FIGURE 9.23 Typical timber plank

FIGURE 9.24 Typical aluminium plank

Saw stools

Saw stools may be made onsite from timber or purchased pre-made with pressed metal legs and a timber head (**Figure 9.25**). They are also available in folding form for easy transport and storage.

Saw stools may be used to create a bench by placing one or two planks on top to cut timber and other materials.

Clamps

Clamps are available in a range of types and sizes, from the traditional heavy duty 'G' clamps, 'F' clamps to quick action clamps, long sash clamps and the very small and lightweight spring clamps (**Figure 9.26**).

FIGURE 9.25 Typical timber saw stool

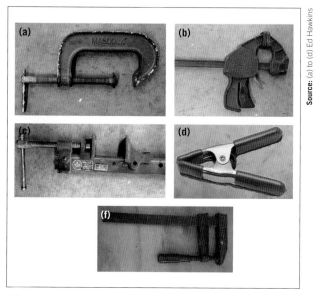

FIGURE 9.26 (a) G clamp, (b) quick action clamp, (c) sash clamp, (d) spring clamp, (e) F clamp

Clamps are used to hold the material being worked, allowing the operator of the hand or power tool to use both hands.

Metalworking tools

The tools listed below are generally used on construction sites that involve working with metals.

Spanners

Spanners are used to tighten or loosen nuts, bolts and hex head screws. They are available in several styles, materials and sizes, including imperial and metric. Styles include single open-end, double open-end, podger (for scaffolding, rigging and formwork centre adjustments), double-ring end, ring and open-end, and a variety of shaft shapes ranging from straight to almost semicircular. They may be made from drop-forged steel or chrome vanadium steel and are available in a host of socket sizes and types. Adjustable or shifting spanners are also very common. Spanners are useful to carpenters for fitting and adjusting surface and/or cavity sliding doors (**Figures 9.27** to **9.34**).

FIGURE 9.27 Drop-forged single open-end spanner

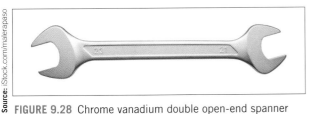

FIGURE 9.28 Chrome vanadium double open-end spanner

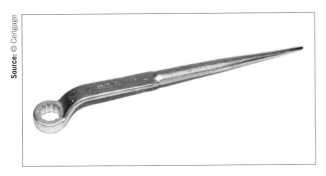

FIGURE 9.29 Podger for scaffolding and formwork centre adjustments – may be used for levering

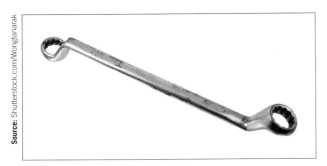

FIGURE 9.30 Double-end ring spanner

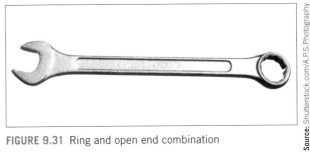

FIGURE 9.31 Ring and open end combination

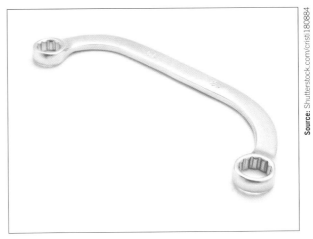

FIGURE 9.32 Half-moon ring spanner

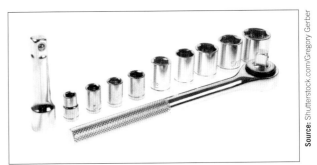

FIGURE 9.33 Square-drive ratchet handle and socket

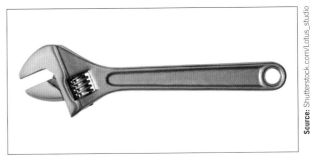

FIGURE 9.34 Adjustable shifting spanner

Pliers

Pliers have a variety of shapes and sizes for particular jobs. They can be used for holding, cutting, banding, twisting and stripping wire. Circlip pliers are specialist pliers used to open or close spring clips (Figure 9.35).

Tin snips

Tin snips are used for cutting straight or curved lines in thin sheet metal such as barges, cappings, corrugated iron, and for cutting banding straps found on slings of timber, pallets of bricks, etc. (Figure 9.36). Small tin snips, also called jewellers' snips, are useful for

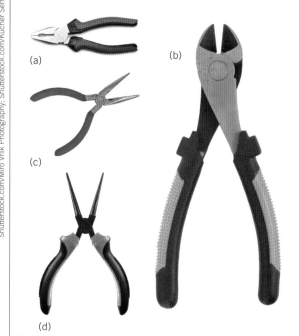

FIGURE 9.35 (a) Insulated combination pliers, (b) insulated diagonal cutters, (c) needle-nose pliers and (d) external straight circlip pliers

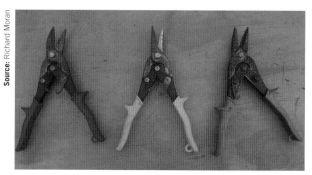

FIGURE 9.36 Red, yellow and green aviation snips

FIGURE 9.37 Jewellers' snips for curved work

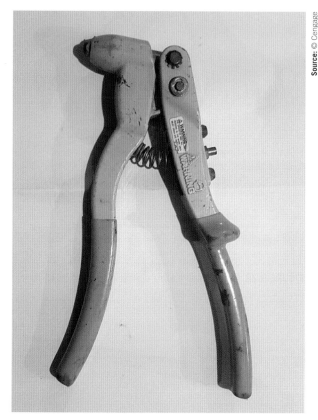

FIGURE 9.38 Hand pop rivet gun

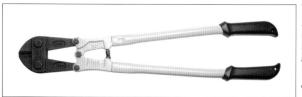

FIGURE 9.39 Long handles for leverage with centre cut jaws for general cutting

trimming thin sheet material and cutting small-radius curves, arcs and circles (**Figure 9.37**).

Aviation snips

Aviation snips (also called tin snips) are the best hand tools for cutting sheets of metal. When choosing a snip for a particular task your choice will depend on whether the waste will be on the right or left hand side: red (cuts left), yellow (cuts straight or left and right) and green (cuts right) (**Figure 9.36**).

Pop rivet guns

Hand pop rivet guns are used with the aid of a rivet to secure two pieces of sheet metal together such as gutters and flashings (**Figure 9.38**).

Bolt cutters

Bolt cutters are used for heavy cutting of rods, bars and thick-gauge wire, the main use being to cut reinforcement bars and steel fabric in preparation for concrete work (**Figure 9.39**).

Bolsters

A bolster is a steel-bodied tool with a broad flat face that forms a thin cutting edge (**Figure 9.40**). It is used with a lump hammer by bricklayers and stone masons to provide a shear cut through masonry materials; for example, cutting dry-pressed bricks to length.

Concreting tools

Listed below are some of the tools that are generally used when working with concrete.

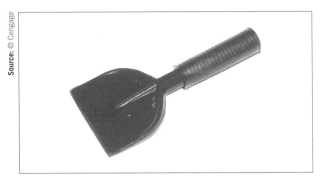

FIGURE 9.40 Bolster – available with a 65 mm, 100 mm and 110 mm blade

Trowels and floats

Wooden floats are used for smoothing and finishing wet concrete, cement render, bedding for ceramic tiling, etc. to give a smooth, textured or non-slip finish (Figure 9.41). They are also used to apply the cement render to masonry walls. Steel trowel finishes are usually smooth and dense to provide a hard-wearing, easy-to-clean surface (Figure 9.42). Wood floats are also useful for floating up the surface of sand bedding prior to laying pavers, to ensure that all hollows are filled and the surface is even.

FIGURE 9.41 Wood float – available in various lengths and widths

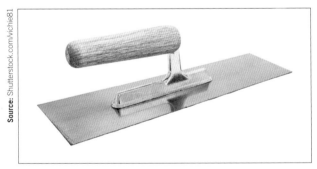

FIGURE 9.42 Steel float – available in a variety of shapes and sizes

Concrete screeds

Concrete screeds are used to screed concrete to ensure a flat, level concrete surface. Available in various lengths (Figure 9.43).

Bull floats

Bull floats are used to smooth concrete surface, after the concrete has been screeded and before the concrete has been finished (Figure 9.44).

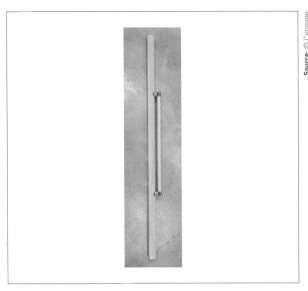

FIGURE 9.43 Concrete screed

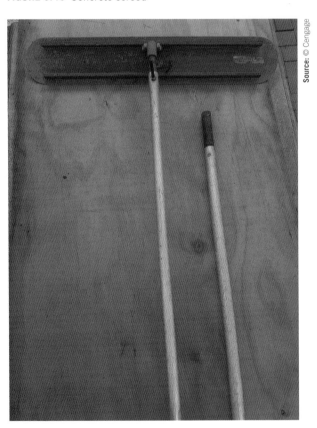

FIGURE 9.44 Bull float and extension handle

Edging tools

Edging tools are used by concreters to give a neat finish around the edges of concrete (Figure 9.45).

Hoses

Hoses are used for a variety of purposes on a building site. They may be used to deliver oxygen, acetylene, abrasive for sand blasting, compressed air, or simply to supply or remove water. Hoses are flexible enough to coil or roll up but have the strength to withstand rough

FIGURE 9.45 Concrete edging tools

treatment. They are normally made of a reinforced plastic or rubber material, or from a tightly woven fabric, as used for fire hoses (**Figure 9.46**).

Painting tools

Putty knives

Putty knife blades vary in length from 100 mm to 150 mm and have a variety of shapes (**Figure 9.47**). They are used to fill nail holes, cracks and surface imperfections. They may also be used to patch or reglaze windows.

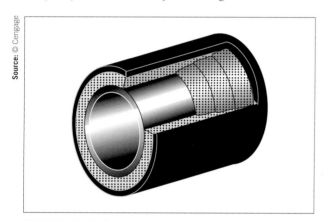

FIGURE 9.46 Typical section through a reinforced water hose

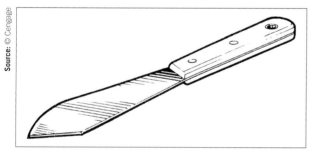

FIGURE 9.47 Putty knife

Filling knives

Filling knife blades vary in width from 50 mm to 150 mm and may be used to apply oil- or water-based fillers to open-grained timbers or shallow holes in the surface of a variety of materials. A broad knife may be used to fill or patch wider cracks and areas with a flat, smooth surface (e.g. plasterboard) (**Figure 9.48**).

FIGURE 9.48 A broad knife and a filling knife

Hacking knives

Hacking knife blades are usually 100 mm to 125 mm long with tapering sides and a thick edge on one side to allow them to be hit with a hammer (**Figure 9.49**). They are used to remove old, hard putty from a window sash to enable removal and replacement of glass.

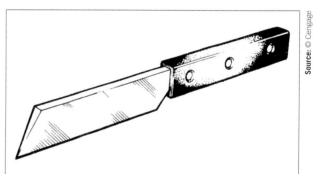

FIGURE 9.49 Hacking knife

Shave hooks

Shave hooks are available in a variety of head sizes to suit the surface they are used on (**Figure 9.50**). They are used in conjunction with a blow torch or liquid paint removers to scrape old paint from ornamental beadings or mouldings and to take out cracks in cornices, etc. prior to filling.

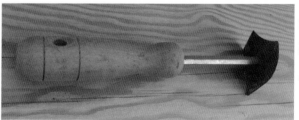

FIGURE 9.50 Shave hook

Paint brushes

Paint brushes are made up of a handle (usually hardwood); the stock, which holds the filling or bristles; a setting of epoxy resin or vulcanised rubber to bind the filling together at the end; and the filling

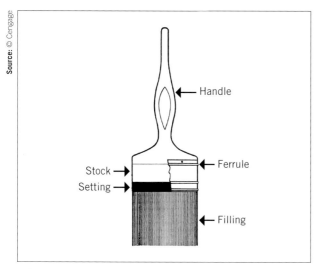

FIGURE 9.51 Standard-type brush

itself (Figure 9.51). The filling is made from pure bristle or animal hair; a synthetic fibre like nylon; a natural fibre such as grass or straw; or a mixture of bristles and fibres. Paint brushes may be used to apply oil- and water-based paints to a variety of surfaces. The size and type of filling, together with the length of the handle, will determine the specific task the brush is designed for; for example, staining, cutting in, flat surfaces, applying a textured finish and even applying diluted acid to brickwork to remove mortar smears.

Rollers

Rollers consist of a central core of heavy-duty cardboard tube impregnated with phenolic resin to resist solvents and water, and are covered with a selection of fabrics. This cover, called the nap or pile, may be made from natural fibres such as wool, mohair or cotton, or synthetic fibres such as acrylic, polyester and nylon. Rollers are used to paint a wide variety of internal and external surfaces such as walls, ceilings, furniture, corrugated shapes, pipes, textured surfaces and wire fences. The roller tube is fitted over a wire frame with a handle, and paint is applied to the roller by rolling it in a ribbed rolling tray to evenly distribute the paint on the nap (Figure 9.52). A variety of fittings, extension arms and cleaning devices is available.

Abrasive papers

A surface-coated abrasive may be defined simply as an abrading medium (grain) bonded to a flexible backing. A common size of abrasive paper sheet is 275 × 225 mm. This is cut into six pieces for use with a cork rubbing block. When using abrasive paper on timber surfaces by hand, care should be taken to work only along the direction of the grain, using a uniform pressure throughout.

The grade of paper used will vary according to the original condition of the timber and the fineness

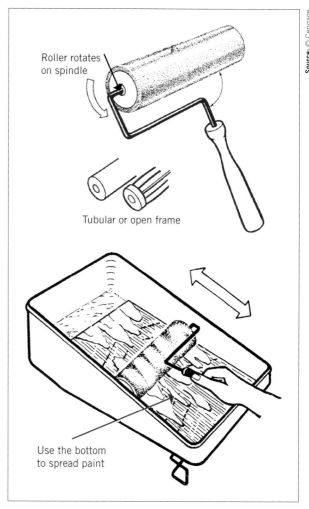

FIGURE 9.52 Roller, roller frame and metal or plastic tray

of finish required. Papers of 60- to 80-grade would generally be suitable to remove marks left by machines or hand tools, and finishing work can be done with grades of 100 to 180. The higher the number, the finer the paper's grit (see Figure 9.53).

FIGURE 9.53 Abrasive papers

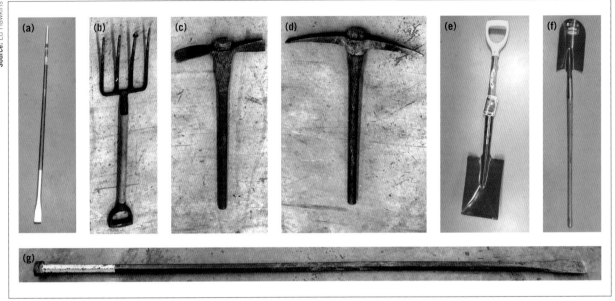

FIGURE 9.54 Hand tools for breaking, cutting and grubbing. (a) Crowbar, (b) Fork, (c) Mattock, (d) Pick, (e) Spade, (f) Long-handled round-mouth shovel, (g) Spud bar

As well, there are different grit materials that may be adhered to the paper or cloth backing.

- Natural abrasive grains include:
  - flint (cream to grey)
  - emery (black)
  - garnet (red).
- Manufactured abrasive grains include:
  - aluminium oxide (reddish-brown)
  - silicon carbide (blue-black).

Excavation tools

Crowbars, picks, shovels, mattocks, etc. (Figure 9.54) may be used by carpenters from time to time in the preparation of the site to allow site set-out activities to proceed. They may also be used to clean up, dig holes for fence posts, move rubbish, dig and trim trenches, grub tree roots or break up old concrete paving.

Light digging, shovelling, cleaning out and spreading tools

After breaking compacted soils or materials into manageable sizes, shovelling or cleaning-out tools are employed to move loose materials clear of the work area or load the spoil directly into a barrow or transporter for disposal.

The most suitable hand tool for this purpose is the long-handled shovel, of either round-nose or square-mouth design, depending on the type of material being moved. The square-mouth shovel is best suited to granular materials such as sand or gravel, while the round-nose is a universal design and performs well in materials of uniform or irregular shape and size. Long-handled shovels allow for a more upright posture. This helps to prevent back strain and enables extra body weight to be applied with the assistance of the more efficient leg muscles. A short-handled, square-mouth

shovel allows more control when cleaning out loose materials from corners and edges of work (see Figure 9.55).

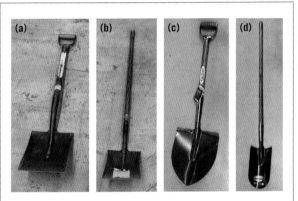

FIGURE 9.55 Light digging, shovelling, cleaning-out and spreading tools. (a) Short-handled square-mouth, (b) Long-handled square-mouth, (c) Short-handled round-mouth, (d) Long-handled round-mouth

Trimming, detailing and finishing tools

As noted previously, a number of the basic breaking and cutting tools and some shovelling and cleaning-out tools are suitable for trimming and finishing operations. For example:

- A cutter end or grubbing mattock is an ideal tool for trimming a flat surface such as a pavement base or trench bottom. A square-mouth, short-handled shovel is a good trimming and detailing tool for medium-density moist soils.
- A crowbar or spud bar is suitable for trimming the sides of trenches or footing pads.

A spade will also give good results when trimming or squaring an excavation, although it is not efficient for shovelling or cleaning out.

Although trimming and detailing are considered finishing operations in the excavation or filling process, the extent of this work will vary according to the degree of accuracy required for each project. For example, a stout steel garden rake of about 500 mm width is sometimes used to spread and level out granular filling and sand base course or screed bed under a slab. The only group of tools not included in the main use groups are special-purpose tools, such as trenching shovels, post-hole shovels and intermediate or narrowed-down long-handled, round-nose shovels (Figure 9.56).

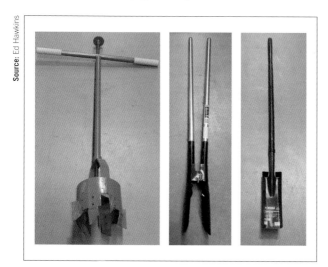

FIGURE 9.56 Special-purpose digging/shovelling tools

Levelling tools and equipment

Before looking at levelling tools and equipment it is important to become familiar with some of the terms used in the levelling process.

These terms include:

- levelling
- level line
- plumb
- datum
- traverse.

Refer to the Glossary for definitions for these terms. You may also want to refer to Chapter 11 for more details.

Spirit levels

Spirit levels are usually made from timber or aluminium, and have a number of glass or plastic tubes containing a liquid with a trapped air bubble inside. The spirit levels are available in lengths of around 250 mm and up to 2.4 m, with a standard length of around 800 mm, which is suitable for checking door heads are level. A straight edge may also be used. Straight edges may be metal (aluminium) or well-seasoned timber (Western red cedar is suitable), have

two straight, parallel edges, and are commonly 2.4 to 3.0 m in length. Care should be taken to prevent damage to the edges, and they should not be left exposed to the weather or stored in a manner that allows them to bow or twist.

Used together, levelling may be carried out by spirit levels and straight-edges in the horizontal plane (Figure 9.57).

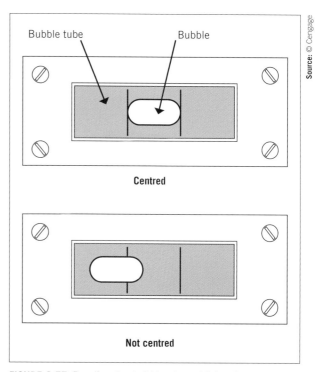

FIGURE 9.57 Reading the bubble of a spirit level

Power and pneumatic tools

The first thing to consider is what is meant by 'power' and by 'pneumatic'. Power means the type of energy that drives or moves the tool or machine. There are now five 'power' sources in relation to hand-operated 'power tools', these include:

- 240-volt AC electrical power
- 12–36-volt DC electrical power
- gas powered – explosive gas (butane or propane) with a battery for ignition purposes
- nitrogen/battery – this tool has a sealed tank of compressed nitrogen gas with a battery
- explosive powered tool (EPT) – uses a 0.22 rifle cartridge.

There is also another form of 'power', which is 'pneumatic' power. Pneumatic simply means compressed air. For example, car and truck tyres are filled with compressed air. So any tool that uses compressed air, not an explosive gas, is in fact a pneumatic tool. This simply means that a tool which is directly or indirectly connected to an air compressor is a pneumatic tool.

240-volt AC electrical power

It is important from a safety aspect to be very aware that 240-volt AC can kill an unsuspecting person. So the first line of defence against electrical shock from the mains service, electrical leads or powered tools, is the installation of an *earth leakage device* or *residual current device (RCD)* or core balance unit. This unit is either fixed to the temporary power board, or is a portable unit which plugs into the general-purpose outlet (GPO) on the mains board or generator and thus becomes a mini-distribution board with four or six outlets (see **Figure 9.58**). This device senses the smallest differential in supply and demand, and then isolates the power in milliseconds.

Source: Arlec Australia

FIGURE 9.58 An RCD protected portable power board

All power from a supply source, whether a temporary builder's service or a permanent installation, should pass through one of these units installed immediately adjacent to the power source.

As an additional measure, users of the temporary power supply should wear heavy rubber-soled shoes, which will give maximum protection against electrocution.

Extension leads

Each state and territory has its own code of practice or guidelines for electrical installation.

These are all based on the national code of practice 'Managing electrical risks in the workplace', which requires portable electrical equipment and leads to be regularly inspected by a licensed electrician. The purpose of inspection is to identify any defects in the leads or equipment that are potential hazards, and to reject or advise on repair or replacement where

possible. The electrician must tag the equipment as evidence that the inspection has been carried out.

These tags (**Figure 9.59**) are clamped to the item after inspection and approval and contain the following information:

- name and licence number of the inspecting electrician
- date of inspection
- name of plant inspected (e.g. power drill)
- due date for re-inspection.

Source: Ed Hawkins

FIGURE 9.59 Electrical tags

All extension leads should be heavy-duty, sheathed for construction and have a rating of 10 A. An extension lead in constant use and drawing near to its full current rating will generate heat and suffer a voltage drop. To minimise damage to the lead and possible hazard to users in this situation, any extension lead of 15 A rating must not be longer than 40 m if it has 2.5 mm² conductors. Every lead should be unwound from its spool or uncoiled fully when in use as, when a lead is coiled up, any heat generated cannot be exchanged efficiently and risks damage or personal hazard to workers.

All leads used on building and construction work sites must have plugs and sockets of clear plastic, to aid inspection for the correct termination of conductors and to identify loose or burnt terminals and wires. Moulded, non-rewireable plugs are also satisfactory.

Leads running from the temporary power supply or from any permanent source should be elevated clear of the ground, over the distance from power source to work site. This is achieved by the use of stands with weighted or broad-based frames and vertical masts with hooks or slots that support the lead without damage.

The minimum height for the lead is 2.1 m from the ground. If the lead can be run adjacent to a wall or other rigid structure, it can be attached at the required height to serve the same purpose.

Any damage to an extension lead, flexible lead to a power tool, or other piece of equipment means it must be taken out of service as soon as the damage is detected. Patching leads, repairing plugs or sockets,

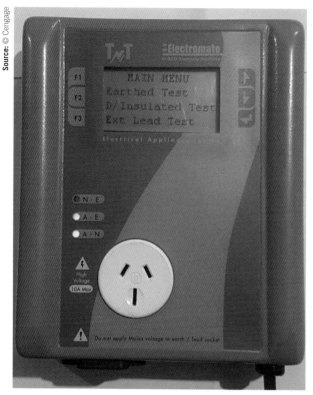

FIGURE 9.60 Tagging machine

applying bandages or trying to camouflage a fault or damage to a lead is not acceptable and breaches national, state and territory WHS legislation.

Circular saw

One of the most widely used electrical tools on the job is the circular saw (Figure 9.61). The circular saw is identified by the diameter of the blade fitted. There are five sizes available:

- 160 mm (6¼″) – battery saw available in this size
- 185 mm (7¼″)
- 210 mm (8¼″)
- 235 mm (9¼″) – most used and preferred saw for carpenters
- 270 mm (10½″).

FIGURE 9.61 Circular saw

The motor of the machine is generally universal and the power source is generally 240-volt AC supply, although there are now good quality 36-volt battery-powered saws available.

These saws may be used for:

- ripping
- grooving
- trenching
- cross-cutting
- rebating
- compound cutting.

Routine maintenance and safety

Routine maintenance:

- Always check the motor housing and the blade for damage prior to using the saw.
- Most blades today are single-use replaceable items, but blades can be sharpened or touched up for reuse many times. Tungsten-tipped blades may be sent to a saw doctor for special sharpening if required.
- Dust, vacuum or blow the body of the saw and around the motor housing so that any build-up of dust doesn't inhibit air flow into the motor.
- Check for worn carbon brushes and replace them as required. *Note:* many new saws are now using brushless motors.
- Store saws in a dry, dust-free box when not in use, to ensure long life.

Safety:

- Always wear eye and hearing protection while operating saws.
- Never operate these tools with loose clothing, long loose hair or necklaces, as these may be caught by the blade.
- The material to be cut must be firmly secured before cutting.
- Ensure that the extension lead and lead of the saw are well behind the saw during cutting operations.

Drop saw/compound mitre saw

This saw is another widely used electrical tool. It allows for a high degree of accuracy in cutting. The motor housing and the blade is set on a hinge or pivot directly above a turn-table to enable the saw to be turned and allows for cuts from 0 degrees (square across the timber) to generally 45 degrees.

If the saw only has this action it is referred to as a drop saw (Figure 9.62), but if the saw incorporates a slide allowing for the motor housing of the saw and the blade to be pulled forward, this is referred as a sliding saw.

Finally, if the motor housing and the blade can be tilted to 45 degrees this is referred to as a *sliding compound mitre saw* (Figure 9.63).

FIGURE 9.62 Drop saw

FIGURE 9.63 Compound mitre saw set to cut a compound mitre cut

 For safety reasons this saw should always be set on a stable work bench or purpose-built saw stand, as using this saw placed on the floor is *extremely dangerous.*

Other safety considerations when using this tool are:

- **Never** try to cut timber *along* its length (ripping the timber) as the blades of this saw are designed for 'cross cutting' only; if the operator attempts to rip timber, the blade may pull the timber through the saw and also pull the operator's hand beneath the spinning saw blade.
- **Never** try to cut a piece of timber that is not secured against the back fence or stop of the turntable. As the timber cuts, the blade will pull the timber back away from the operator and slam it against the fence, possibly jamming the saw blade in the timber.

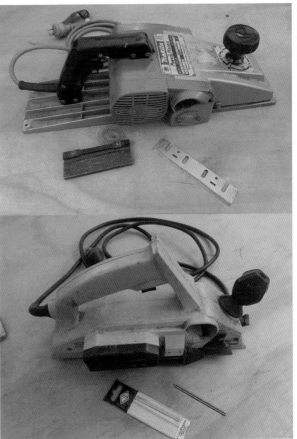

FIGURE 9.64 82 mm and 155 mm electrical planers and blades

- **Never** cut wide pieces of timber by cutting into the timber and pulling the blade towards you. *Always* start the cut on the edge of the timber closest to you and push the motor housing and the blade away from you.

Routine maintenance and safety
See the checklist for **circular saws** on page 341.

Electric planers
Planers are used to remove excess timber while leaving a smooth finish. The depth of cut can be adjusted to a maximum depth of 3 mm. Planers can range from 82 mm common to 155 mm wide timber. Edges can be chamfered using the groove on the bottom of the planer. There are two types of blades: double edge disposable or those that can be resharpened multiple times.

Routine maintenance and safety
Routine maintenance:

- Always check the motor housing, planer lead and the blade for damage prior to using the planer.
- Most blades today are single-use replaceable items, but some blades can be sharpened or touched up for reuse many times.
- Dust, vacuum or blow the body of the planer and around the motor housing so that any build-up of dust does not inhibit airflow into the motor.

- Store in a dry, dust-free environment to prolong the life of the planer.

Safety:

- Always wear eye and hearing protection when using the planer.
- Always unplug planer from power source before adjusting and replacing blades.
- Never put fingers near the blades.
- Never hold the piece of timber being planed – ensure the material is firmly secured instead.

Jig saw

The portable jig saw is used to cut concave and convex shapes in thin materials (**Figures 9.65** and **9.66**). It is useful onsite due to its portability where access to a band saw is not possible. These saws are available in a wide range of sizes and styles, from light-duty single-speed to heavy-duty variable-speed. They may be used to cut curved shapes or along straight lines in materials such as ply, hardboard, soft and hard metals and cardboard. Care should be taken when cutting brittle material, as the blade cuts on the up-stroke, which may cause surface chipping of the material.

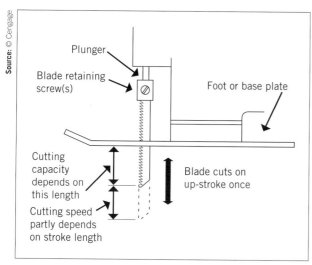

FIGURE 9.65 Cutting action of a jig saw

FIGURE 9.66 Jig saw cutting a piece of ply

Routine maintenance and safety

Routine maintenance:

- Replace blades if excessive pressure has to be applied when cutting or if blades are burnt or bluing.
- Ensure that air vents in the machine casing are not blocked, as this will cause overheating.

Safety:

- Never put fingers or hands under the material while cutting.
- Make sure sufficient clearance is available under the piece being cut.
- Drill a hole in the material first so that the blade may be inserted. Do not use the blade to start a cut in hard materials, as this may shatter the blade. Plunge cutting is permissible only in soft materials.

Sabre saw

Another saw with a similar action to the jig saw is the sabre saw (**Figure 9.67**). This is designed for two-hand operation to cut steel pipe, steel plate and floor panels, or for awkward setting positions.

Portable router/trimmer

The portable electric router is a high-speed spindle moulder and shaper, which may be used in the workshop or onsite. The versatility of the router lies in the variety of bits and cutters that are designed for its use, and unlike the joinery shop fixed bench-type spindle moulder, it is portable (**Figure 9.68**).

FIGURE 9.67 Cordless and powered sabre saws

FIGURE 9.68 (From left to right) Plunge router, standard router and trimmer router

It has a high-speed revolving cutter (2000–27 000 rpm) that gives a very neat, clean cut. There are a wide range of routers available, with differing power and speed ratings to suit a variety of work. They are usually purchased with accessories such as an adjustable fence, template guide, etc. for cutting grooves and rebates, trimming, edging, and also for joints.

Routine maintenance and safety

Routine maintenance:
- Always check the motor housing, router/trimmer lead and the blade for damage prior to using the router.
- Check for damaged cutters.
- Dust, vacuum or blow the body of the router and around the motor housing so that any build-up of dust doesn't inhibit airflow into the motor.
- Store in a dry, dust free environment to prolong the life of the router.

Safety
- Always wear eye and hearing protection.
- Unplug router/trimmer from power source before replacing cutters.
- Hold router firmly when switching on, as torque may twist machine out of hands.
- Never put fingers near the cutters.
- Never start the router with the cutter in contact with the job as it will kick and cause damage to the job and possibly injure the operator.
- Never hold the piece of timber – ensure material is firmly secured and make sure two hands are on router during use.

Electric and pneumatic drills

Drills cover a range of tools that provide rotation for boring, cutting and fastening operations. Most drills are electrically operated (240-volt or rechargeable battery type); however, they can also be operated by air or by flexible drive from a petrol motor.

Drills are classified according to their size, use, power source, speed, or the type of boring or cutting 'bit' they drive and the material they bore into. Sizes range from palm-size pistol-grip through to fixed machine tools for workshop use. Some drills deliver medium- to high-frequency impact or hammer forces, as well as rotating for boring into concrete, masonry or rock. Manufacturers' information charts will assist in selecting the right drill for a specific task.

The cutting tools, bits or drills that actually do the boring are as specialised as the drills that drive them with regard to speed of rotation, diameter of the hole and best sharpening angle for the cutting edge. Common drill types are:
- *pistol-grip drill* – can be used one-handed for small-diameter holes in soft materials like timber and aluminium

- *pistol-grip with side handle* – designed to be used with both hands to prevent the drill from twisting if it jams. Used for larger holes or heavy drilling (Figure 9.69)
- *side handles with breast plate* – used mainly for drilling steel or large-diameter holes when more pressure needs to be applied
- *angle-head drill* – used for drilling in confined spaces or awkward positions
- *impact drill/driver* – used for drilling holes in masonry materials with a tungsten carbide-tipped bit (Figure 9.70). The impact as well as the rotation allows for easy drilling into these materials. Impact drivers are fitted with a hexagonal chuck that allows the fitting of screwdriver and hex-head bits quickly and easily to the driver. The impact action with the turning action drives screws in timber (Figure 9.71).

Source: Ed Hawkins

FIGURE 9.69 Pistol-grip drill with side handle drilling steel

Source: Ed Hawkins

FIGURE 9.70 Tungsten carbide-tipped drill bit (masonry bit)

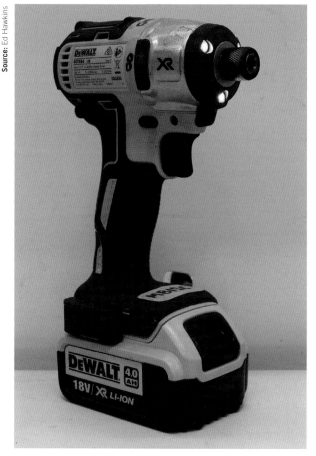

FIGURE 9.71 Battery-powered impact driver

- *pneumatic drills* – used for very heavy-duty work, as found on a large building site or mine. They may be used to drill holes in concrete for reinforcing steel starter bars, into solid rock to receive long rock bolts, or even underwater where the use of electricity is not possible. They are powered by compressed air.
- *power screwdriver* – this operates on similar principles to drills, and some drills may actually be used as screwdrivers due to their variable speed and reverse functions. It comes with a variety of bits and fittings to accommodate most head shapes.

Drill bits must be securely held by the chuck on the drill. Chucks may be key operated (Figure 9.72) or simply tightened by hand if a keyless chuck is fitted (Figure 9.73).

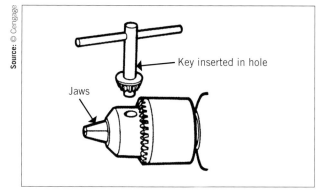

FIGURE 9.72 Chuck operated with a key

Key inserted in hole

Jaws

FIGURE 9.73 Keyless chuck type

Routine maintenance and safety

Routine maintenance:

- Any electrical problem, including replacement of leads, must be dealt with by a licensed electrician.
- Internal greasing of gears, internal cleaning and so on should be carried out by a qualified service person.
- Brushes should be replaced as necessary and checked for contact.
- Pneumatic hoses should be neatly coiled when not in use, with ends capped to prevent foreign matter from entering. Store hoses off the floor away from acids, oils, solvents and sharp objects.

Safety:

- Always wear eye protection as material can be thrown from the drill bit into the operator's eyes.
- Never attempt to open or close a keyed-chuck with the power still connected.

Angle grinders

Angle grinders are used to cut bricks, pavers, concrete, stone, blocks, terrazzo, slate, marble and steel (Figure 9.74). The blade is made from a reinforced abrasive designed for specific materials. It turns at very high speeds and has been known to shatter under load. Angle grinders are available in sizes to carry blades of 125 mm, 230 mm and 300 mm diameter.

FIGURE 9.74 125 mm and 100 mm grinders

Routine maintenance and safety

Routine maintenance:

- Regular cleaning and dusting of the air vents around the motor housing is essential to prevent blockages.
- Check and replace carbon brushes as required.

Safety:

- Abrasive discs generate a large volume of dust, so it is essential to wear a respirator.
- The guard must be in place at all times during use.
- Hot sparks are thrown out during the cutting process, so appropriate clothing, especially safety goggles, should be worn.
- The grinder should be laid on its back after it is switched off, as the blade takes some time to stop turning.

Portable electric sanders

The main types of sander in use are the drum, belt, disc and orbital sanders (Figures 9.75–9.77). Drum sanders are traditionally used for sanding timber floors to a smooth finish. Belt sanders are used for removing large amounts of timber very quickly and, if used correctly, are capable of producing a much flatter surface than a disc sander. Belt sanders are also used for sanding timber floors along the wall line where drum sanders are unable to go. Disc sanders are generally used for removing material quickly – usually paint from timber or steel. Care should be taken as these sanders can easily gouge timber. Finally, orbital sanders, which come in a range of shapes and sizes, are designed to finish the surface of the timber or base material prior to painting or finishing.

Papers for machine use are available as sheets, rolls, discs and belts. Abrasive grains or grits used for coating the paper may be of natural stone, or are manufactured in an electric furnace under temperature conditions up to 2300 °C.

Sanding should always be in the direction of the grain, as cross-grain sanding will tear the timber fibres and leave marks. The dust bag should be fitted at all times to reduce dust levels in confined spaces.

Orbital and reciprocating sanders are not designed to remove large quantities of timber like the belt

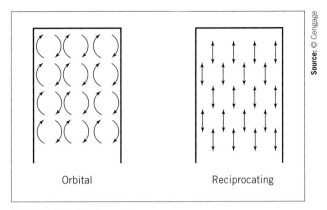

FIGURE 9.76 Sander actions

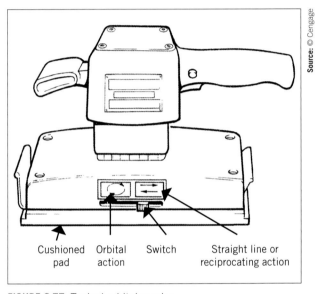

FIGURE 9.77 Typical orbital sander

sander, but are mainly used to finish off prior to painting or staining. They are both fitted with a cushioned base pad and may also be fitted with a dust bag. Both machines are safe to use as the movement of the base is limited to little more than a vibration.

Routine maintenance and safety

Routine maintenance:

- For belt sanders – always fit the belt with the indicator arrow facing the correct direction, otherwise the join in the belt might come apart.
- For belt sanders – adjust the travel of the belt sideways with the adjustable tracking knob.
- For belt sanders – keep a firm grip with both hands when in use, as most belt sanders exert a high level of forward thrust.
- For all sanders – keep housing vents free of dust and check and replace carbon brushes as required.
- For all sanders – keep vents on the motor housing free of dust to prevent clogging and overheating.
- Check and replace carbon brushes as required.

Safety:

- *Always* wear a respirator when using sanders in a confined location.

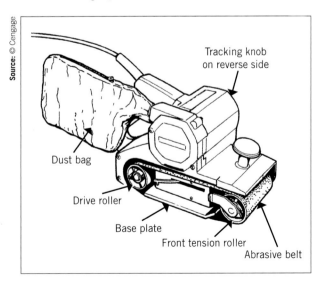

FIGURE 9.75 Belt sander showing main parts

When operating a belt sander, keep the moving belt away from clothing or the body. Loose clothing may be dragged into the rollers and an injury could result.

- Dust extraction bags should be used with all sanders (where provided) or alternatively connect the sander to a vacuum cleaner to reduce the amount of fine dust, which creates an unhealthy work environment.

Pneumatic, gas-powered, nitrogen/battery-powered nailers and explosive-powered tools (EPT)

Construction workers onsite are required to fasten different materials together, such as putting together a timber frame, securing skirting boards to a wall or even securing a timber batten to a steel or concrete column. All these actions may be done using a powered device that drives in a nail.

The first and most common device is what is commonly known as a 'nail gun'. This may be powered in the following ways:

- pneumatic – compressed air
- gas-powered – often referred to as a gas gun
- nitrogen/battery-powered – some manufacturers refer to these devices as nitrogen-powered and others simply call it a battery gun.

All of these devices have a common factor – they are designed with all the same features. Some will even accept the same nails, which are specially manufactured to load into a clip (Figure 9.78) on the nailer rather than being loaded individually.

FIGURE 9.78 Nails being loaded into nailer

The purpose of a nailer is to drive nails quickly and efficiently into and through timber to fasten materials together.

All fasteners work on the same basic principle – a sealed chamber with a piston (similar to a car motor) is filled with compressed air or an exploding or compressed nitrogen gas, driving a piston that pushes the nail out of the nailer and into the material (Figures 9.79 and 9.80).

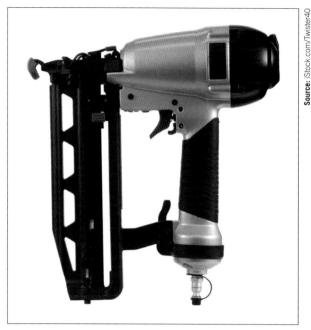

FIGURE 9.79 Compressed air nailer

Source: Ed Hawkins

FIGURE 9.80 Gas nailer, generally referred to as a gas nail gun

Explosive-powered tools (EPTs) are often referred to by their manufacturer's name; for example, Ramset™ gun or Hilti gun. These nailers are specialty tools and are designed to nail timber to steel or masonry; they are not to be used to nail timber to timber.

The most common tool is now the low velocity or *indirect-acting* explosive-powered tool (EPT). This uses a 0.22 cartridge to drive a piston down, pushing a hardened nail down the barrel of the tool, through the timber and into the steel or masonry (Figure 9.81).

The high velocity or *direct acting* explosive-powered tool (EPT) uses a 0.22 cartridge to directly push a hardened nail down the barrel of the tool, through the timber and into the steel or masonry (Figure 9.82). This tool is being phased out and many building sites will not permit their use because of the danger involved. These tools in untrained hands *have killed and will kill.*

FIGURE 9.81 A stripped down indirect-acting EPT

FIGURE 9.82 A direct acting EPT

EPTs in general are considered dangerous and it is a requirement that danger warning signs be placed to notify others on the site that an EPT is being used (**Figure 9.83**).

FIGURE 9.83 Explosive-powered tool danger warning sign

Routine maintenance and safety

Routine maintenance:

- For all nailers – regularly inspect for damage to the casing.
- Some compressed-air-type nailers require a few drops of oil to be put into the air intake to lubricate the piston seals. (*Note:* read the manufacturer's booklet, as putting oil into synthetic-sealed nailers will damage these seals.)
- Never exceed the maximum allowable pressure for a compressed air nailer as set by the manufacturer – generally around 120 psi.

- Always place a nailer in a secure, water-resistant box when you have finished using it.

Safety:

- *Always* wear eye and ear protection when using this tool.
- *Never* point these tools at anyone at any time – loaded or unloaded.
- *Never* walk around with your finger on or over the trigger.
- *Always* disconnect the power source of the tool if leaving it for a period of time.
- *Always* lock away an EPT when not in use.

LEARNING TASK 9.2

SAFE AND EFFECTIVE USE OF HAND TOOLS

Undertake a practical exercise to demonstrate safe and effective use of hand tools. Here is a list of potential projects that you can complete using a variety of carpentry hand tools:
- carry-all tool box
- carry-all fastener box
- timber mallet
- cross-legged stool.

Identifying, selecting and using plant and equipment

On any construction site today in Australia, workers are operating **plant** and equipment. For the purpose of this book and for simplicity, the plant and equipment covered will only include:

- 240-volt power supply – generators
- compressors
- pneumatic and electric jackhammers and breakers
- concrete mixers
- concrete vibrators
- wheelbarrows/brick barrows
- industrial vacuum cleaners
- industrial work platforms
- ladders and trestles.

Refuelling petrol compressors and generators: Safety

When using petrol-driven power tools:
- only refuel after the power tool has been switched off and has stopped running
- always use a suitable funnel when refuelling
- avoid splashing fuel on hot parts and electrical components.

240-volt power supply – generators

A builder has two choices when it comes to obtaining temporary electrical power onsite. The builder may use the mains (240-volt) supply to a single board mounted on a pole (**Figure 9.84**) containing one or more general-purpose outlets, or on larger sites the supply is likely to be both 240 volts (single-phase), and 415 volts (three-phase) for heavy equipment. The power may be reticulated

LEARNING TASK 9.3 HAND AND PORTABLE POWER TOOLS – QUIZ

Download a copy of the CFMMEU's *OH&S Bulletin – Hand and Portable Power Tools* from https://vic.cfmeu.org.au/sites/vic.cfmeu.org.au/files/uploads/OHS/Alerts/power%20tools%20CFMEU%20OHS%20Alert.pdf.

Use this document to complete the following questions.

1 Tools found to be defective should be reported and replaced immediately. True or false? ....

2 What are the two most common injuries that can happen when using power tools?

_____

_____

3 Before servicing a tool you should always have the power source connected to the tool.
True or false? ....

4 How should a tool be pulled up or lowered while working on an elevated work platform?

_____

_____

5 What is the best way to carry knives or sharp objects?

_____

to a number of points around the site and involve main and sub-boards. All work associated with the application, installation of pole and board and reticulation of power, is the domain of the licensed electrical contractor. The builder simply engages the contractor, signs the agreement for supply on the application and plugs into the service when it is installed, inspected and connected by the supply authority.

For short-term power supply onsite, a generator may be used. It is normally a petrol-powered, air-cooled portable unit used to supply 240-volt power to operate electric-powered tools (**Figure 9.85**).

These are generally available as small, extremely quiet and light-weight 1-kVA generators ranging to large, onsite 8-kVA generators.

Most carpenters using a generator will usually have a 2.5-kVA, 3.5-kVA or a 5.0-kVA generator, depending upon the type of work and the tools that need to be powered by it.

For example, a carpenter might want to buy a generator to operate *all* these tools simultaneously: a circular saw, impact drill and angle grinder.

Circular portable saw	4 amps
Impact drill	2 amps
Angle grinder	4 amps
	= 10 amps

Formula: $kVa = \dfrac{amps \times volts}{1000}$

$10 \text{ amps} \times 240 \text{ volts} = \dfrac{10 \times 240}{1000}$

$= 2.4 \text{ kVa}$

In this case, the minimum capacity of the generator would be 2.4 kVA. A reserve of 1–2 kVA is advisable for start-up of tools; therefore, the nearest standard size would be 3.5 kVA.

Note: 1 kVA = 1 kilovolt-ampere or 1000 volt/
amperes per hour : 1 amp
= 1 ampere × 240 volts
= 240 watts

Therefore, a tool rated at 960 watts = 960 W ÷ 240 V = 4 amperes.

Portable site generators are an essential unit on new, large housing estates where mains power is not yet available and in country or isolated areas where access to electricity is not possible. Most units are compact enough to fit into the back of a utility or trailer.

In some instances large diesel-powered generating units for operating heavy equipment with high current demand are available, mounted on the back of a truck or purpose-made trailer (**Figure 9.86**).

Routine maintenance

With any mechanical equipment, maintenance is vital for the continued running of the machine as well as the safety of those who are operating it. Listed below are some of the potential checks that need to be done to keep the machine running.

- Check the oil level before use each day when using a four-stroke engine.
- Check petrol level each day and make sure the correct oil/petrol mix is used in two-stroke engines.
- If the generator fails to start, check the spark plug. Clean and reset the plug or replace it as required.
- Clean the air filter regularly to prevent a build-up of oil choking the air flow.
- Replace the pull start cord when it becomes frayed or breaks.
- Check power-point outlets for wear or damage.
- Make sure equipment is cleaned after use and before storage.

FIGURE 9.84 Typical builder's temporary power pole and board

Compressors

A site compressor is usually portable and is used to operate nail guns, paint spraying equipment, needle guns for paint removal, small sand-blasting guns, small air tools generally and to supply air for respiratory masks via an air purifier (Figure 9.87).

FIGURE 9.85 Portable site generator

FIGURE 9.86 Trailer-mounted diesel generator

FIGURE 9.87 Portable site compressor

Just like generators, compressors come in a range of sizes and configurations. The differences are dependent on the following:

■ *L/s (litres per second)* or *cfm (cubic feet per minute)* is the amount of air being brought from the compressor to the tool. Tanks with higher cfm ratings deliver more air. This is the measure of capacity, for example:

- 1.2 L/s (2.5 cfm) or (70 L/minute)
- 4.6 L/s (10 cfm) or (283.2 L/minute)

■ *kPa (kilopascals)* or *psi (pounds per square inch)* is the pressure exerted by the compressed air on the walls of its container. Generally, the higher the number the higher the pressure, resulting in greater force pushing the nail.

Trailer-mounted site compressors are available for use with a breaker or rock drill on demolition work and for breaking up firm soil (**Figure 9.88**).

FIGURE 9.88 Trailer-mounted site compressor

Larger units are available to operate a range of air tools and sand blasters. Jobs include large excavation, breaking concrete, demolition and as a stand-by in case a factory compressor breaks down or there are power restrictions.

Petrol- or diesel-powered site compressors supply:

■ 47 L/s;
■ 118 L/s.

Routine maintenance

■ An operating compressor unit should be set as level as possible in a clear area, where cool air free of dust is available.
■ Prior to starting, oil, water and fuel levels must be checked to avoid the possibility of engine wear.
■ When starting up the plant, the air cocks should all be opened. After the engine has been started and had sufficient time to warm up, the air cocks may be closed.
■ When stopping the plant, the air in the receiver should first be released by opening the cocks; then the engine itself should be stopped by allowing it to lose speed gradually.

■ Pneumatic tools should be handled carefully and serviced frequently to help ensure trouble-free use of the plant.
■ Prior to coupling up the tools to the air line and putting them into service, any moisture or dirt in the line should be blown out to prevent it entering the working parts of the tools.
■ All bolts should be tightened up regularly and the shanks of chisels and points should be of the correct length and cross-section.
■ Air lines should be checked to ensure all joints are secure, to ensure a minimum of air loss at these points.
■ Careful maintenance and attention to these points will greatly lengthen the life of equipment and ensure the best working efficiency.

Pneumatic and electric jackhammers and breakers

Electric demolition hammers (**Figure 9.89**), commonly called 'Kango hammers', are efficient tools for light demolition of brickwork, concrete and hard-packed clay. They are also available with a rotary hammer drill facility for boring holes into rock or reinforced concrete.

These electric hammers are available in sizes of 8.3 kg, 10 kg, 15 kg and 33 kg weights. They may use moil points, chisels or clay spader attachments and/or 13 mm to 38 mm diameter drill bits.

FIGURE 9.89 Electric demolition hammer with moil point

Pneumatic jackhammers or breakers (Figure 9.90) are available to carry out similar work to the smaller demolition hammers. They are used for chipping, drilling and scabbling. Small points, chisels and drill bits ranging from 12 mm to 25 mm may be used with this 7-kg hammer.

FIGURE 9.91 Tilting drum mixer

is mounted on a steel trunnion, which is tipped by a hand-operated lever.

Being small, this mixer can be moved easily from one building site to another; and as stated above is the mixer preferred by bricklayers.

Horizontal drum mixer

The horizontal drum mixer has a larger capacity than the tilting drum type, and is of much heavier construction (Figure 9.92). These machines have two openings; one is for loading and the other for discharging or emptying. The machine is fitted with a hopper into which the dry materials are placed. The hopper is elevated to load the dry materials into the drum, water is added and the material is then mixed as it slowly works through the machine and out the other end.

FIGURE 9.90 Air breaker used for heavy work

Routine maintenance

- Safety goggles and ear muffs should be used, as well as a silencer, to cut excessive noise.
- Moil points, chisels and other attachments need grinding and hardening on a regular basis to allow efficient operation.
- Loosely coil air hoses when not in use and cap ends to prevent foreign matter from entering.

Concrete mixers

There are four types of concrete mixers available; they range in size and shape – each has its advantages and all are used in the construction industry.

Tilting mixer

This small-sized mixer is generally used by bricklayers on building sites to mix mortar (Figure 9.91). They can be powered by one of three sources – electric motor, petrol or diesel engines – though the smallest mixers are not usually available with diesel engines. These mixers are mounted on a steel chassis with the power unit at one end. The power is usually transferred from the engine to the pinion shaft by a roller chain and sprocket or a V-belt drive, and the drum revolves at about 20 rpm. The whole steel mixing drum and drive

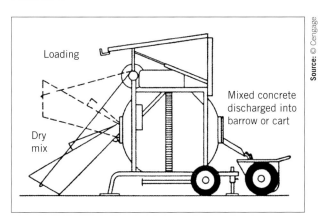

FIGURE 9.92 Mobile horizontal drum mixer

The advantage of this type of mixer is that it is continually mixing concrete; there is no stopping and starting as it is one process.

Inclined drum mixer

The inclined drum is the conical-shaped drum that is usually fitted onto a truck (Figure 9.93). The drum has only one opening and the mixing blades are arranged so that the drum will discharge (or empty) when the rotation is reversed.

FIGURE 9.93 Inclined drum mixer with a capacity of up to 8.0 m³

Source: iStock.com/Andyqwe

Pan mixer

Pan mixers are not often seen on construction job sites as they are not really transportable (Figure 9.94). They are more often used in batching plants, as they are designed to produce the highest-quality concrete in the shortest possible time. Whereas most other types of mixer will take up to 120 seconds in normal circumstances to produce a mix, the pan mixer should not require more than 30 seconds. They are fed with material through a chute in the top of the pan and discharge through an opening in the base. The larger models have up to three openings, which have a sliding gate cover to prevent loss during mixing.

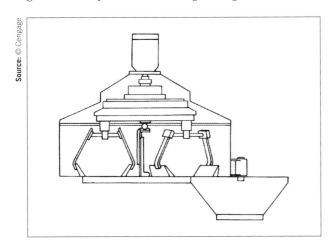

FIGURE 9.94 Mobile pan mixer with capacities of 0.2 m³ to 1.5 m³

Source: © Cengage

Routine maintenance

- Check electrical connections for proper and secure attachment. No bare wire or coloured cables should be visible at any connection point.
- Make sure all electrical connections are waterproof.
- If petrol motors are used, check oil levels as required and clean, adjust and/or replace spark plugs as required.
- Cleaning should take place as soon as possible after final use of the mixer. On very hot days this may need to be carried out as a quick wash between mixes.

- Use a stiff brush and water to clean off remaining slurry from the outside and inside of the barrel.
- If the slurry or leftover mix has started to harden inside the barrel and on the mixing blades, pieces of broken brick or dry blue metal may be placed in the barrel with water – this allows the mixing action to dislodge any stubborn remains.

Note: Do not allow leftover slurry in the barrel to completely set prior to cleaning as this will build up with future use and prevent effective mixing action.

Concrete vibrators

The poker or immersion vibrator is the most common of all the concrete vibrators; it consists of a steel poker head and a flexible drive shaft which connects the poker head to the power unit. The poker head has an internal impeller which causes the poker to vibrate (Figure 9.95).

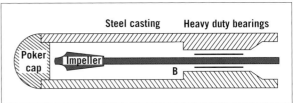

FIGURE 9.95 Section through the vibrating head of the poker vibrator

Source: © Cengage

The impeller generates vibrations as it is thrown against the casing; the casing often needs to be tapped against something to put the impeller off centre and to start it vibrating. The frequency of the vibrations varies a lot between different models and can be as low as 9000 vibrations per minute. The poker cap may be made of polyurethane material to protect the formwork. The flexible drive shaft is usually between 3.0 m and 6.0 m, but can be up to 12.0 m. The power unit is usually fuelled by petrol, but may also be diesel-powered, electric or pneumatic (Figure 9.96).

The purpose of the vibrating action is to enable trapped air and water in the concrete mix to rise to the surface. The removal of air and excess water allows the voids to be filled with a sand and cement paste, making the concrete very dense and durable.

Poker vibrators have a variety of casing diameters and should be inserted at relative spacings into the concrete so as not to create air pockets or voids. Table 9.1 gives a guide to sizes and relative spacings for insertion.

Other vibrators available include the twin screedboard vibrator, which is used for paths, roads, runways, etc. (Figure 9.97), and the vibrating table, which is normally used for pre-cast concrete units such as step treads, concrete manhole covers, etc. The external form of the vibrator is bolted to the

FIGURE 9.96 Portable immersion vibrator (petrol-driven)

TABLE 9.1 Diameters and relative spacings

Diameter of poker (mm)	Spacing when inserted (mm)
26	150
30	200
38	300
50	450
60	600
65	750
76	900
105	1000

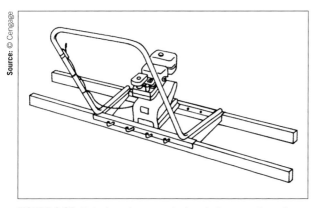

FIGURE 9.97 Petrol engine mounted on twin screedboards

outside of steel or timber forms and used where access is limited for other types of vibrator. Single-engine-mounted screedboard vibrators are available for use by two persons for screeding wide concrete driveways and pavement slabs.

Routine maintenance

- Check all electrical connections for proper operation and water tightness.
- Check oil levels and clean filters regularly on petrol-driven models.
- Always clean off any concrete or cement paste with running water and a stiff brush after every use.

GREEN TIP

When working with concrete, your tools will need cleaning at the end of the job. Make sure you dispose of any washdowns from cleaning these tools in an appropriate space. The best solution may be to collect concrete waste until it dries and then it becomes 'clean hard fill'.

Wheelbarrows/brick barrows

Barrows are the traditional carriers of material on building sites. They are used to carry both wet and dry material to places that heavier and larger-capacity machines cannot reach. The single, central-wheeled type usually consists of a frame constructed of tubular steel with a bolt-on heavy steel tray (Figure 9.98). Wheelbarrows are sometimes galvanised with strong bracing at points of stress. The wheel is either a 355 × 75 mm solid rubber or 410 × 100 mm pneumatic tyre.

FIGURE 9.98 Standard builder's barrow

Wheelbarrows have a range of capacities, varying from light duty 60-litre barrows (0.060 m³) up to 125-litre (0.125 m³) heavy duty builder's barrows. A recent development is the ball barrow, which in place of the conventional wheel has a 355 mm diameter pneumatic ball and a steel body with a moulded rubber bin (Figure 9.99). The ball barrow is lighter than a conventional barrow, weighing only 10 kg.

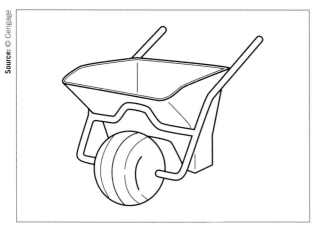

FIGURE 9.99 The ball barrow

Manufacturers claim that the ball rides lightly over rough and muddy sites where wheels would sink in. All parts are snap-fit, enabling simple onsite replacement. The body is smooth and easily cleaned, with a capacity of up to 110 litres (0.11 m³).

The two-wheeled barrow has a tubular steel chassis with a detachable steel body that tips forward (Figure 9.100). It is equipped with a pram handle and is available with alternative bodies.

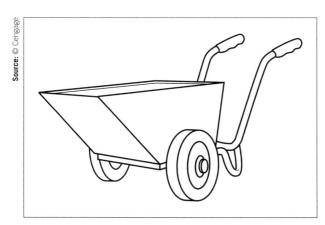

FIGURE 9.100 The two-wheeled barrow with tilt-forward tray

Routine maintenance and safety

Routine maintenance:

- Cleaning should be carried out as soon as possible after use. A stiff brush should be used with water to dislodge any leftover concrete slurry as a build-up of dried slurry makes future cleaning and smooth discharging of mixes difficult.
- Regularly oil the wheel bearing.
- Regularly inflate the pneumatic tyre to the correct pressure as a tyre with low pressure is more difficult to push and to control.

Safety:

- Use your legs and not your back when lifting the handles of a wheelbarrow.
- Always try to wheel over flat, rubbish-free surfaces, as any sudden jolt may injure your back.

- When constructing barrow ramps, make them the width of the barrow tray. Single planks are dangerous, as they tend to flex excessively and don't provide a surface wide enough to walk and wheel on safely. Also nail or screw low cleats across the width of the ramp to stop your boot slipping.

 Tipping should be done gradually. Don't discharge the load all at once, as this allows the barrow to catapult forward, which may cause injury.

Industrial vacuum cleaners

Industrial vacuum cleaners (Figure 9.101) are designed for wet and dry vacuuming of floors in factories, shopping centres and all commercial premises as well as building sites.

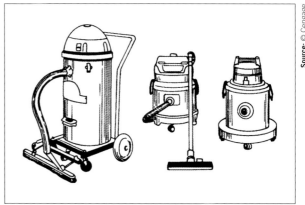

FIGURE 9.101 Various types of industrial vacuum cleaners

Specialist HEPA (high efficiency particulate air) vacuum cleaners must be used for asbestos dust removal, as well as other hazardous material. Bag removal/replacement and cleaning of these machines must be carried out within strict safety guidelines.

Industrial vacuum cleaners are suited to the removal of surface water from flooded floors, cleaning ceilings and removal of water from waterlogged carpets. Capacities range from 23 litres, with a 1200-watt motor, to 210 litres, with dual electric motors.

Industrial vacuum cleaners are also available in smaller back-pack models.

Routine maintenance and safety

- Always check machine leads for bare wires, cracks or cuts before use. Do not use the machine if any are found.
- Check that the switch is in the 'off' position before plugging the machine into the power point.
- Ensure that the lead is always behind the machine when operating.
- Never use extension leads.
- Never stretch the lead to its limit – use the closest power point to proceed.
- Switch the machine off at the power point before unplugging.

- Never remove the plug from the power point by jerking the cord.
- Always empty the bag after use.
- Clean filters regularly.
- Keeping the machine clean extends the useful life of the unit and also reflects well on the operator.
- Debris around the base of the machine and wheels should be removed when noticed or at the end of each use.
- Always wipe the machine over after use with a clean damp cloth and thoroughly dry it prior to storage.

Industrial work platforms

Industrial work platforms can be divided into two main groups:
- elevated work platforms (EWPs)
- temporary structures, such as scaffolding.

Elevated work platforms

Trailer lifts or cherry pickers (**Figure 9.102**), scissor lifts and man lifts are all designed to allow easy access to elevated work areas. They may be used for tree lopping, factory maintenance, changing light globes/fittings in factories, painting, sign writing or reaching awkward places.
- Trailer lifts or cherry pickers will slew 360° and are powered by petrol engines. They have hydraulic outriggers for stability and the controls are mounted in the platform basket, which has a capacity of 150 kg and can lift to a height of 14.0 m.
- Scissor lifts are self-propelled and electrically operated or battery-powered, lift vertically only and to a height of 7.5 m.
- Man lifts are petrol-powered hydraulic lifts for one person and light equipment. They have a safe working height up to 9.0 m, a capacity of 136 kg, and come with adjustable outriggers.

Routine maintenance and safety
- All petrol or diesel engines must have oil levels checked regularly.
- Refer to the machine's 'log book' for precise maintenance requirements.
- A safety harness must be worn at all times when working in the basket or on the platform (refer to your state or territory's code of practice for requirements).
- Platforms must not be used near overhead electrical cables.
- The operator of the trailer lift/cherry picker is required to have a certificate of competency.

Temporary structures – scaffolding

Scaffolding can be divided into different types; however, this book will look solely at light-weight mobile scaffolding (**Figure 9.103**). Scaffolding allows workers to operate on a safe platform above ground level. As with all work at heights, there are strict guidelines set out in the national WHS Regulations 2011, as well as the national code of practice 'Managing the risk of falls at workplaces'.

Where work is performed using mobile scaffolds, workers should be trained to ensure the scaffold:
- remains level and plumb at all times
- is kept well clear of power lines, open floor edges and penetrations
- is not accessed until the castors are locked, to prevent movement
- is never moved while anyone is on it
- is only accessed using an internal ladder.

Ladders and trestles

Ladders are usually constructed from aluminium, steel, timber or fibreglass. Selecting the correct ladder for the job is an important safety consideration (**Figures 9.104 to 9.109**). As a Safe Work Australia report of work-related fatalities in Australia from 1 July 2003 to 30 June 2011 revealed:

half of the falls that resulted in a fatality involved distances of three metres or less in the eight years 2003–11. Falls from ladders accounted for the greatest number of fatalities (37 fatalities: 16%).

FIGURE 9.102 Mobile cherry picker work platform for safe access to difficult locations

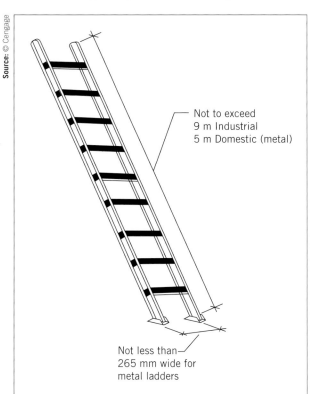

FIGURE 9.103 Light-weight mobile scaffolding

Source: Ed Hawkins

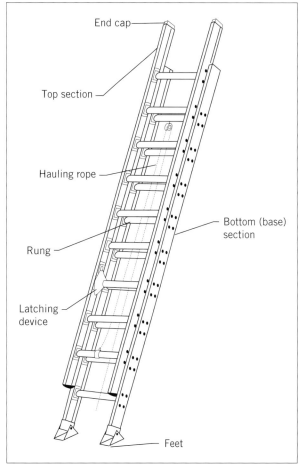

FIGURE 9.105 Extension ladder

End cap

Top section

Hauling rope

Rung

Bottom (base) section

Latching device

Feet

FIGURE 9.104 Single ladder

Not to exceed
9 m Industrial
5 m Domestic (metal)

Not less than
265 mm wide for
metal ladders

Source: © Cengage

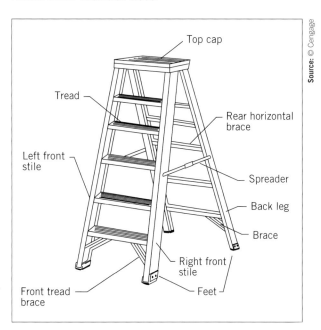

FIGURE 9.106 Step ladder

Top cap

Tread

Rear horizontal brace

Left front stile

Spreader

Back leg

Brace

Right front stile

Front tread brace

Feet

Source: © Cengage

PART 3

9

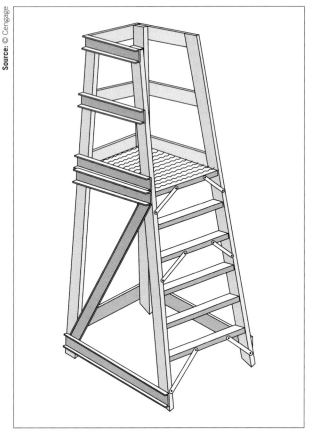

FIGURE 9.107 Platform-type step ladder

Lengths range from
1.5 m step, up to
4.5 m extension

FIGURE 9.108 Dual-purpose step/extension

It is easier for workers to select the correct ladder when the following information is permanently marked or labelled in a prominent position:

■ the name of the manufacturer
■ the duty rating, in the largest lettering possible, i.e. industrial use or domestic use, and the load rating in kilograms; always select industrial use

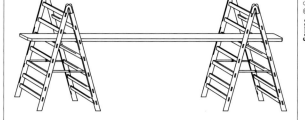

FIGURE 9.109 Trestles – used with a plank to create a working platform

■ the working length of the ladder (closed and maximum working lengths for extension types)
■ hazard warnings, in the largest lettering possible (particularly for metal ladders; for example, do not use where electrical hazards exist
■ instructions for use; for example, double-sided step ladders should display the words 'to be used in a fully open position'.

Safe use

■ Ladders must be in good condition, free from splits, knots and broken or loose rungs.
■ Follow the 1:4 (base:height) ratio rule with ladders. For example, the foot of a 4-metre ladder should be at least 1 metre away from the wall against which the ladder is leaning. Make sure the top of the ladder extends at least 1 metre above the landing or platform.
■ The ladder should be securely fixed at the top and bottom and footed securely on a firm, level foundation.
■ Never put ladders in front of doorways, or closer than 4.6 metres to bare electrical conductors (sometimes it is safe to put timber ladders closer than this; e.g. 1.5 metres). An electrical current can jump from a conductor to an aluminium ladder without direct contact.
■ When working with or on electrical equipment, use only wooden ladders. Do not use metal or wire-reinforced ladders when working near exposed power lines.
■ One person only should be on a ladder at a time and tools should be pulled up with a rope. Workers ascending or descending should face the ladder.
■ Two ladders must never be joined together to form a longer ladder.
■ Ladders should not be placed against a window.
■ Timber ladders should never be painted, as this could cover faults in the timber.

2.2 Use equipment and hand, power and pneumatic tools following WHS requirements and manufacturers' recommendations

New and inexperienced operators should be given extensive instruction on the safe use of each of the common power tools before using them.

This instruction should be provided individually or at an appropriate ratio between learners and instructor. Close supervision is also extremely important as the new and inexperienced operator gains knowledge. Remember, a 'learner' car driver does not have a single session of instruction and become totally competent; it takes time and practice.

Power tool safety checklist

Portable electric-powered tools can be a source of both mechanical and electrical hazards unless they are maintained and used correctly. The checklist in Table 9.2 contains points of general safety and ways of avoiding damage to tools and equipment or injury to the operator.

TABLE 9.2 Power tool safety checklist

Power tool safety checklist	Yes	No
General lead safety		
Are extension leads neatly coiled when not in use to allow ease of future use?	☐	☐
Are *all* leads stored away from oils, solvents, acids, heat and sharp objects?	☐	☐
Are the plugs clear plastic? (Extension leads are required by law to have *clear* plastic plugs so that broken or exposed wires can easily be detected.)	☐	☐
Are leads carried rather than dragged? Are copper wires exposed because of dragging?	☐	☐
Are tools carried or lowered by a lanyard or the handle; not by the lead?	☐	☐
Are plugs disconnected by pulling the plug out directly; not by the lead?	☐	☐
Are leads fully rolled out; *not* left partially rolled up when using high-amp tools? (e.g. welder)	☐	☐
Is the outer casing of the lead intact, with *no* coloured cords visible along the length of the lead or at the ends?	☐	☐
Are leads free of patches or repairs with insulation tape?	☐	☐
Are leads clear of the operator's feet and the tool's cutting edge while being used?	☐	☐
Are the leads protected when they are laid across a driveway or barrow runs?	☐	☐
Are extension leads supported at a vertical height of 2.1 m above the work area and passageways up to 4 m from where the tool is to be used?	☐	☐
Are the pins firm, not loose? (Loose connections cause the plug to overheat.)	☐	☐
Tool safety		
Is the work area tidy before, during and after using a power tool? (An untidy work area is potentially dangerous.)	☐	☐
Is the work area floor damp / wet? (DO NOT MIX ELECTRICITY AND WATER!)	☐	☐
Is the operator wearing appropriate safety glasses/goggles?	☐	☐
Are the operator and nearby workers wearing appropriate hearing protection?	☐	☐
Is the operator wearing loose clothing or jewellery? (Jewellery may become entangled in the moving parts of the tool.)	☐	☐
Is the material being worked on correctly secured? (The tool operator should have both hands free to control the tool.)	☐	☐
Is any part of the operator or assistant in a position to be hit or cut by a runaway tool?	☐	☐
Does the operator carry the tool with their finger on or over the trigger while it is still plugged in?	☐	☐

If the answer is 'Yes' to any of these questions, the relevant workers must review work practices before using any power tools.

All power tools are potentially dangerous if not used correctly. This includes wearing or using the correct PPE (personal protective equipment).

Every time a tool is used, you will need to conduct checks for tool damage, missing guards, damaged leads, faulty triggers, poor maintenance, etc. (see Figure 9.110). These are just some of the items to be addressed prior to operating power tools. This step is especially important if the tool is unfamiliar to the operator; for example, where the tool has been borrowed, hired or is brand new.

The following details outline the safety precautions required for power tool operation to allow preparation for safe use of the tools.

The list is not comprehensive, but it does outline the main safety requirements and checks to be carried out for the safe operation of a portable power saw.

- Check all adjustments and adjustable fittings to ensure they are tightened before use.
- Ensure that the saw is fitted with a hood guard and a returning spring guard.

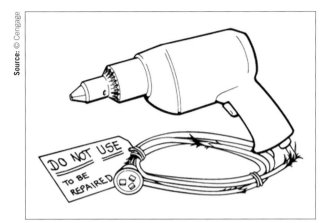

FIGURE 9.110 Labelling faulty equipment

- Never tie the guard back.
- Always operate the saw with both hands.
- After all cutting operations, check to see that the guard has returned before placing the saw on the ground or other surface.
- Rest the saw on a timber block when not in use to prevent damage to the guard.
- Use the correct type of blade for the work being undertaken.
- Switch off the power and remove the plug before making any adjustments to the saw.
- Keep the work area clean and clear of off-cuts.
- Always wear safety glasses and hearing protection when using the saw.

To use a power tool correctly means obtaining training and instruction. It is highly recommended that you gain personal training from an experienced operator in how to operate power tools. However, you can also access supplementary training on video or DVD that will assist with the personal training. There are also video clips on YouTube that will give you a good basis for learning how to use tools correctly.

LEARNING TASK 9.4 POWER TOOL INSTRUCTION

Undertake safe power tool instruction with a teacher or instructor in relation to the following basic portable power tools:
- circular saw
- compound mitre saw
- planer
- impact drill
- compressed air nailer and compressor.

Note: For safety and compliance with WHS requirements, it is recommended that power tool instruction should be carried out at a ratio of no greater than three students per teacher/instructor for approximately four hours per session. It is also recommended that safe instruction and use be conducted as required for AQF (Australian Qualifications Framework) level 3 competencies.

LEARNING TASK 9.5 POWER TOOL USE

Use basic portable power tools, as directed by your teacher/instructor, to assist in the construction of various practical projects and/or site activities.

Note: For safety and compliance with WHS requirements, it is recommended that only one or two tools be used at any one time by a class or training group. The tools should be set up and used in a designated area clearly visible to the teacher/instructor at all times.

2.3 Sharpen and maintain tools

When sharpening and maintaining construction tools it's important to read the user manual for each of the tools and closely follow the instructions for that tool. Some tools have non-serviceable parts, and these should be taken to a specialist service centre or returned to the supplier. Some general things to remember when thinking about sharpening and maintaining tools are that blades will need to be kept sharp and clean, any moving parts will need to be kept clean and not lubricated in such a way that they will attract dust or sawdust, and any electrical components are inspected regularly according to your state or territory's regulatory authority. Proper maintenance of construction tools will aid you in operating a safe, efficient business and you will reduce the risk of accidents or equipment failure. Throughout this chapter there are references for sharpening and maintaining specific tools.

Sharpening blades for chisels, planers, knives, etc. is usually a two-step process. The first step is to grind the blade to the correct angle for the intended use; for example, a little bit sharper for softwood, a little less sharp an angle for hardwood. The second step is to hone a sharp edge on the tool. Generally, a grinding stone that spins at high speed, and has the potential to burn your tool, has been used with a tool rest supporting the tool and set up to give you the correct grinding angle. However, there are a lot of other tools available and you should investigate what works best for you and what is available in your area. For example, wet, slow grinding tools that operate quietly and slowly give you a very sharp finish. It's worth your while to learn how to sharpen different tools so that if you're in a place where you can't spare the tool for someone else to sharpen it at a sharpening shop then at least you can put a good sharp edge on it to continue work, especially in remote areas. This goes for drill bits as well. You should learn how to put a reasonable edge on a drill bit so that you can continue work if you accidentally blunt your drill bit.

3. Clean up

3.1 Clean up, meeting all legislative and workplace requirements for safety, waste disposal and materials handling

At the completion of the job, it is important to clean the work area of any garbage, scraps and material off-cuts, as well as packing away all the tools that have been used. There are several questions to consider in the cleaning up and packing away process, such as:

- Is the work area safe for the next team of workers on the site; this means clean and tidy, and all the off-cuts, shavings and sawdust have been cleaned away?
- Have the tools that were used been cleaned, checked for damage and then correctly packed away?
- Are there tools that need to be maintained to keep them operating properly?
- Are there cutters, bits or blades that need to be replaced before next use?

Cleaning the work site

Brooms are the basic tools used in cleaning up and are available in a variety of sizes for specific uses and may have heads containing bristles of straw, polypropylene or animal hair (Figure 9.111). Stiff-bristle brooms such as the yard broom and straw broom may be used on rough surfaces, and soft-bristle brooms may be used on smooth surfaces, such as lined or coated floors. Brooms may also be used to provide a textured non-slip finish to wet concrete. Broom handles are made from seasoned hardwoods which are straight-grained, strong and flexible.

Brooms form an essential part of the carpenter's equipment kit, as they may be required to clean up a work area before carrying out a task, or to clean up after work has been carried out. It is each tradesperson's responsibility to clean up his or her own mess and to leave a clean, dust-free and safe work surface in line with WHS acts and regulations. It is important for you to learn how to push the dirt with a broom rather than vigorously sweep the dirt and create a lot of dust.

3.2 Check, maintain, store and secure tools and equipment and report any faults

It is important to use hand and power tools safely and correctly, and part of that is the maintenance of your tools. If tools are not maintained properly they will not operate to their full potential and may become a potential hazard and result in wasted time.

FIGURE 9.111 (a) Stiff straw or millet broom, (b) Stiff yard broom of straw or polypropylene, (c) Broad soft-bristle floor broom of animal hair or polypropylene

Below are some points that may assist a tradesperson in maintaining some of the basic tools:

- Hammers and pinch bars should be kept dry and rust-free. Handles should be kept free of oil and grease to prevent slipping during use.
- Spanners, pliers and nips must be kept rust-free, lightly oiled at moving parts and stored in a dry dust-free container.
- Tapes, especially metal-blade retractable tapes, should be kept dry and clean before retracting the blade; otherwise rusting and jamming will occur. Also, the blade should be fed back gently without allowing the hook to hit hard against the tape's body, as it may be snapped off.
- Four-fold rules should have the knuckles oiled and be gently exercised when first used. This will prevent stiff joints from snapping.
- Concreting trowels must be thoroughly cleaned, dried and stored in a toolbox after use. Don't let cement paste harden on these tools, as this will allow a build-up to take place every time they are used.
- Straw or millet brooms should not be stored wet, as this will allow the bristles to rot.
- Hoses should be loosely coiled or wrapped on a reel to prevent kinks and tangles. Protect the hose

with timber battens if it is laid across a traffic road. Take care using hoses in extremely cold weather, as plastic hoses become brittle and may snap during use.

■ Saw stools should be stored out of the weather and all screwed joints should be checked for tightness on a regular basis.

■ Painting tools should be scraped clean after use and shave hooks should be filed to keep the edges keen.

■ Brushes and rollers should be cleaned with the recommended solvent, rinsed with warm, soapy water, dried and stored away. During use they may be wrapped in plastic to prevent drying out.

■ Abrasive papers must be kept dry, especially those with animal glue holding the grit, otherwise they will clog, lose the grit or tear easily during use.

■ Shovels, spades, picks, mattocks, etc. should have excess soil scraped off and then be thoroughly washed with a hose. Lightly oil exposed steel areas when these tools are not being used for extended periods of time.

 COMPLETE WORKSHEET 3

SUMMARY

In Chapter 8 you have learnt how to select and safely use and maintain carpentry tools and equipment, meeting all relevant requirements of the National Construction Code, Australian Standards, work health and safety (WHS), and Commonwealth and state or territory legislation.

You learnt about hand tools, power tools, and pneumatic tools and equipment.

You have also learnt how to:

- plan and prepare
- select, check and use tools and equipment
- clean up.

REFERENCES AND FURTHER READING

Acknowledgement

Reproduction of the following Resource List references from DET, TAFE NSW C&T Division (Karl Dunkel – Program Manager – Housing and Furniture) and the Product Advisory Committee, is acknowledged and appreciated.

Texts

Safe Work Australia (2013), *Work-related injuries and fatalities involving a fall from height, Australia,* Code of Practice, retrieved from www.safeworkaustralia.gov.au

Safe Work Australia (2015), *Managing the risk of falls at workplaces – Code of practice, March 2015,* retrieved from www.safeworkaustralia.gov.au

TAFE Commission/DET (1999 / 2000), *Introduction to trade – Certificate III in General Construction (Carpentry) Housing,* course notes (CARP series), C&T Division, Castle Hill, Sydney

TAFE Commission/DET (1999 / 2000), *Building materials and hand tools – Certificate III in General Construction (Carpentry) Housing,* course notes (CARP series), C&T Division, Castle Hill, Sydney

TAFE Commission/DET (1999 / 2000), *Power tools and equipment – Certificate III in General Construction (Carpentry) Housing,* course notes (CARP series), C&T Division, Castle Hill, Sydney

Web-based resources

Resource tools and VET links

www.training.gov.au, Skills training

www.safeworkaustralia.gov.au, Work-related Injuries and Fatalities Involving a Fall from Height

www.safeworkaustralia.gov.au, Managing Electrical Risks in the Workplace – Code of Practice

www.safeworkaustralia.gov.au, Managing the Risk of Falls at Workplaces – Code of Practice

www.comlaw.gov.au, Work Health and Safety Regulations 2011

www.cfmeuvic.com.au, search safework sites for safety notices on Hand and Portable Power Tools

Search on your favourite tool store's site or your favourite tool manufacturer's website and search for user manuals for various tools.

GET IT RIGHT

In the photo below, the person is changing the planer blade unsafely.

Identify the unsafe practices and provide reasoning for your answer.

Source: Richard Moran

_____

_____

_____

_____

_____

_____

 **WORKSHEET 1**

Student name: _____

Enrolment year: _____

Class code: _____

Competency name/Number: _____

Task: Read through the sections beginning at *Plan and prepare* then complete the following questions.

1   Complete the following statement.

Hand tools have an important role in _____ and
_____ the work to be undertaken.

2   List the five (5) things that need to be considered before even starting the tool or machine.

1   _____

2   _____

3   _____

4   _____

5   _____

**WORKSHEET 2**

Student name: _____

Enrolment year: _____

Class code: _____

Competency name/Number: _____

Task: Read through the sections beginning at *Select, check and use tools and equipment* then complete the following questions.

1 List five (5) hammer types used for general carpentry activities.

1 _____

2 _____

3 _____

4 _____

5 _____

2 List the four (4) main types of planes.

1 _____

2 _____

3 _____

4 _____

3 Briefly describe the main use of pincers.

_____

_____

4 Why should a utility knife only be carried or stored in your nail belt or tool bag?

_____

_____

5 Identify the type of plank in the photos.

Source: Ed Hawkins

Source: Ed Hawkins

6   List the five (5) main types of clamps that can be used.

1 _____

2 _____

3 _____

4 _____

5 _____

7   What is the name of the spanner that a scaffolder or formworker would use?

_____

8   List five (5) actions or tasks that may be done with pliers.

1 _____

2 _____

3 _____

4 _____

5 _____

9   Tin snips are used for cutting a straight line only in thin sheet metal. (Circle the correct answer.)

TRUE   FALSE

10   Bolt cutters are used for cutting reinforcement bars and steel fabric for concrete work. True or false? (Circle the correct answer.)

TRUE   FALSE

11   Apart from applying putty to glaze windows, state three (3) other uses for a putty knife:

1 _____

2 _____

3 _____

12   What is the main use for a hacking knife?

_____

13   What are the three (3) main parts of a paint brush?

1 _____

2 _____

3 _____

14   What is the name given to the soft covering on a paint roller?

_____

15   State the three (3) main types of natural abrasive grains used for abrasive papers:

1 _____

2 _____

3 _____

16 State the two (2) main types of manufactured abrasive grains used for abrasive papers.

1 _____

2 _____

17 Name two (2) tools that may be used for trimming or detailing an excavation.

1 _____

2 _____

18 List the six (6) sources of power that may be used to drive or work a power/pneumatic tool.

1 _____

2 _____

3 _____

4 _____

5 _____

6 _____

19 List four (4) types of concrete mixers available for use on-site.

1 _____

2 _____

3 _____

4 _____

20 What is another name for the poker vibrator?

_____

21 Complete the statement:

Trailer lifts or cherry pickers have a platform basket with a capacity of _____
and are able to extend to a maximum height of _____ m.

22 State the minimum section size of an aluminium plank:

_____ × _____ mm

23 Extension leads used on all building sites need to be inspected and tagged by a licensed electrician. State the four (4) main items that must be included on the clamped electrical tag.

1 _____

2 _____

3 _____

4 _____

24 State the minimum height above ground for an extension lead when used on-site:

_____ m

25 To ensure safe use of powered tools, why should leads not be left partially rolled up during use?

_____

_____

26 State why a portable site generator is essential on large new housing estates.

_____

27 What do the letters HEPA stand for in relation to industrial vacuum cleaners?

_____

28 According to the *Model Code of Practice – Managing the risks of falls in the workplace 2015*, what five (5) things should workers be trained to ensure when using mobile scaffolds?

1 _____

2 _____

3 _____

4 _____

5 _____

29 Fill in the missing information below.

1 _____

2 _____

3 _____

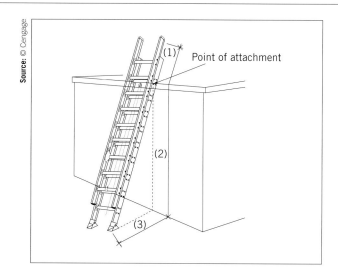

Source: © Cengage

30 What are the two (2) important things that need to happen for a new and inexperienced operator to gain knowledge and experience?

1 _____

2 _____

31 Extension leads are required by law to have *clear* plastic plugs so that broken or exposed wires can easily be detected. (Circle the correct answer.)

TRUE   FALSE

 **WORKSHEET 3**

Student name: _____

Enrolment year: _____

Class code: _____

Competency name/Number: _____

Task: Read through the sections beginning at *Clean up* then complete the following questions.

1 Why is it important to check the tools for serviceability and clean and sharpen them after you finish a job?

_____

_____

2 What might be the problem with vigorously sweeping the dirt up at the end of the job?

_____

3 When planning to service or maintain a tool, what is the first step you would take?

_____

4 How long after you finish using concrete tools should you clean them off?

_____

5 Why would you consider lightly oiling your metal tools from time to time?

_____

PERFORMANCE OF SKILLS EVIDENCE

Student name: _____

Enrolment year: _____

Class code: _____

Competency name/Number: _____

Task: Working with your teacher, prepare to demonstrate your skills by completing the following task. To demonstrate competency, a candidate must satisfy all the elements, performance criteria and foundation skills of this unit by safely and effectively, across three different carpentry tasks, use and maintain all of the listed tools and equipment at least once:

1   hand tools:

retractable tape measure

folding or steel ruler

combination square

string line

chalk line

hand saws

carpenter's hammer / claw hammer

wood chisel

hand plane

trimming knife

clamps

bevels

spirit level

tin snips

2   power/battery/pneumatic tools and equipment:

circular saw

reciprocating saw

sliding compound saw

jig saw

angle grinder

planer

laminate trimmer or router

drill

rotary hammer drill

impact driver

nail gun

bench grinder

extension lead

portable residual current device

air compressor and hoses.

The candidate must also replace blades/cutters/grinding discs in:

a power saw

a powered planer

a router

grinder

The candidate must also:

grind, sharpen and hone a hand plane blade

grind, sharpen and hone a chisel.

PERFORM CONSTRUCTION CALCULATIONS TO DETERMINE CARPENTRY REQUIREMENTS

10

This chapter covers the following topics from the competency CPCCOM3001 Perform construction calculations to determine carpentry material requirements:

- plan and prepare
- calculate area and volume of construction materials for the project
- calculate the requirements for construction for the project
- check and record results.

This unit of competency specifies the skills and knowledge required to plan and perform calculations to determine material requirements for a construction project.

1. Plan and prepare

This chapter covers various materials used in construction and how to estimate the quantities needed.

It will refresh your understanding of how to calculate areas and volumes as well as linear measurements and how they are applied to the different types of construction materials.

It covers the accepted practice of adding a percentage of waste on the various construction materials before giving the final quantities needed.

It follows industry practices of what can be ordered; for example, concrete must be ordered to the nearest 0.2 m^3, timber to the nearest 0.3 m.

There is an explanation of the difference between quantity surveying and builders' quantities.

It is a lot easier to work out various quantities using a calculator, like that shown in Figure 10.1. Buy a good-quality scientific calculator, as you will need its cosine (cos) and tangent (tan) functions.

FIGURE 10.1 A calculator

1.1 Review drawings, specifications and workplace requirements for a construction project

Reviewing drawings, specifications and workplace requirements for a construction project is essential. The reason that we review the drawings and specifications is so we are working on the current version. By the time we quote and win the tender, items might have changed, so always confirm that the plan and specifications are the latest before starting.

If you find that there are anomalies in the plans, such as room sizes have changed or material selection has changed, you need to ask questions and adjust pricing if necessary.

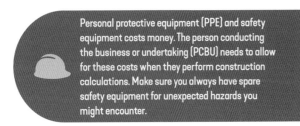

Personal protective equipment (PPE) and safety equipment costs money. The person conducting the business or undertaking (PCBU) needs to allow for these costs when they perform construction calculations. Make sure you always have spare safety equipment for unexpected hazards you might encounter.

1.2 Plan all work to comply with laws and regulations, the National Construction Code (NCC), Australian Standards, work health and safety (WHS) and environmental requirements, manufacturers' specifications, workplace requirements, drawings and specifications

All work must be planned to comply with laws and regulations, including the National Construction Code (NCC), Australian Standards, work health and safety (WHS) and environmental requirements, codes of practice, manufacturers' specifications, workplace requirements, drawings and specifications.

That is a lot to take in, and don't forget that every state is different. The only constants are Australian Standards, and the ones that are particularly relevant to this unit are:

- AS 1684-2010 Residential timber-framed construction
- AS 1288-2006 Glass in buildings – Selection and installation
- AS 1860.2-2006 Particleboard flooring, Part 2: Installation
- AS 2050:2018 Installation of roof tiles
- AS 3600:2018 Concrete structures
- AS 4773.2:2015 Masonry in small buildings, Part 2: Construction.

Even the NCC has some different amendments in certain places for each state. One example of this is side boundaries. Whereas in every other state of Australia the side boundary from the outside wall is 1 m, in New South Wales it is 900 mm, but the caveat is you must have a clearance of 675 mm from the edge of the gutter to the side boundary (NCC – BCA Volume 2).

With WHS, every state will have the various codes of practice, but they may not be the same in each state. Please check with your local WHS regulator. Most states seem to take the lead from Safe Work Australia.

For environmental issues, contact the local council or your state or territory's Environment Protection Authority (EPA).

When installing any product, contact the manufacturer for installation procedures, especially on new products or products that you have not used before.

If working with timber, the Timber Development Association is a good source of information on species to use, how to install and any special treatments that should be applied before installing.

GREEN TIP

The local government authority (local council) is normally the main organisation that construction workers will have to deal with regarding environmental issues in your area of construction. Do some research online and make sure you are aware of their environmental requirements before you start calculating costs of work you plan to do.

LEARNING TASK 10.1

CLASS/GROUP ACTIVITY: COMPARING THE COST OF ONE CONTROL FOR A HAZARD OVER ANOTHER

Pick a hazard from a site you are familiar with and, considering the controls that you can identify based on researching documentation like user manuals, codes of practice and a competent person's experience, make a note of two different controls that could cost different amounts of money to implement. Estimate the two costs and discuss this with your classmates or teacher.

COMPLETE WORKSHEET 1

2. Calculate area and volume of construction materials for the project

This section will introduce you to equipment used to calculate the area and volume of construction materials for a project, and also the methods used to do so.

2.1 Review drawings and specifications to determine dimensions of each type of construction material for the project

When calculating the area or volume of construction materials for the project, you will be required to review drawings and specifications to assist you in correctly calculating the area and quantities of each material required for the project.

In this section you will see all the tools needed except for one crucial element – the plans and specifications.

Figure 10.2 shows a section through a brick veneer building and Figure 10.3 shows a plan view of the ground floor of a house.

Measurements and equipment used

For most construction materials you will be working out the area or the volume of an item then applying that to the quantities needed. Linear measurements are also used, and within those linear measurements other formulas may have to be used, such as $\pi \times d$.

The construction industry only works in metres (m) or millimetres (mm) for linear measurements; centimetres (cm) are never used as they are not recognised as true SI units. SI stands for the International System of Units, which is the metric system of measure that is used globally. Developed by the French, it eliminates the confusion that still exists between imperial and US measuring systems.

The International System of Units is the basis for absolute measuring. It has base units such

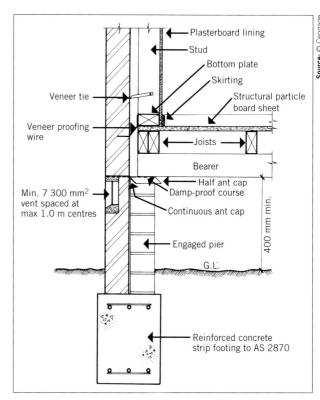

FIGURE 10.2 Vertical section through external wall of brick veneer construction

as millimetres (mm), metres (m), pascals (Pa), kilograms (kg) and seconds (s). For a complete list, visit the Australian Government Style Manual website at https://www.stylemanual.gov.au/grammar-punctuation-and-conventions/numbers-and-measurements/measurement-and-units.

Typical tools used in measuring are shown in Table 10.1.

Making calculations: formulas and processes

What you are trying to quantify will determine what calculation you will use on the type of material that you are measuring.

Area formula = Length × Width (m²) rounded up to two decimal places.

Trades that use this are:

- carpenters
- painters
- carpet layers
- plasterboard fixers
- floor and wall tilers
- bricklayers.

Volume formula = Length × Width × Height (m³) rounded up to three decimal places.

Trades that use this are:

- concreters
- carpenters
- plumbers
- mechanical engineers (air conditioning)
- excavators.

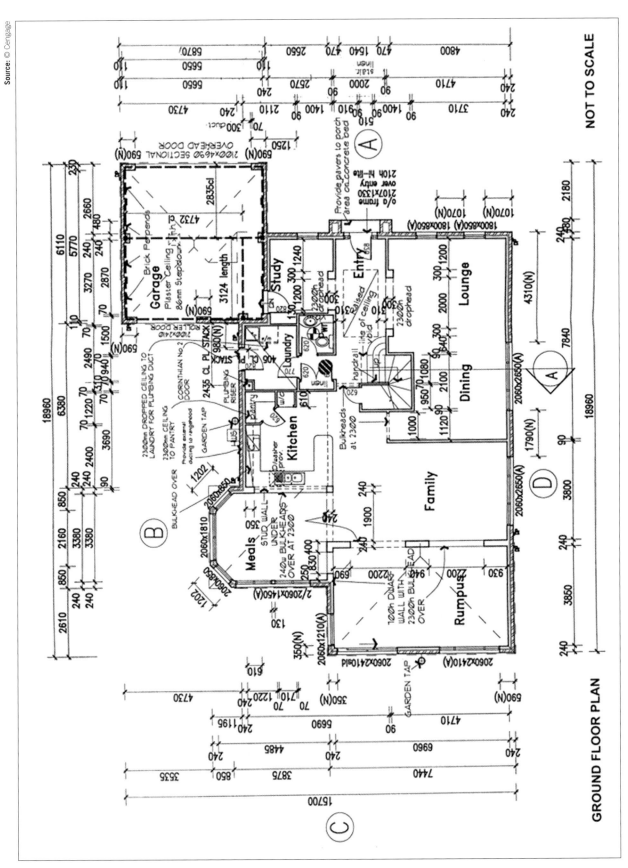

FIGURE 10.3 Ground floor plan

TABLE 10.1 Typical tools used in measuring

1-metre four-fold rule If you use one with a bevel edge, there is less likelihood of parallax error. Parallax error is when an object or point of measurement appears to shift or change position as a result of a change in the position of the observer.	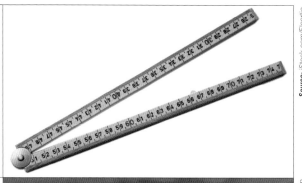 Source: iStock.com/Floortje
Retractable tape measure This is the most common type of measuring tool for the building industry. Retractable tape measures come in a variety of lengths, with 8 metres as the usual choice. The wider the blade the better, and the more rivets holding the hook on the better. The hook adjusts depending on whether the measurement to be taken or marked out is internal or external. Some have a magnetised hook.	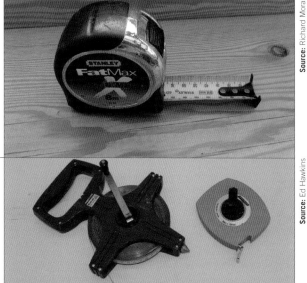 Source: Richard Moran
Open reel and closed-case long tapes These tape measures can be made from fibreglass or steel, with fibreglass preferred over long distances as it is not affected by heat, which can expand metal and give a false reading. The general range is 30 metres up to 100 metres. The hook fits over a nail when setting out.	 Source: Ed Hawkins
Scale rules These are important when drawing items to scale or taking measurements off plans where there are no dimensions given. Warning: always use dimensions given and do not rely on the scale rule as some of those plans might have been photocopied, which may not represent the true size. A good all-round scale rule is a Kent 62M, which has the best basic set of scales on the rule. The rule may be triangular in shape as well as the usual flat shape.	 Source: Ed Hawkins

>>

Trundle wheel

A trundle wheel is a lightweight wheel constructed from timber or plastic which has a circumference of exactly 1 metre. It has a handle attached to allow a person to walk it along easily, and a counter so that every rotation (1 metre) is counted. These tools are not accurate and are mainly used to obtain a quick estimate of distances over 100 metres up to several kilometres.

Laser measuring devices

These devices have been improved over the years and offer many features other than basic distances, with most devices including areas and volumes while others add angles for roofing. In particular, Bosch has an accessory where the device clips into a straight edge to lay on timber or steel to get precise angles or to test levels. These measurements can be Bluetoothed to a smartphone after a photo is taken of the room and superimposed onto the photo, including dimension lines.

Linear measurement is length in metres in increments of 0.3 m when buying timber.

To make sure that you cover all aspects when calculating material, you need to have a system that you stick with and not miss anything.

- Always start in the top left-hand corner and work your way around the building or room.
- Use a highlighter marker on the plan to indicate you have covered that part.
- Use centreline measurements for excavation of footings, concrete for the footings and brickwork.
- Use dimension lines and avoid scaling off the plan.
- Break up the quantities into the various trades.
- Don't rush.

2.2 Calculate the area of each type of lining material

The most common type of lining material a construction worker may typically encounter would be plasterboard (Gyprock – brand name from CSR).

Other types of lining material can be 3D panels, plywood panels, formed concrete, corrugated Colorbond sheets and timber boards. All these products use the same method as the one demonstrated except for timber boards. These are calculated as demonstrated in calculating timber flooring later in this chapter. Use the area formula to calculate linings in the rumpus room from the plan in Figure 10.3 (see the 'How to' box below).

HOW TO

CALCULATE AREA OF PLASTERBOARD NEEDED FOR RUMPUS ROOM FROM THE PLAN IN FIGURE 10.3 (AREA = LENGTH × HEIGHT) – ANY DEDUCTIONS OVER 1 M²; FLOOR TO CEILING HEIGHT = 2400 MM; DOOR HEIGHT 2100 MM

Area to be covered in 10 mm plasterboard as per manufacturing requirements; allow 10% wastage.

- → 3850
- ↑ 6960
- ← 3850
- ↓ 6960

Total length 21 620 × height

| Area before deductions | = 21.62 m × 2.4 m |
| | = 51.89 m² |

Deductions:

Windows 2 × (2060 × 2410)	= 12.53 m²
Door into meal room 2040 × 830	= 1.74 m²
Openings to family 2 (2400 × 2200)	= 10.56 m²

Goes to full height of ceiling

Total deductions	= 12.53 + 1.74 + 10.56
	= 24.83 m²
Area to be sheeted	= total area minus deductions
	= 51.89 – 24.83
	= 27.06 m² + wastage 10%
	= 27.06 × 1.1
	= 29.77 m²

How many sheets? Use 4800 × 1200 sheets

Area of sheets	= 4.8 × 1.2
	= 5.76 m²
No. of sheets	= Total area – deductions / area of sheet
	= 29.77 / 5.76
	= 5.17 sheets, say 6 sheets

2.3 Calculate the total area of the building wrap and of each type of external cladding material

Building wrap is an insulation material that wraps the entire outside of a building before putting on a cladding. How does a construction worker calculate the total area of the building wrap? Use the area formula to calculate how many rolls of wrap are needed.

CALCULATE AREA OF WRAP NEEDED FOR ENTIRE HOME FROM THE PLAN IN FIGURE 10.3 (AREA = LENGTH × HEIGHT); DO NOT DEDUCT ANY OPENINGS; BOTTOM PLATE TO TOP PLATE HEIGHT = 2550

Area to be covered in TYVEK wrap and installed as per manufacturer's instructions.

Do not deduct for any openings.

Use the overall measurements of house plan for area. Disregard the various floor shapes.

Size of wrap is 30 m × 0.9 m per roll = 27 m²

Area of building

↔   = 18 960

↕   = 15 700

Total area	$= 34.66 \times 2 \times 2.55$
	$= 176.77$ m²
Number of rolls	$=$ total area / coverage of 1 roll
	$= 179.77 / 27$
	$= 6.55$ rolls, say 7 rolls of wrap

Calculating building external cladding

Calculating the cladding of the building will depend on the type of cladding that is being used. For many years in Australia the most common type of cladding has been brick veneer. It is unique to the Australasian region, including New Zealand. It looks like a full brick home but is only one skin wide.

Brick veneer is not a structural component, but its job is to keep out the weather and environmental elements and noise.

There are many other types of cladding, such as weatherboards, concrete panels, full brick and metal sheeting. All of these will rely on using area calculations at some point. For this exercise, brick veneer and Hardie weatherboard will be used from the plan in Figure 10.3.

CALCULATE NUMBER OF BRICKS USING THE PLAN IN FIGURE 10.3

Formula for bricks is: Area of wall minus any deductions (any openings such as doors and windows more than 1 m²; if under 1 m², no deduction is made as per mode of measurement.)
× number of skins of brickwork × 50 bricks m²
Hume Access cream face bricks
Allow 5% wastage
Example: Western wall – rumpus, family, dining, lounge
→ Height of brickwork from top of footing to underside of eaves = 2525 or 2.525 m
→ Rumpus + family + dining + lounge = 18 960 or 18.96 m² long

Area	$= 18.96 \times 2.525$
	$= 47.87$ m² total area
Deductions	$= 2$ windows 2060×2650
	$= 2.06 \times 2.65 \times 2$
	$= 10.92$ m²
	$=$ total area minus deductions
	$= 47.87 - 10.92$
	$= 36.95$ m²
No. of bricks	$=$ area × no. of skins × 50
	$= 36.95 \times 1$ skin (brick veneer) × 50
	$= 1848 \times 1.05$ wastage
	$= 1940$ bricks (500 bricks per pallet), say 2000 bricks

Calculating building external cladding

The next example is for 'weatherboards' using the same information as the brick calculation.

HOW TO

CALCULATE OTHER CLADDING INSTEAD OF BRICKS USING THE PLAN IN FIGURE 10.3

Formula for cladding is: Area of wall minus any deductions (any openings such as doors and windows more than 1 m²; if under 1 m², no deduction is made as per mode of measurement).

Linear metres of boards per 1 m² = 1000/effective cover

$$= 1000/278$$

$$= 3.6 \text{ m per m}^2$$

Use James Hardie PrimeLine Weatherboard Heritage range 300 wide boards

Effective cover 278 mm, length 4.2 m

Use bricklaying calculations for area to

be covered = 36.95 m²

Linear metres = 36.95 × 3.6

$$= 133.02 \text{ m allow 10\% waste}$$

$$= 133.02 × 1.1$$

$$= 146 \text{ m}$$

Length of boards is 4.2 m

No. of boards = 146 / 4.2

$$= 34.8 \text{ boards, say 35 boards}$$

$$\text{@ 4.2 m long}$$

2.4 Calculate the total area of each type of roofing material

Common roofing materials for a domestic builder include the following three main choices:

1. Concrete tiles: These tiles are made using a cement mixture like mortar. Under great pressure and heat the tiles are formed up in moulds and coated with a special concrete paint or glaze. Colour can also be added to the mortar mix so it goes right through instead of the colour just coating the surface of the tile.

2. Terracotta tiles: As the name suggests, these are clay-based tiles. The clay is pressed into a mould, usually covered in a glaze, and fired in a kiln. These tiles are expensive.

3. Colorbond, metal, tin, corrugated iron, Custom Orb. This type of roofing material is making a huge comeback, especially in regional and country towns.

No matter what type of roof covering you choose, the method of calculating the area of roof is the same. The difference is the method of preparation and fixing techniques needed for the various products.

The way to calculate the area of a roof is to treat the roof shape as a gable roof no matter what the roof shape is. Break the roof areas up into manageable sections, e.g. roof over dining, lounge.

Find what the pitch of the roof is and the rafter length needed. Don't forget to add the overhang and calculate the half-span or run of the roof. To find the length of the rafter, use 1/pitch cos.

This will give you the rafter length over 1 m of half-span or run.

HOW TO

CALCULATE ROOFING TILES USING THE PLAN IN FIGURE 10.3

Method

1. Treat all hip shapes or broken hip and valleys as gable roofs.
2. Divide sections of house into different areas, i.e. roof area A, B and so on.
3. Disregard 2nd storey as the plan for that is not provided.
4. Allow 60 tiles per m².
5. Allow 5% for broken tiles and cutting in.
6. Use Boral contour concrete tile.
7. Do not allow for ridge and hip capping or valleys.
8. Use 25° pitch.
9. Calculate length of rafter including overhang.
10. Allow 450 overhang.

Length of rafter	= 1/25 cos × half-span or run include o/hang (you will need a scientific calculator)
	= 1/25° cos = 1.1034 per half-span or per m run
Roof A will be over rumpus, family, dining and lounge	
Half-span	= 7440 / 2
	= 3720 plus o/hang
	= 3720 + 450
	= 4170 or 4.17
Length of rafter for roof A	= 4.17 × 1.1034
Area of roof A	= length of fascia plus o/hang both ends × length of rafter
	= (18 960 + 900) × 4.601 × 2 sides
	= 19 860 or 19.86 × 4.601 × 2
	= 182.75 m²

>>

Area of roof B will be over kitchen, laundry and study

Half-span	= 4485 plus o/hang 450
	= 4935 (rafters would be pitched to a pitching plate, so do not divide by 2)
	or 4.935 m
Length of rafter	= 1.1034 × 4.935
	= 5.45 m
Area of roof B	= length of fascia × length of rafter
	= 6.380 × 5.45 × 1 side
	= 34.74 m² × 1
	= 34.74 m²

Area of roof C will be garage

Half-span	= 6110 / 2 = 3055 plus o/hang 450
	= 3505 or 3.505
Length of rafter	= 1.1034 × 3.505
	= 3.88 m
Area of roof C	= length of fascia plus o/hang both ends × length of rafter
	= (6110 + 900) × 3.88 × 2 sides
	= 7010 or 7.010 × 3.88 × 2
	= 54.4 m²

Area of roof D will be meals (treat as a gable, not as a half-hexagon)

Half-span	= 240 + 3380 + 240 / 2 = (3860 / 2) plus o/hang 450
	= 3860 / 2 = 1930 plus 450 o/hang
	= 2380 or 2.38 m
Length of rafter	= 1.1034 × 2.38
	= 2.626 m
Area of roof D	= length of fascia plus o/hang both ends × length of rafter
	= (3860 + 900) × 2.626 × 2 sides
	= 4760 or 4.760 × 2.626 × 2
	= 25 m²
Total area of roof	= area A + B + C + D
	= 182.75 + 34.74 + 54.4 + 25
	= 296.89 m² × wastage 5%
	= 296.89 × 1.05
	= 311.73 m²
Total area of roof	= say 312 m²

2.5 Calculate the quantity of materials that are measured by volume

When working on the volume and quantities of material needed, especially when you can use that same calculation in a variety of materials, it is best to have a systematic approach. The areas in which volumes are used on a domestic construction site are volume of soil to be removed for slabs etc., for cutting and filling sites to make them level to work on, detailed excavation for footings, the amount of concrete that is needed for those footings and any other concreting

work. When working out the various parts of the house for concrete, some may be added all in together and at other times it will be step by step as each job is different. The formula never changes; it will always be l × w × d.

Sometimes you might need to find the area first, such as in slabs, then apply the thickness of the slab to get the volume.

One of the best ways of calculating foundation to be dug, footings to be poured, and even centreline of double brick, is to use the centreline method. It's a systematic approach and saves work.

CALCULATE CENTRELINE MEASUREMENT TO USE FOR VOLUME OF SOIL AND VOLUME OF CONCRETE
USING THE PLAN IN FIGURE 10.3

Using the plan, calculate the volume of soil to be taken out and concrete needed for footings.

Measure the perimeter, starting at the top left-hand corner. Use arrows to notate the direction. For centreline measurements, take off the width of the footing, which is 450 mm. Depth of cut in foundation material is 750 mm. Height of footing is 600 mm. Internal measurements don't change.

→	5870 – 450	= 5420 for centreline
↓	2660	= 2660 internal measurement, no change
→	10030	= 10030 internal measurement, no change
↓	16300 – 450	= 15850
←	7920 – 450	= 7470
↑	2610	= 2610 internal measurement, no change
←	4725	= 4725 internal measurement, no change; disregard semi-hexagon
↑	3860 – 450	= 3410
→	1195	= 1195 internal measurement, no change; disregard semi-hexagon
↑	6380	= 6380 internal measurement, no change
→	4730	= 4730 internal measurement, no change
↑	6110 – 450	= 65415
	Total length	= 70 140 or 70.14 m
	Volume of spoil / excavation	= 70.14 × 0.45 × 0.75 (footing depth plus soil cover of 150 mm)
		= 23.67 m³ soil
	Volume of concrete	= 70.14 × 0.45 × 0.60
		= 18.939 m³ of concrete; allow 5% waste due to not a perfect trench
		= 18.939 × 1.05 (times 1 nothing changes; include the 5% or in money terms 5 cents)
		= 19.885 m³
	Concrete order:	19.885, round up to nearest volume of 0.2 m³ for ordering
		Therefore, 19.885 goes to 20 m³

Assume 80 mm slump, strength 20 MPa, so order 20 m³ 20 MPa 80 mm slump

---

Make sure you separate the waste that comes from the various products you use in your construction work into different locations on site, so it is easy to dispose of according to your local council requirements.

 COMPLETE WORKSHEET 2

3. Calculate the requirements for construction for the project

There are many components that make up the construction of any building, and each component could have different ways of calculating those components from framing to flooring and anything in between.

CLASS/GROUP ACTIVITY: COMPARING THE COST OF ONE CONTROL FOR A HAZARD OVER ANOTHER

Pick a small job that you may have completed at work lately (or imagine a small pergola, for example) and list the materials needed to do the job. Then, do some research online for costs of the materials, and make up a table to present the costs to classmates or your teacher so you can demonstrate that you know how to cost, and neatly present the costings and material lists, for a small job.

3.1 Calculate the quantity of wall and roof framing materials

Wall framing materials

Wall framing materials are usually a count of studs, plates, noggings, lintels and bracing material, then put in orderable lengths. There are a few rules, such as allowing extra studs for clusters at wall junctions, allowance for trimmers and double top plates to take trusses if that is the method of roof framing used.

HOW TO

CALCULATE FRAMING MATERIAL USING THE PLAN IN FIGURE 10.4

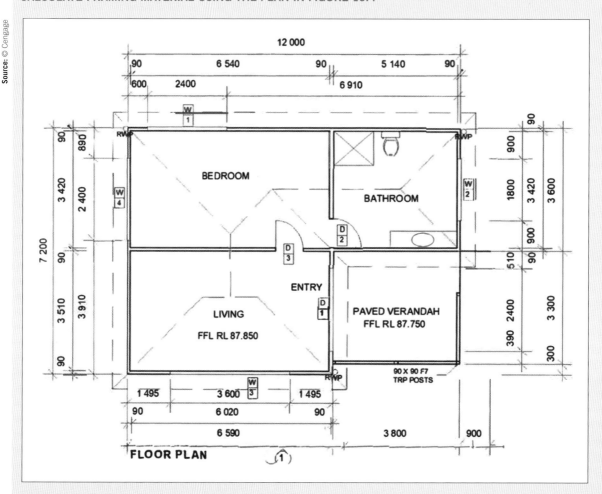

FIGURE 10.4 Floor plan

Specifications

Window and door schedule (height of windows and doors is always the first measurement)

W1, W4	1500 × 2400	Aluminium sliding with meranti reveals
W2	1500 × 1800	Aluminium sliding with meranti reveals
W3	1500 × 3600	Aluminium sliding with meranti reveals
D1	2040 × 2400	Aluminium sliding with meranti reveals
D2, D3	2040 × 820	Hume Hampton 5 prime coat door with timber jambs

Frame specifications

Use 90 × 35 MGP10 (machine-graded pine) radiata H2 blue pine for all plates, studs, trimmers and noggings
Use double top plates to take trusses
Spacing of studs @ 450 c/c

>>

Noggings @ 1350 c/c max
Lintels: W3 and D1 use 2/240 × 45 MGP12 cypress pine
W1 and W2 use 2/225 × 45 MGP12 cypress pine
All plates to be scarfed at all junctions
Maximum length of timber available 7.2 m
Finish ceiling level is 2400

Method

1 Treat each room separately.
2 Start at the top right-hand corner.
3 Don't deduct for any openings for studs, as the excess can be used for all trimmers.
4 Include 1 extra stud for all internal junctions.
5 Calculate the number of studs = length of plate divided by max. spacing + 1 + 1 each junction and external corner.

Bedroom

→	90 + 6590 + 90	= 6770
↓	90 + 3420 + 90	= 3600
		= 10 370 × (2 walls the same)
		= 20 740 or 20.74 m / 0.45
		= 46 + 1 + 1 + 4
		= 52 studs

Bathroom

→	90 + 5140 + 90	= 5320
↓	90 + 3420 + 90	= 3600
		= 8920 × 2
		= 17 840 or 17.84 / 0.45
		= 40 + 1 + 1 + 4
		= 46 studs

Living and entry

→	6590	6590
↓	90 + 3510 + 90	= 3690
		= 10 280 × 2
		= 20 560 or 20.56 / 0.45
		= 47 + 1 + 1 + 4
		= 53 studs

Total studs
= 52 + 46 + 53
= 151 studs order 90 × 35 MGP10 radiata H2 blue pine 151 @ 2.4 m
= 362.4 m

Roof framing materials

Calculating roof framing material is only needed if it is a 'stick' or 'cut' roof – in other words, not a truss roof. For truss roofs, you would need to see a truss manufacturer and they would price the trusses and associated fixings as well as give you a plan on how to install the trusses, including the bracing of the trusses.

HOW TO

CALCULATE ROOF FRAMING MATERIAL USING THE PLAN IN FIGURE 10.4 ROOF FRAMING SPECIFICATION FOR A CUT ROOF

Rafters: 100 × 50 MGP10 blue pine @ 450 c/c
Cripple creepers, valley creepers, jack rafters, crown end and centring rafters: 100 × 50 MGP10 blue pine
Ridge, valleys and hips 150 × 19 MGP15 blue pine

Disregard all ceiling frames, purlins, struts, hanging beams, strutting beams, scissors or half-scissors.
Roof pitch 25 degrees
Overhang 450 mm
Treat all roofs as a gable.

Process
Treat each roof separately.

Roof A

1   Find length of rafter per m run

$$= 1/25 \cos$$
$$= 1.104 \text{ m per run}$$

2   Find half-span over roof area A – bedroom & living

$$= 6590 + (2 \text{ o/hang}) / 2$$
$$= (6590 + 900) / 2$$
$$= 7490 \text{ or } 7.49 / 2$$
$$= 3.745 = \text{half-span or run}$$

3   True length of rafter = rafter length × run

$$= 1.104 \times 3.745$$
$$= 3.8 \text{ m orderable length} = 4.1 \text{ m to allow for plumb cuts}$$

4   Number of rafters roof A

$$= \text{overall length} + \text{overhangs} / 450 + 1 \times 2 \text{ sides}$$
$$= (7200 + 900) / 450$$
$$= 8100 \text{ or } 8.1 / 0.045$$
$$= 18 + 1$$
$$= 19 \times 2 \text{ sides}$$
$$= 36 \text{ rafters @ } 3.9 \text{ m}$$

Timber order would be 100 × 50 MGP10 blue pine 36/3.9

Roof B – Bathroom

1   Find length of rafter per m run

$$= 1/25 \cos$$
$$= 1.104 \text{ m per run}$$

2   Find half-span over roof area B – bathroom = width of room + overhangs / 2

$$= 3600 + 900$$
$$= 4500/2$$
$$= 2250 \text{ or } 2.25 \text{ half-span or run}$$

3   True length of rafter = rafter length × run

$$= 2.25 \times 1.104$$
$$= 2.484 \text{ orderable length} = 2.7 \text{ m should be long enough for plumb cut}$$

4   Number of rafters roof B = overall length + 2 walls + overhangs / 450 + 1 × 2 sides

$$= 5140 + 90 + 90 + 450$$
$$= [(5770 \text{ or } 5.77) / 450] + 1 \times 2$$
$$= 12.8, \text{ say } 13 + 1$$
$$= 14 \times 2 \text{ sides}$$
$$= 28 \text{ rafters @ } 2.7$$

Timber order would be 100 × 50 MGP10 blue pine 28/2.7

>>

You now have how many rafters needed which will be used to frame up the roof. Next come the hips, valleys and ridges (assume this roof is a hip roof and no Dutch gable in it as shown).

Roof A ridge length	= length of rooms + overhangs – span
	= 7200 + 900 – span
	= 8100 – 7490 (use span from calcs in rafters)
	= 610 mm or 0.610 m
Roof B ridge length	= length of room + overhang – half-span (in this case)
	= 5770 + 450 – half-span
	= 6220 or 6.22 m – 2.25 (use span from calcs in rafters)
	= 3.97 m or 3970 mm
Hip length	= plan length hip, rise per run to find true length hip

The plan length of a hip is 1.414 per metre run (how do we know this) – if you have a 1 m × 1 m roof and you apply Pythagoras' theorem of $A^2 + B^2 = C^2$

Therefore, 1 m^2 = 1 × 1 = 1; do that to both A and B you have 1 + 1 = 2 m^2, and the $\sqrt{}$ of 2 = 1.414 m; to find the rise, use the pitch (25 degrees) tan = 0.466 m (see Figure 10.5).

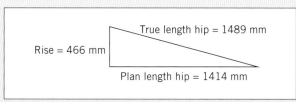

Rise = 466 mm

True length hip = 1489 mm

Plan length hip = 1414 mm

FIGURE 10.5 Hip length

Source: © Cengage

Using Pythagoras again	= $\sqrt{}$ (Rise m^2 + PL Hip m^2)
	= $\sqrt{}$ (0.466 m^2 + 1.414 m^2)
	= $\sqrt{}$ (0.217 + 2.0 m^2)
	= $\sqrt{}$ (2.217)
	= 1.489 m per run or (half-span + overhang)
Hips and over bathroom	= (Run or half-span + oh) × TL hip
	= (3600 + 900) / 2 = 2.25 × 1.489 = 3.35 m, say 3.6 m for plumb cuts
Using Pythagoras again	= $\sqrt{}$ (Rise m^2 + PL Hip m^2)
	= $\sqrt{}$ (0.466 m^2 + 1.414 m^2)
	= $\sqrt{}$ (0.217 + 2.0 m^2)
	= $\sqrt{}$ (2.217)
	= 1.489 m per run or (half-span + overhang)
Hips and over bed & living	= (Run or half-span + oh) × TL hip
	= (6590 + 900) / 2 = 3.745 × 1.489 = 5.6 m, say 6 m for plumb cuts
Timber order for hips & valley	= 150 × 25 MGP15 blue pine 4/6 m, (the 4th one is used for construction purposes), 3/3.6 (third one used for valley as shown)

For clarity 150 × 25 MGP15 blue pine 4/6 m, 3/3.6 m

Total 150 × 25 MGP15 = 32.4 m

Be very careful when working at height on construction jobs because falling is the major cause of injury in the construction industry each year in Australia. A fall can have major implications on your ability to work and do leisure things. Always get the latest information on how to perform a task at height safely.

3.2 Calculate the dimensions and quantity of sheets of each type of flooring and lining material, ensuring that the most economical layout is employed

Bearers and joists are used when a concrete slab is not. It can allow for storage underneath if on a sloping site.

Sizes and grades of timber are determined by AS 1684 Residential timber-framed construction. Manufactured timber products such as LVLs have made internal piers redundant (LVLs = Laminated veneer Lumber and Glulam beams. For termite-prone areas, the use of metal sub-floors has increased, with different companies using different systems. The type of floor system you use is up to the consumer, but most floors are constructed on a platform system instead of a cut-in system. You can use particle board, plywood or tongue-and-groove (T&G) boards in a platform system, but generally only T&G boards are used in a cut-in floor system.

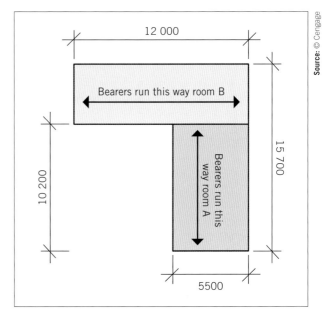

FIGURE 10.6 Floor plan

HOW TO

CALCULATE BEARERS AND JOISTS

Method
Bearers generally run the longest length of the room. Disregard placement of piers.

Specification: F11 100 × 75 hwd bearers @ 1800 c/c

F11 100 × 50 hwd joists @ 450 c/c. These run in the opposite direction to the bearers.

Formula for bearers = (width of room / max spacing) + 1 (why do we add the extra 1?)

If you start on the left-hand side of the room and you have worked out the number of bearers using the formula, you will still need one more to sit on the outside wall).

Calculation: room A, 5.5 / 1.8 = 3.06; must go to a whole number, so 4 + 1

No. of bearers room A = 5

Maximum length of bearer is 7.2 m

Room A is 10.2 m long; therefore, any combination of linear lengths to make up 10.2 is OK
I have chosen 5 @ 5.7 and 5 @ 4.5; stagger these along piers

Timber order room A	= 100 × 75 F14 hwd 5/5.7, 5/4.5
	= 51 m
No. of bearers room B	= 5.5 / 1.8 = 3.06, so 4 + 1 = 5 bearers
Room B is 12.5 m long	= I have chosen 6.6 m lengths
	= 6.6 × 2
	= 13.2 m
Timber order for room B bearers	= 100 × 75 F14 hwd 10 @ 6.6 m
	= 66 m

Room A joists

Formula = length of room / max spacing + 1 + 1 for every load-bearing wall

No. of joists room A	= 10 200 / 450 + 1 + 2 (outside walls)
	= 22.7 joists, round up to 23 + 1 + 2
	= 26

>>

>>

Timber order for room A joists	= 100 × 50 F14 hwd 26 @ 5.7 m (nearest orderable length)
	= 26 × 5.7
	= 148.2 m

Room B joists

Formula = length of room / max spacing + 1 + 1 for every load-bearing wall

No. of joists room B	= 12 500 / 450 + 1 + 2 (outside walls)
	= 27.8 joists, round up to 28 + 1 + 2
	= 31
Timber order for room B joists	= 100 × 50 F11 hwd 31 @ 5.7 m (nearest orderable length)
	= 31 × 5.7
	= 176.7 m

Calculate flooring: Platform floor (means it covers the entire floor space to make it easier to work on and all frames can be placed with safety)

Using particle board flooring moisture-resistant (MR) 3600 × 900 18 mm thick sheets, allow 5% wastage for cutting.

Method

1 Calculate the m² areas A and B.
2 Calculate m² of a sheet.
3 Divide m² into the room areas.
4 Include waste.

Room A	= 10.2 × 5.5
	= 56.1 m²
Room B	= 12.5 × 5.5
	= 68.75 m²
Total area	= A + B
	= 124.85 m²
Area of 1 sheet	= 3.6 × 0.9
	= 3.24 m²
No. of sheets	= 124.85 / 3.24
	= 38.53 × wastage
	= 38.53 × 1.05
	= 40.46 sheets, say 41 sheets

3.3 Calculate the length of linear flooring and lining material, ensuring that the most economical layout is employed

Calculating the length of linear, or strip, flooring and lining material is best done by finding out the width of the flooring or lining material you are going to be using and determining the effective cover width of that material. The effective cover width means the width of the material that covers the surface area of the wall or floor that you are installing the material onto. It is the width of the material minus any tongues or overlaps the material may have moulded into its edge. The effective cover is determined by this simple formula:

- The number of linear metres required of your material to cover each square metre of wall or floor area = 1 metre divided by the effective cover of the material. For example:
  - Flooring that is 60 mm wide may look like this:

Source: © Cengage

  - Therefore: 1.0 m ÷ 0.06 m = 16.7 m required to cover 1 m² of floor area.

You must know how many linear metres of material you require to cover every square metre of wall or floor. This means you can simply divide the wall or floor area that needs covering by the linear metres per square metre for your cover material. The result will be the number of linear metres of your floor or wall strip cover material you require to cover your job. The following 'How to' box shows some examples.

HOW TO

Calculate the same floor with T&G boards

Not end-matched boards
Secret nail profile
Brushbox species
Effective cover (EC) 60 mm × 19 mm thick
Allow 10% wastage for cutting over joists

Method

1 Use same areas as sheet flooring.
2 Linear metres per square metres.
3 Include the waste.

Linear metres per m²	= 1000 / effective cover
	= 1000 / 60
	= 16.7 m per m²
Area A	= 56.1 m² × 16.7
	= 937 m
Area B	= 68.72 m² × 16.7
	= 1148 m
Total linear metres	= (937 + 1148) × 1.1 for waste
	= 2085 × 1.1
	= 2294 m of brushbox T&G boards, secret nail profile 60 EC × 19 mm thick

---

3.4 Calculate the dimensions and quantity of sheets of external cladding material, ensuring that the most economical layout is employed

For sheet cladding there are several products on the market that are used instead of bricks for an outside cladding. One example is Koolwall. Panels made of a polystyrene sheet material are fixed to the studs, and a mesh-like fabric is fixed to the panels. After this, a special render is applied to make the cladding waterproof and fire-resistant. Other examples include James Hardie's Scyon range of products. Also, there are thin brick or stone claddings that come in sheets to be fixed to the studs. Whatever system you use, it is the same method of calculation. However, the method of fixing is different for all types; follow the manufacturer's installation techniques.

Step 1. Calculate the length of the outside perimeter.

Step 2. Find the height from damp-proof course (DPC) to underside of eave.

Step 3. Multiply those two figures together for total area.

Step 4. Calculate any deduction for doors, windows or other openings; disregard any opening less than 1 m².

Step 5. Subtract deduction areas from total areas.

Step 6. That will give you an area to cover.

Step 7. Calculate the area of one (1) sheet.

Step 8. Divide that sheet area into the area to cover to give you the number of sheets.

Step 9. Add the nominated wastage.

Step 10. Round up to full sheets.

Use the plan in **Figure 10.7** for calculation.

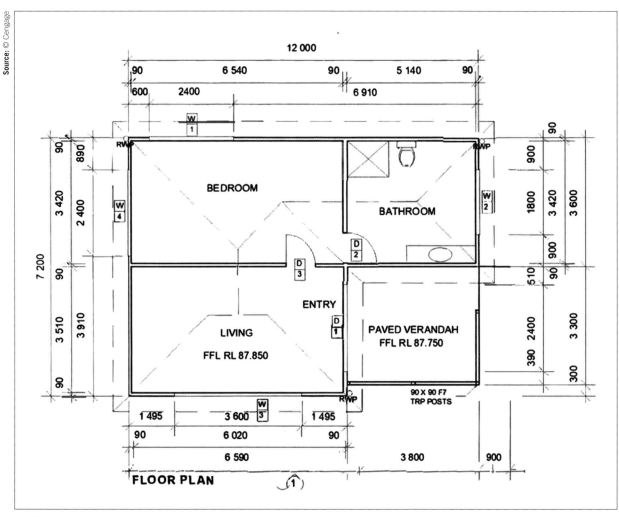

FIGURE 10.7 Floor plan

HOW TO

EXTERNAL CLADDING SHEETS

Step 1.

→	= 12 000
↓	= 3600
← = 5140 + 90	= 5230
↓ = 3510 + 90	= 3600
← = 6540 + 180	= 6720
↑	= 7200

Total length   = 38 350 or 38.35 m

Step 2. Height wall from DPC to underside of eaves = 2525 or 2.525 m

Step 3. Total area   = 38.35 × 2.525

                                  = 96.83 m²

\>>

Step 4. Deductions from openings W1 = 2.1 × 2.4, W2 = 2.1 × 1.8, W3 = 2.1 × 3.6, W4 = 2.1 × 2.4, D1 = 2.1 × 2.4

W1, W4, D1	= 2.1 × 2.4 × 3
	= 15.12 m²
W2	= 2.1 × 1.8
	= 3.78 m²
W3	= 2.1 × 3.6
	= 7.56 m²
Total deductions	= 15.12 + 3.78 + 7.56
	= 26.46 m²
Steps 5 & 6. Final area to be covered	= 96.83 − 26.46
	= 70.37 m²
Step 7. Area of 1 sheet, use Scyon Matrix cladding 2990 × 1190 × 8 sheets	
	= 2.99 × 1.19
	= 3.56 m²
Step 8. Divide that sheet area into the area to cover to give you the number of sheets	
	= 70.37 / 3.56
Step 9. Allow for waste	= 19.77, allow 5% waste
	= 19.77 × 1.05
Step 10. Number of sheets	= 20.76 sheets, round up to 21 sheets

3.5 Calculate the length of linear external cladding material, ensuring that the most economical layout is employed

There are many varieties of external linear cladding, from various species of timber and profiles such as Shiplap to fibre cement products that look similar but are impervious to termite attack and are fire-resistant. Another product is Weathertex, which is easier to cut and you don't need special tools compared to fibre cement products. Again, whatever product you use, follow the manufacturer's instructions and fixing techniques. Any of these products use the same principles as the sheet material discussed in the previous section, except for one thing: the effective cover of the product, which is the same as flooring products.

HOW TO

CALCULATE EXTERNAL LINEAR CLADDING (WEATHERBOARDS)

Use calculations from the previous section.
Step 1. Calculate the length of the outside perimeter.
Step 2. Find the height from DPC to underside of eave.
Step 3. Multiply those two figures together for total area.
Step 4. Calculate any deduction for doors, windows or other openings; disregard any opening less than 1 m².
Step 5. Subtract deduction areas from total areas.
Step 6. That will give you an area to cover.
Step 7. Calculate the linear metres per m² by dividing 1000 by effective cover of board.
Step 8. Multiply the linear metres by the area to be covered.
Step 9. Allow for waste.
Step 10. Divide length of board into total length.
Step 11. Final number of boards to buy.

Use figures from the previous task on sheet material.

Step 1 from previous outside perimeter	= 38.35 m
Step 2 from previous	= 2.525 m
Step 3 from previous	= 96.83 m²

Step 4 from previous deductions = 26.46 m²

Steps 5 & 6 from previous = 70.37 m²

Step 7. Calculating linear metres per m². Use James Hardie PrimeLine Heritage weatherboard 300 wide cladding

Effective cover (EC)	= 278 mm, length of boards 4.2 m
	= 1000 / 278
	= 3.6 m per m²
Step 8. Linear metres	= Final area × m per m²
	= 70.37 × 3.6
	= 253 m
Step 9. Allow for waste 10%	= 253 × 1.1
	= 278 m³
Step 10. Number of boards needed	= 278.3 / 4.2
	= 66.26 boards
Step 11. Final total number of boards	66.26, round up to 67 boards of James Hardie PrimeLine Heritage weatherboard 300 wide cladding

3.6 Calculate the dimensions and quantity of sheets or units of roofing material, ensuring that the most economical layout is employed and allowing for overlaps

When calculating roof sheeting, the choice of what you can use is vast, with different profiles, colours and methods of fixing. The most common type is corrugated Colorbond sheeting. These sheets can be cut to any length. Some companies even have trucks that come to site and manufacture the sheets onsite, so there are no issues about correct lengths or colour matching. Before screwing down sheets a vapour barrier must be installed. Take the barrier about 20 mm into the gutter. Sheets are laid with 50 mm into the gutter. The overlap of sheets is usually 1½ ribs, and always lay sheets on the lee side of prevailing winds first (see Figure 10.8).

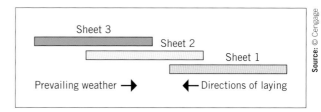

Source: © Cengage

FIGURE 10.8 Laying roof sheets

At the ridge line, make sure you have turned up the valleys of the rib to stop water blowing back under the sheet. Minimum pitch of roof is 5 degrees. When measuring, treat each roof as a gable and break up roof sections. When laying hip roofs, what you cut off can be used for the opposite hip.

HOW TO

CALCULATE QUANTITY OF SHEETS FOR METAL ROOFING USING THE PLAN IN FIGURE 10.7

Effective cover (EC) is 762 mm
Pitch 25 degrees
Overhang 450 mm
Allow 50 mm extra into gutter

Step 1. Rafter length roof A	Find length of rafter per m run
	= 1/25 cos
	= 1.104 m per run
Step 2. True length (TL) of rafter	= span of roof + 900 / 2
	= 7490 / 2
	= 3745 or 3.745 × 1.104
	= 4.135 + 50 into gutter
	= 4185 mm = length of sheet roof A
Step 3. Find area of roof	= (length of roof + overhang) × TL rafter
	= 7200 + 900

$$= 8100 \times 3.745$$
$$= 30.33 \text{ m}^2$$

Step 4. Find area of 1 sheet $= \text{length of sheet} \times \text{EC}$
$$= 4.185 \times 0.762$$
$$= 3.2 \text{ m}^2$$

Step 5. Find number of sheets for roof A
$$= \text{area of roof} \div \text{area of 1 sheet}$$
$$= 30.33 \div 3.2$$
$$= 9.48 \text{ sheets, round up to } 10 \times 2 \text{ sides}$$
$$= 20 \text{ sheets 4185 mm long}$$

Step 1. Rafter length roof B   Find length of rafter per m run
$$= 1/25 \cos$$
$$= 1.104 \text{ m per run}$$

Step 2. True length of rafter $= \text{span of roof} + 900 / 2$
$$= 3600 + 900 / 2$$
$$= 2250 \text{ or } 2.25 \times 1.104$$
$$= 2.484 + 50 \text{ into gutter}$$
$$= 2534 \text{ mm} = \text{length of sheet roof B}$$

Step 3. Find area of roof $= (\text{length of roof} + \text{overhang}) \times \text{TL rafter}$
$$= 5140 + 180 + 450 \text{ (only 1 overhang)}$$
$$= 5770 \text{ or } (5.77 \times 2.534)$$
$$= 14.62 \text{ m}^2$$

Step 4. Find area of 1 sheet $= \text{length of sheet} \times \text{EC}$
$$= 2.534 \times 0.762$$
$$= 1.93 \text{ m}^2$$

Step 5. Find number of sheets for roof A
$$= \text{area of roof} \div \text{area of 1 sheet}$$
$$= 14.62 \div 1.93$$
$$= 7.56 \text{ sheets, round up to } 8 \times 2 \text{ sides}$$
$$= 16 \text{ sheets 2534 mm long}$$

LEARNING TASK 10.3

CLASS/GROUP ACTIVITY: MATERIALS COSTING FOR CONSTRUCTION JOBS

Contact a local supplier, either directly or online, and obtain costs and details on how they like the materials listed below to be ordered:

- particle board sheet flooring
- linear flooring
- external cladding sheets
- corrugated iron roofing sheets.

 COMPLETE WORKSHEET 3

4. Check and record results

Review all quantities and make sure you have the correct answer and have used the right formula for what you are quantifying. Is it in m² or m³? Does it ask for total linear metres or number of sheets or other material as it is sold at trade suppliers? Have you followed industry practice in ordering, and rounded up units as required?

4.1 Record workings and review calculations for accuracy

As you have worked through each element you would have an answer for every component of the building that has been asked for. When quantifying materials, it is best to refresh and review those answers in case you missed something.

One famous quote – when a building company won a tender, the director said, 'What did we miss?'

Always check and double-check your findings. Make sure you have read the specifications and have used the appropriate method for finding area, volume or linear metres.

Have you broken the answer down to orderable material sizes or acceptable building practice, such as ordering concrete in 0.2 m³ or timber in industry lengths? Have you checked the effective cover per m²?

4.2 Record results of calculations as required for costing and ordering materials

You have done the hard work. Now we have to put that information into typical ordering procedures for the different types of material as well as adding the cost associated with those materials. All prices have been sourced from local hardware businesses as of 15 November 2021. Prices may be different in other locations.

A good way to check how to write up your order for all the various building materials you may use on a construction site is to contact the supplier and ask them how they like the orders placed. You can sometimes get a copy of their latest prices and this will give you a guide for your ordering methods as well.

Calculating area of lining material

Plasterboard linings

10 mm plasterboard, sheet sizes 4.8 × 1.2 – 6 sheets @ $28.00 sheet (no allowance for glue, fasteners, battens, setting or labour) = $168.00

Calculating building wrap and external cladding

Building wrap

Tyvek wrap or similar to wrap whole of frame
Roll size 30 m × 0.9 m = 7 rolls @ $91.00 per roll = $637.00

Bricks for external cladding

Hume Access cream face bricks (no allowance for scaffolding, mortar, veneer ties, vent holes or labour)
= 4 pallets or 2000 bricks @ $670.00 / thousand bricks
= $2680.00

Cladding other than bricks

James Hardie PrimeLine Weatherboard Heritage range 300 wide boards, length of boards 4.2 m (no allowance for fixings, Tyvek wrap [insulation], scaffolding or labour)
= 35 boards @ $43.79/ board
= $1532.65

Calculating roofing material

Roofing tiles

Boral Contour concrete tiles (no allowance for sarking, anti-ponding board, battens, fixings, ridge or hip tiles, pointing up mortar, tile hoist, edge protection or labour to lay.
= 312 m^2 tiles @ $12.75 m^2
= $3978.00

Calculating the quantity of materials by volume

Volume of spoil in trench and amount of concrete to go into trench

= 20 m^3 MPa 80 mm slump @ $273 per m^3 (no allowance for reinforcing steel, concrete pumps, formwork or labour to lay)
= $5460.00

Calculating requirements for construction

Wall framing

= 100 × 38 MGP10 H2 blue radiata pine – 151 studs or 362.4 m @ $4.65 per m
= $1685.16

Roof framing

= 100 × 50 MGP10 H2 blue radiata pine for all rafters, centring rafters, crown end rafters, jack rafters and cripple creepers. Roof area A 36 / 3.9 m = 140.4 m
Roof B 28 / 2.7 m = 75.6 m
Total = 216 m @ $5.85 m
= $1263.60 150 × 25 MGP15 H2 radiata pine ridges and hips
Ridge A = 0.610 m; ridge B = 3.970 m; hips 4/6 m, 3/3.6
Total 150 × 25 = 39.38 m, say 40 m @ $8.20 m
= $328.00

Flooring particle board sheets

3.6 × 0.9 × 19 mm yellow tongue practical board flooring 41 sheets @ $36.50 per sheet
= $1496.50

Tongue-and-groove (T&G) flooring, bearers and joists

100 × 75 hwd bearers room A 5/5.7, 5/4.5 = 51 m
100 × 75 hwd bearers room B 10/6.6 = 66 m
Total 100 × 75 hwd bearers 117 m @ $18.50 per m
= $2164.50
100 × 50 hwd joists room A 26/2.7 = 148.2 m
100 × 50 hwd joists room B 31/5.7 = 176.7 m
Total 100 × 50 joists = 324.9, say 325 m @ $9.90 per m
= $3217.50
75 × 19 Brushbox T&G flooring 2294 m @ $2.35 per m
= $5390.90

External cladding sheets

Scyon Matrix cladding 2990 × 1190 × 8 – 21 sheets @ $132.49 per sheet
= $2782.29

External linear cladding

James Hardie PrimeLine Weatherboard Heritage range 300 wide boards, length of boards 4.2 m (no allowance for fixings, Tyvek wrap [insulation], scaffolding or labour)

67 boards @ $43.79 per board
 = $2933.93

Roofing material

Colorbond Custom Orb roofing Monument colour, effective cover 762 mm

Roof A 20 / 4185 mm = 30.33 m²
Roof B 16 / 2534 mm = 14.62 m²
Total m² = 44.95 m² @ $22.00 m²
 = $988.90

 COMPLETE WORKSHEET 4

Calculating other areas and volumes – different geometrical shapes

TABLE 10.2 Formulas commonly used in the construction industry

Any figure/shape	Perimeter formula	Unit of measurement
© Cengage	$P = S_1 + S_2 + S_3 + S_4$	m or mm
Square ('L' shaped, or complex shaped area calculations are done by breaking the shapes down into simple shapes, and then adding the singular areas up to get a total)	**Area formula** $A = S^2$ $A = S_1 \times S_2$ $A = L \times W$ $A = L \times H$	**Unit of measurement** m²
Rectangle ('L' shaped, or complex shaped area calculations are done by breaking the shapes down into simple shapes, and then adding the singular areas up to get a total)	**Area formula** $A = L \times W$ $A = L \times H$	**Unit of measurement** m²
Triangle	**Area formula** $A = \dfrac{B \times H}{2}$ $A = \frac{1}{2} \times B \times H$	**Unit of measurement** m²
Parallelogram	**Area formula** $A = B \times PH$ $A = L \times W$	**Unit of measurement** m²
Trapezoid	**Area formula** $A = \dfrac{(B1 + B2)}{2} \times PH$ $A = \frac{1}{2} \times (B1 + B2) \times PH$	**Unit of measurement** m²
Circle	**Area formula** $A = \pi\, r^2$ $\pi = 3.142$	**Unit of measurement** m²

>>

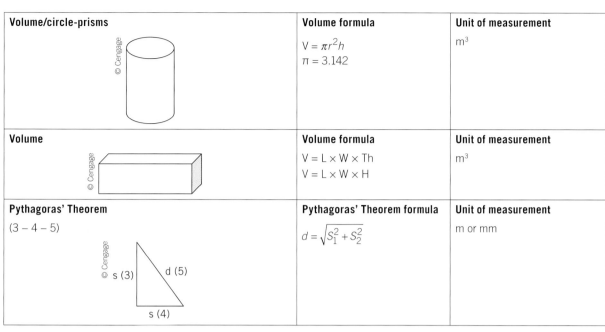

Volume/circle-prisms	Volume formula	Unit of measurement
	$V = \pi r^2 h$ $\pi = 3.142$	m^3
Volume	Volume formula	Unit of measurement
	$V = L \times W \times Th$ $V = L \times W \times H$	m^3
Pythagoras' Theorem (3 – 4 – 5) s (3) d (5) s (4)	Pythagoras' Theorem formula $d = \sqrt{S_1^2 + S_2^2}$	Unit of measurement m or mm

Legend: A: Area; B: Base; C: Circumference; D: Diameter; L: Length; H: Height; Th: Thickness; P: Perimeter; r: Radius; S: Side; W: Width; PH: Perpendicular Height; π(pi) = 3.142

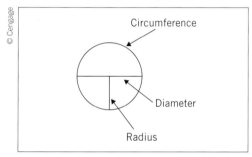

Refer to Chapter 6 for more information on areas and volumes.

Difference between quantity surveying and builders' quantities

Throughout these exercises in quantities and costing we have used builders' quantities or typical sizes and methods of buying the products. Whether they be in sheets, lengths or litres, it is how a builder would have to first work out the amount of material then apply a cost appropriately.

A quantity surveyor must use a set of rules as per the *Australian and New Zealand Standard Method of Measurement of Building Works* current edition 2018

(often referred to as 'the mode'). For example, when working out reinforcement steel it is to be costed by the tonne, even though the workings would be there to calculate how many lengths of trench mesh or sheets of fabric are required. If you go on to the Certificate IV Building and Construction or the Diploma of Building and Construction, you will be expected to quantify buildings this way. If you ever come across a bill of quantities on the work site, have a look to see what it is like. Most domestic buildings would not have a bill of quantities as they are mainly used for commercial works and sent out for tender as part of the contract conditions.

SUMMARY

In Chapter 10 you have learnt how to plan and perform calculations to determine material requirements for a construction project. We have covered how to break down the components of various materials using formulas to find areas, volumes or linear metres depending on the building product being calculated. We have also covered how to:

- plan and prepare
- calculate area and volume of construction materials for the project
- calculate the requirements for construction for the project
- check and record results.

REFERENCES AND FURTHER READING

Laws, A. (2020). *Site Establishment, Formwork and Framing* (4th edn), Cengage Learning Australia, Melbourne, Vic.

Staines, A. (2012). *The Roof Building Manual* (5th edn), Pinedale Press, Caloundra, Qld.

Useful weblinks

www.australbricks.com.au, 2021 Brick and Paver Collection, Victoria, Search for The Gyprock Red Book

WoodSolutions, Timber Flooring, Design Guide 09, **www.woodsolutions.com.au/publications**

Australian Standards

AS 1684-2010 Residential timber-framed construction

AS 1288-2006 Glass in buildings – Selection and installation

AS 1860.2-2006 Particleboard flooring, Part 2: Installation

AS 2050:2018 Installation of roof tiles

AS 3600:2018 Concrete structures

AS 4773.2:2015 Masonry in small buildings, Part 2: Construction

NCC – BCA Volume 2

Apps to download

Quick Chippy app, Quick Chippy, **https://www.quickchippy.com.au/quick-chippy-app**

Quick Decks app, Quick Chippy, **https://www.quickchippy.com.au/quick-decks-app**

Quick Walls app, Quick Chippy, **https://www.quickchippy.com.au/quick-walls-app**

Search for 'carpentry calculator' in your app store

www.blocklayer.com.au has lots of construction calculators

GET IT RIGHT

In the photo below, the workers are doing their measuring and marking out task with some incorrect practices. Identify these incorrect methods and provide reasoning for your answer.

Source: Richard Moran

_____

_____

_____

_____

_____

_____

 **WORKSHEET 1**

Student name: _____

Enrolment year: _____

Class code: _____

Competency name/Number: _____

Task: Read through the sections beginning at *Plan and prepare* then complete the following questions.

1 List five (5) Australian Standards or documents that could be relevant to this unit.

_____

1 _____

2 _____

3 _____

4 _____

5 _____

2 List four (4) hazards that may not be obvious if you don't stop and think through your work processes.

_____

1 _____

2 _____

3 _____

4 _____

3 Where can installation instructions be obtained for installing a product?

_____

4 When creating a SWMS, how long should the document be?

_____

5 Does the NCC contain information about side boundary distances in housing?

_____

6 What is the crucial element needed for calculating quantities on a building site?

_____

7 Carry out your own research. What is the Australian Standard for reinforced concrete footings?

_____

Student name: _____

Enrolment year: _____

Class code: _____

Competency name/Number: _____

Task: Read through the sections beginning at *Calculate area and volume of construction materials for the project* then complete the following questions.

1   Do you need access to the plans and documents for a project to work on area and volume of construction materials?

  YES   NO

2   For most construction projects you will be working out the area or volume of an item and then applying that to the quantities needed.

  TRUE   FALSE

3   List two (2) units of measurement that are used in the construction industry.

  _____

4   List three (3) International System of Units used in construction:

  1   _____

  2   _____

  3   _____

5   What would a 30 m tape be used for in construction?

  _____

6   What is the formula for volume?

  _____

7   What is the formula for area?

  _____

8   When using the formula for area, to how many decimal places do you round up?

  _____

9   When using the formula for volume, to how many decimal places do you round up?

  _____

10   Name four (4) trades that would use an area formula in calculations.

_____

11   Name three (3) trades that would use volume in calculations.

_____

12   What does linear mean?

_____

 **WORKSHEET 3**

Student name: _____

Enrolment year: _____

Class code: _____

Competency name/Number: _____

Task: Read through the sections beginning at *Calculate the requirements for construction for the project* then complete the following questions.

1   List five (5) items to be measured when working out the quantities for wall framing.

  1   _____

  2   _____

  3   _____

  4   _____

  5   _____

2   List four (4) logical methods to apply when calculating wall framing studs.

  _____

  _____

3   List a formula to use when calculating wall framing studs.

  _____

4   Calculating the rafters needed in a roof is required in a truss roof as well as a pitched roof.

  TRUE   FALSE

5   List a formula to use when calculating sheet flooring.

  _____

  _____

6   List a formula to use when calculating strip flooring.

  _____

  _____

7   When calculating quantities, there are some basic rules or systems you should follow so that you don't miss anything. What are they?

  i   _____

  ii   _____

  iii _____

iv _____

v _____

vi _____

8   Calculate the volume of soil/spoil to be removed from a trench 15 m long, 750 mm deep, 600 mm wide. Answer in m^3.

_____

_____

9   Calculate the volume of concrete to go into the trench if the footing is 600 mm deep. Allow 5% wastage.

_____

_____

_____

_____

_____

10 The following questions relate to the figure Q1 in the workbook. Calculate the m^2 of 13 mm plasterboard for the ceiling. Max. sheet size 4800 × 1200.

   Area of plasterboard = 11.7 × 7.3 = 85.41 m^2
   m^2 of one sheet   = 4.8 × 1.2 = 5.76 m^2
   No. of sheets   = 85.41 / 5.76
                          = 14.8 sheets, say 15 sheets

11 Calculate the litres of paint if painting a ceiling with Dulux Ceiling White by 2 coats. Average coverage = 16 m^2 per litre of paint. Minimum paint can size 4 L or 10 L. Allow 5% wastage.

   Litres of paint for ceiling = 85.14 m^2 / 16
                               = 5.3 × 2 coats
                               = 10.7 L × wastage 5%
                               = 10.75 × 1.05
                               = 11.24 L, use 3 × 4 L cans or 1 × 4 L + 1 × 10 L cans

12   Calculate the m² of floor tiles for the same room using 50 × 50 mosaics on mat size 300 × 300 coverage.

Number of tile mats:

Coverage of 300 × 300 mat = 0.3 x 0.3

= 0.09 m²

Area = 85.41 / 0.09

= 949 mats of 50 × 50 mosaic tiles

13   Calculate the amount of bricks in Q1 if 50 bricks per m² 230 walls (2 skins), garage door opening 2100 × 3000, height of walls 2550. Allow 5% wastage.

Number of bricks:

⟶ 11.7 + 0.46 = 12.16 m

↓ 7.3 + 0.46   = 7.76 m

Total   = 19.92 m × 2

= 39.84 m × 2.55 m (height)

= 101.59 m² × 1.05 wastage

= 106.67 m²

Deductions   = 2.1 × 3 = 6.3 m²

Total   = 106.67 – 6.3

= 100.37 m² × 2 skins × 50

= 10 037 bricks, say 11 000 bricks

 **WORKSHEET 4**

Student name: _____

Enrolment year: _____

Class code: _____

Competency name/Number: _____

Task: Read through the sections beginning at *Check and record results* then complete the following questions.

1 Do you need to double check your calculations when working out the quantities for a building materials order?

    YES   NO

2 Provide an example of how you might order some different building materials listed in the following table. One answer has been provided for you.

Material	Order procedure answers (depending on supplier's requirements)
Lengths of timber	Type of timber and section size, number of lengths/lengths of pieces
Plasterboard	
Building wrap	
Bricks	
Cladding	
Roof tiles	
Sand	
Sheet flooring	

PERFORMANCE OF SKILLS EVIDENCE

Student name: _____

Enrolment year: _____

Class code: _____

Competency name/Number: _____

Task: Working with your teacher, prepare to demonstrate your skills by completing the following task. To demonstrate competency, a candidate must satisfy all the elements, performance criteria and foundation skills of this unit by analysing the drawings and specifications for a building with a minimum of six (6) rooms, including a kitchen and bathroom and linear external lining, and preparing a detailed list of materials and calculated quantities of each material for:

wall and roof framing

internal lining and flooring

external cladding and roofing.

1   Calculate the area of the roof: Pitch is 20 degrees, overhang is 450 mm.

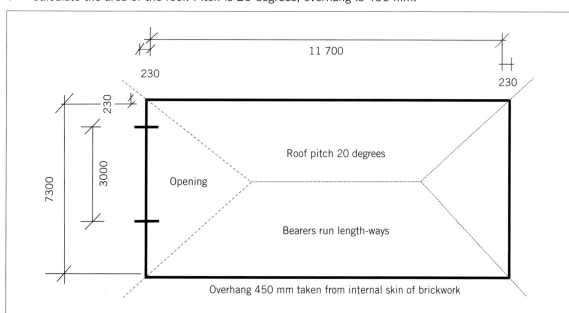

Area of roof:

Length of rafter	= 1/20 cos = 1.064 m
Half span or run include o/h	= 3.65 + 0.45 = 4.1 m
Total length of rafter	= 4.1 × 1.064 = 4.36 m
Area of roof	= length of building + o/hang × rafter × 2 sides
	= 11.7 + 0.46 + 0.9
	= 13.06 × 4.36 × 2
	= 113.88, say 114 m²

2   Using the same plan, calculate bearers and joists if: F11 100 75 hwd bearers @ 1600 c/c F11 100 50 Hwd joist @ 450 c/c, Max. length of timber is 6.0 m.

Bearer timbers	= width of room / max spacing + 1
	= 7.3 / 1.6 = 4.56, round up to 5 + 1
	= 6 bearers order
	= F11 100 × 75 hwd bearers 6/6.0 m, 6/5.7
Joist timbers	= length of room / max spacing +1 + 1 for external walls
	= 11.7 / 0.45
	= 26 + 1 + 2 = 29 joists
Order	= F11 hwd 100 × 50 joists 29/5.4, 29/2.1

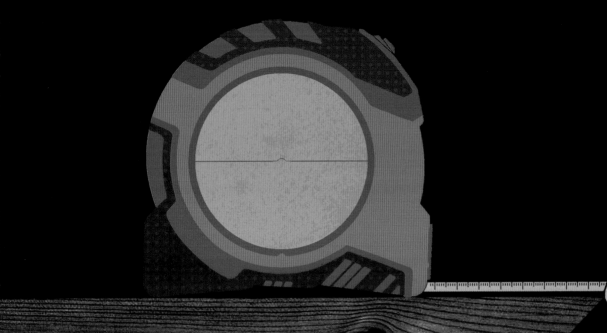

11 APPLY BASIC LEVELLING PROCEDURES

This chapter covers the following topics from the competency 'Apply basic levelling procedures':
* Plan and prepare
* Set up and use levelling device
* Clean up.

This unit of competency specifies the outcomes required to carry out levelling in a single plane for the purpose of establishing correct and accurate set-out of building components. It includes the set-up, testing and use of levelling devices, and establishing and transferring heights using a range of levelling equipment.

The unit supports workers in the construction industry who use a variety of common methods and equipment when working with others and as a member of a team. It applies to levelling work on residential and commercial work sites.

No licensing, legislative, regulatory, or certification requirements apply to this unit of competency at the time of endorsement.

Source: © Commonwealth of Australia, 2022

1. Plan and prepare

1.1 Job requirements are obtained, confirmed with relevant personnel, and applied to planning

Before any job starts, it is important to gather all the relevant materials and information needed and to work out what all the members of your work team will be doing. You need to be familiar with the plans and specifications for the job so you know the important points to mark out.

In New South Wales, the *Surveying and Spatial Information Act 2002* (NSW) and a Surveying and Spatial Information Regulation control the official measuring of land. For example, in the Act it states that there are penalties for disturbing any declared Survey Mark.

Before starting to use levelling devices, it is necessary to know what it is you want to achieve. Some of the options are:

1   If the heights you want to measure are close and the job is small, a spirit level may be sufficient; alternatively, you could use a spirit level and a straight edge.
2   If the job involves transferring heights over many metres and working by yourself, it would be beneficial to use a laser level.
3   For measuring heights around corners and out of line of sight, you will need to move the line-of-sight instruments or use a water (hydrostatic) level.

If you only require a transfer of height from one point to another, there is no need to set up booking sheets (a written way to calculate height differences) and log the measurements. However, if different heights are to be marked and calculated and positions measured in different places on the job site, you will need an accurate system of recording and processing the heights. By doing this, you can also build in checks so you eliminate error.

You will need to obtain permission from site owners (and approval to start earth work from the local government authority) before going onto a site and commencing any work.

Once you have learnt how to measure vertical height and horizontal measurements over small-scale jobs, you will learn how to measure heights over the distances needed on a building site.

1.2 Inspect work site for hazards and report them according to workplace procedures

Safety considerations could be as simple as looking up to ensure you don't make contact with overhead power lines with a metal measuring staff or ensuring that you avoid holes or obstacles on the site where you are working with instruments. Pegs can be especially hazardous on-site because of the sharp ends protruding above the ground. Protect these sharp ends with caps or some other safety method. Safe Work Australia has a

Traffic Management: Guide for Construction Work, an Excavation Code of Practice and a Safe Work Method Statement for High Risk Construction Work Information Sheet, which give good advice on how to operate safely. In the SafeWork NSW Code of Practice for Work Near Overhead Power Lines, it states that workers must not go any closer than 3 m to mains power lines and no closer than 0.5 m to low voltage insulated supply power lines.

For further information on safety and the risk assessments that should be undertaken before work commences, including a Job Safety Analysis and Safe Work Method Statement, please take the time to review the chapters *CPCWHS3001 Identify construction work hazards and select risk control strategies* and *CPCCWHS2001 Apply WHS requirements, policies and procedures in the construction industry*.

Always use a site safety plan and a safe work method statement.

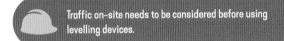
Traffic on-site needs to be considered before using levelling devices.

1.3 Health and safety requirements for levelling procedures are confirmed and applied to planning

Levelling is not too dangerous on-site, but some consideration needs to be given to making sure you are safe and efficient. Some things you need to be familiar with are:

- the site safety plan (which will alert you to any dangers on the site)
- the safe work method statements (SWMSs), which show the plan for how the levelling job will be done and how you will handle any potential dangers
- the operator manuals for the tools and equipment you may use; this will help you to be efficient, to look after the tools, and to use them properly and safely
- if safety signs are needed, what signs will be used and where they will be used. For example, laser tools have dangers associated with their laser beam. AS 2397 and AS/NZS IEC 60825.14 state that all laser equipment used in construction must either have labelling or documentation stating what class of laser it is. Generally, Class 1 is safe for normal use and Class 2 normally safe because a person's 'blink' reaction will offer protection except in low-light environments. In all cases, laser light should not be viewed through binoculars or similar.

Remember that you have good safety information in the user manuals for all your tools, and this should be followed. Stay alert for hazards in your vicinity while you concentrate your focus on the levelling instruments and be aware of your path of travel for trip hazards.

1.4 Check levelling tools and equipment and report or fix faults

The tools used to measure heights on a site are:

- spirit level
- water level
- laser level
- optical level
- tripod.

The tools used to record and/or position height measurements are:

- levelling **staff** (extendable measuring stick, normally just called a 'staff')
- tape measure
- string lines
- pegs.

The testing, maintenance and minor servicing of each tool will be discussed in this chapter, as well as planning for the work, cleaning up the work site and making the site safe after this part of the work is completed.

There are a number of tools that can be used to measure heights. The choice of tools will depend on how accurate the measurements need to be and over what distance the measuring has to be done.

Spirit level

A **spirit level** consists of a small tube filled with fluid and an air bubble; it is used to determine a **plumb** (or level) position and is suitable for transferring a height a short distance. Lines (commonly called 'gates', shown in **Figure 11.1**) are marked around the tube and the bubble of air will sit within these lines when the tube is level. The tubes have been inserted with one

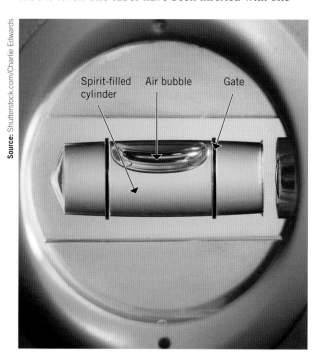

FIGURE 11.1 Marks on the spirit level (called 'gates') help to locate the position of the bubble

tube at 90° and the other tube parallel to the length of the level. This means that when the level is held up to a wall (see **Figure 11.2**) or set on a peg, it is possible to determine a plumb position. Some different types of spirit levels are shown in **Figure 11.3**.

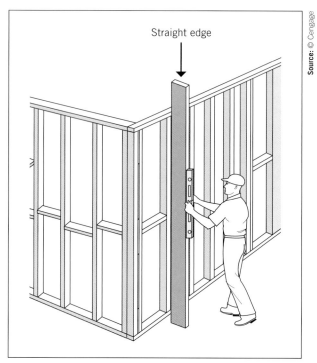

FIGURE 11.2 Using a spirit level and straight edge to check a wall for plumb

The distance a height can be transferred will depend on the length of the level or of a straight-edged piece of material used with the level (see below), which is usually limited to about 2 m to 3 m. This is about the length of an item that can be comfortably carried around in a tool trailer or ute. A level should be checked as it can become inaccurate through wear or damage. This is done easily by checking the bubble position when each end of the level is set on two solid objects (e.g. two screw heads fixed into a bench); then turn the level end-for-end and recheck the bubble position. An accurate level will display the bubble in the same position in relation to the gates with each check. An inaccurate level will display the bubble in different positions. The solution is to adjust the bubble part of the level if possible or replace the level. The correct level or plumb position with the inaccurate level is halfway between the two readings.

Transferring heights using an accurate spirit level will usually involve the use of a straight edge. This may require two people but it is a very quick way to transfer positions. Simply hold the level on the straight edge and line up the mark on one end of it. When the level reads plumb, mark the position where the transferred height is required from the other end of the straight edge, as shown in **Figure 11.4**.

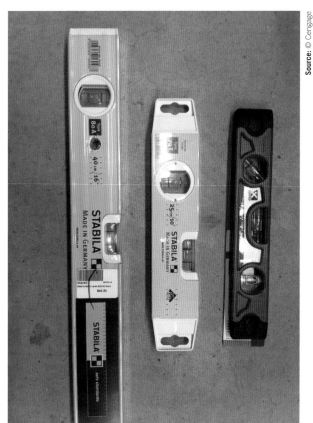

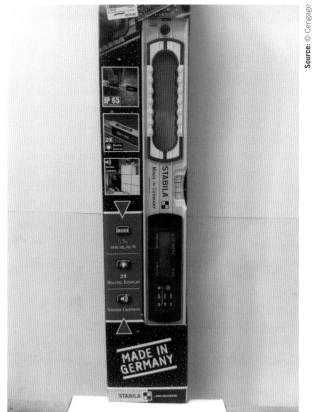

FIGURE 11.3 Some different types of spirit levels

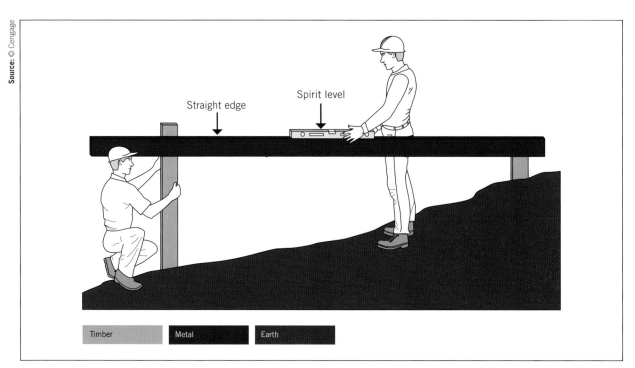

FIGURE 11.4 Using a level and straight edge to transfer or measure heights

Using a spirit level and a straight edge is only practical if you are transferring a **position** over short distances with no obstructions between the points. It is important to store spirit levels on their surface so they don't develop a bow in them and to store them where they won't be damaged.

Straight edge

A **straight edge** is a piece of material that is straight and can be used in levelling to extend the effective distance that a level can measure from and to. If you have a 1.2 m spirit level and a 2.1 m straight edge, then by holding them together as shown in Figure 11.4 points can be effectively transferred over a longer distance. The straight edge must be made out of material that will not warp and twist. Usually, they are made from aluminium, which is strong, light and very stable. It is important to store them flat and protected.

Line level

A **line level** (see Figure 11.5) is a small tool consisting of a tube filled with liquid and air and hooks to enable it to be hung on a string line. It is set up by hanging it in the centre of the string line so that it gives an average reading and allows for sag in the line. The string line can be made fairly level by moving one end of the line up or down until the bubble is in the centre of the gates. This tool is quite useful for home handy people because it is small and portable; however, its accuracy can be compromised due to factors such as wind and the small size of the bubble compared to the distance being

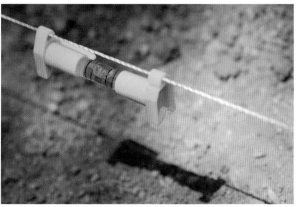

FIGURE 11.5 A line level

levelled. The string line used for this method must be pulled very tight and if the bubble is not in the centre of the line then it will not give an accurate reading.

Hydrostatic (water) level

A **hydrostatic (water) level** consists of clear plastic tubes filled with water (see Figures 11.6 and 11.7). The plastic tube may be clear or coloured, with clear tubing at each end. The water level works on the principle that water will always find its own level if it is left open to the air. The two open ends of the tube are held up near the heights to be measured and the water finds its level in the two ends. A hydrostatic level can be used by one or two people. For one person, temporarily fix the tube near the first height, then use the other end of the tube to locate the level position. Or have one person hold

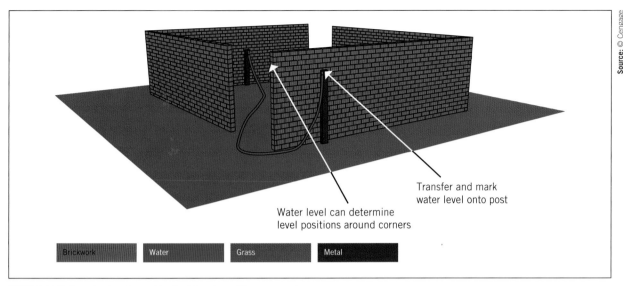

Transfer and mark
water level onto post

Water level can determine
level positions around corners

| Brickwork | Water | Grass | Metal |

FIGURE 11.6 A water level used to find level positions around corners

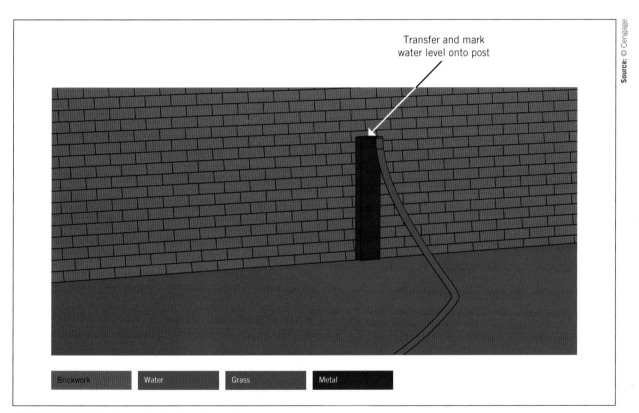

Transfer and mark
water level onto post

| Brickwork | Water | Grass | Metal |

FIGURE 11.7 Marking the water level onto a post

the end of the tube up near the initial height point that you want to transfer from, while the other person holds the other end in the position that you want to transfer to (see Figure 11.8). The water will settle (providing there are no air bubbles in the line) at the same level at both ends of the tube. Simply raise or lower the end of the tube used to find the second point until the water is at the same height as the initial height that you want

to transfer from. When the water has settled, mark the water position onto the point that you want to transfer to and both points will be the same height. This method works well for 'out of sight' applications such as around corners and for multiple transfers of one height to many positions (see Figure 11.9). Obviously, the distance that a height can be transferred is limited by the length of tubing used.

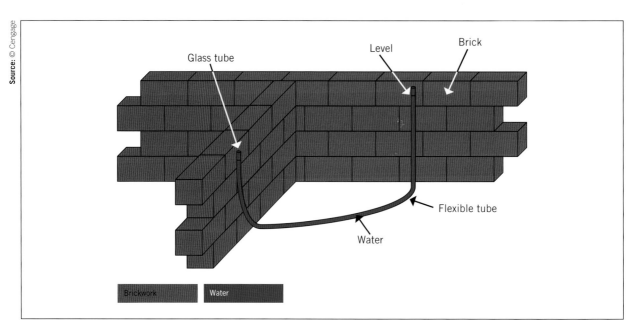

FIGURE 11.8 Using a water level for height transfer

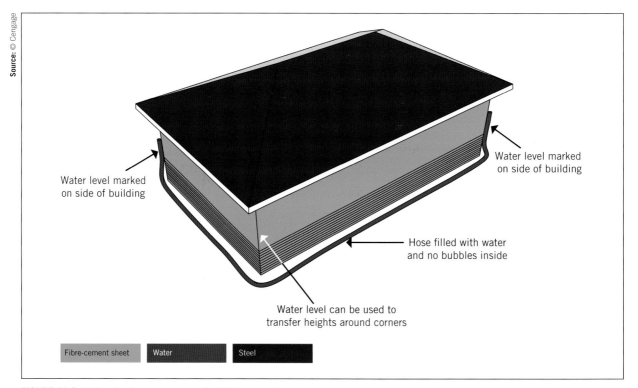

FIGURE 11.9 Water levels can measure heights around corners

Boning rods

Boning rods are 'T'-shaped tools that can be used to sight lines over reasonably long distances (see Figure 11.10). They are a simple way to determine straight surface lines for trenches, roads and landscaping (see the 'How to' box). See Figure 11.11 for an example of how this may work.

HOW TO

1 Set up one boning rod at the desired height and position.
2 Set up another boning rod at the other end of the work area in the required height and position.
3 Use a third boning rod to establish spot heights between the first and second rods.

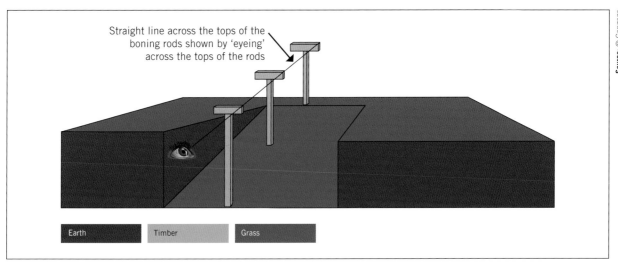

FIGURE 11.10 Boning rods in use: example 1

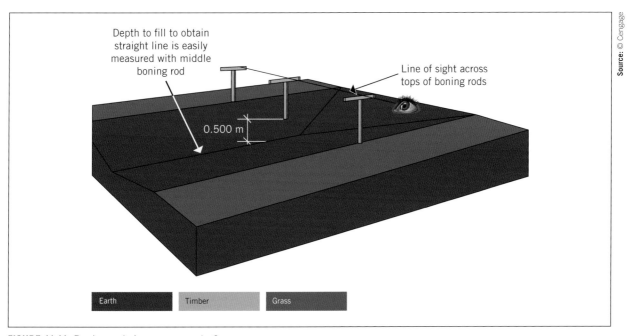

FIGURE 11.11 Boning rods in use: example 2

If the surface needs to have a uniform curve on it, as in a roadway surface, then the middle boning rod can be cut slightly shorter than the two end rods. When sighting is done over the three rods, this has the effect of curving the surface in the middle of the road (see Figure 11.12).

Boning rods have the further advantage that the line of the surface indicated by the rods doesn't need to be level; it may be sloped. If you were using a levelling device such as a laser level or an optical level (see discussion of these below), the line of the surface would be level. This enables boning rods to be used for work that is not intended to be level but needs to be straight – for example, driveways, drainage trenches or landscaping.

Tripods

The next two levelling instruments to be discussed – *laser levels* and *optical levels* – rely on a tripod to hold them in a steady position. A **tripod** is a frame with three adjustable legs and a top (stage), which allows an instrument to be fastened to it (see Figure 11.13).

Setting up a tripod involves the simple steps outlined in the 'How to' box on page 14.

The two main instruments that are used on a tripod are laser and optical levels. These two instruments are particularly good at determining the difference in height between long distances with points a long way apart (up to approximately 30 m). Laser levels can easily be operated by one person, whereas an optical level requires at least two people. Both of these levels can be

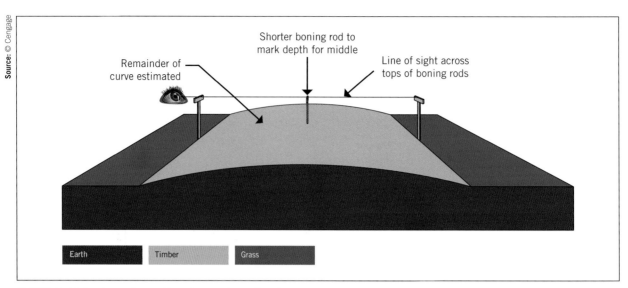

Remainder of
curve estimated

Shorter boning rod to
mark depth for middle

Line of sight across
tops of boning rods

| Earth | Timber | Grass |

FIGURE 11.12 Boning rods used to set curve in road surface

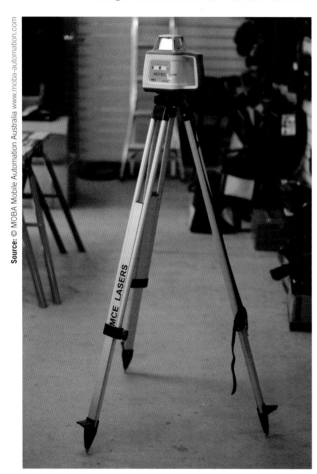

FIGURE 11.13 Tripod holding a laser level

FIGURE 11.14 Laser level with spinning beam emitter shown on top

used for complex measurements with many points over a large area or for simple tasks such as measuring how much fill is required for a slab or for measuring footings. We will now look at each of these levels in turn.

Laser levels

Laser levels are electronic instruments that you sight through, like a telescope. They emit a signal in such a way that the beam is at the same height in a 360° circle (see Figure 11.14).

When used for setting out, the laser level is attached to a tripod and it sends out a level beam of light or a signal. In most cases, a sensor (Figure 11.15) will pick up the beam and a mark can be transferred to any number of positions, all being the same height.

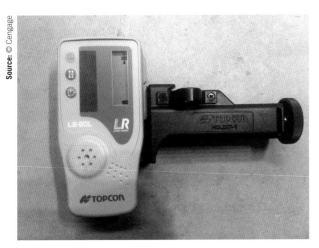

FIGURE 11.15 A laser receiver/sensor

Sometimes the laser emits a beam of light and this can be used to mark positions with a pencil. A mark can be made where the light beam shines on parts of the building. Because the light beam is in line and level wherever it shines, this line will indicate level points. (Laser levels can also be used to transfer vertical positions; however, this is outside the scope of this book.) The beam emitted from light-beam emitting lasers can be hard to see outdoors, and tinted glasses may be necessary to see the beam. The sensor can be attached to a staff (an extendable measuring stick), a purpose-built holder that has an inbuilt tape measure or even just clamped to a light stick of timber on which you have made your own markings.

Some individual instruments may have their own set-up requirements, which will be detailed in the instruction book for that level. Follow these instructions closely! Laser instruments can be dangerous if used incorrectly, and any warnings in the instructions must be taken seriously (see Figure 11.16).

FIGURE 11.16 Warning label on a laser instrument

Laser levels have three screws on the base of the instrument that can be adjusted, one at a time, so the circle bubble can be located within the centre of the circle. This means the level is within the limits of its compensating mechanism and it will be able to give level sightings. A mistake beginners commonly make when setting up instruments on a tripod is to 'chase' the bubble around the circle by adjusting screws too quickly and not waiting for the bubble to settle. Avoid this by only screwing one or two screws at a time and then waiting to see the resultant movement of the bubble. Alternatively, some laser levels have a reset button to automatically set them at level.

Once the instrument is set up and the laser is rotating, you can transfer a height around the job site. Simply take the sensor to the first height point (job datum or benchmark) and sit the staff on that height point while moving the sensor up or down until it receives the beam either audibly on its receiver or on the screen, or on both. Clamp the sensor on that point and move to the next desired point. When the sensor is receiving the beam correctly, simply transfer a mark from the bottom of the staff or stick onto this new position. (This is why the laser level is a one-person instrument! Only one person is required to carry the staff with the sensor attached around the site, marking the position of the heights, while the laser unit sends out a beam for the receiver.)

If you need to read and calculate various heights on the site and work out the differences between each height, the staff will need to be set on each of the points to be measured and the sensor will only be moved up or down on the staff until it is receiving the signal correctly. At this point, a measurement is recorded from the sensor position on the staff and all the measurements are 'booked' for calculation of the different heights. This is discussed further after the optical level is considered.

Optical levels

An optical level is a delicate instrument with sights and lenses similar to those in a telescope (see Figure 11.17). At one end there are some fine lines, called crosshairs, along with stadia lines (see Figure 11.18). (Stadia lines and crosshairs are set inside the level and can be focused to an individual's eyesight. They provide a reference for sighting/reading positions on a staff.) The crosshairs allow you to sight through the level

FIGURE 11.17 An automatic level

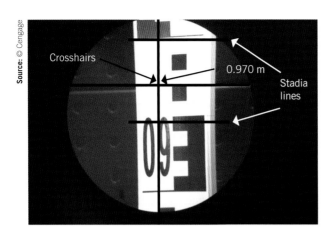

FIGURE 11.18 Crosshairs and stadia lines on a staff

instrument and line up the measurements on a staff or the marks on a stick. As the level is rotated, the crosshairs will always be the same height if the level is set up correctly. An **automatic level** works by having a swinging stage with a prism on it that compensates for minor differences in out-of-level set-up. This gives accuracy without having to check for the instrument being set up level each time you swivel the instrument around.

There are two main types of optical levels commonly used on building sites; the automatic level with a damper on the swinging prism stage and an automatic level without a damper. The automatic level with the damper has a spring mechanism that holds the swinging pendulum stage from being damaged during movement. When the instrument is first set up, a button needs to be pushed on the instrument so that the mechanism can swing free and give you an accurate reading. The automatic level without the damper has nothing to dampen the movements inside the instrument during set-up and carrying.

Setting up the optical level is very similar to setting up the laser level, with a round bubble to locate within a circle. There are a number of optical level types – for example, the dumpy level, the tilting level and the automatic level. However, in this book we will look only at the automatic level. For set-up of other types, refer to that level's instruction manual. Once set up, their use is basically the same.

The optical level also has three adjusting screws on the base of the instrument. By adjusting them one at a time, in the same way as the laser level screws, the circle bubble can be located on the centre of the circle mark. For automatic levels, this means the swinging mirror mechanism inside the level is within the limits of its compensating mechanism and this will give level sightings through the optical sight. You can hear the swinging mechanism by gently swivelling the instrument from side to side.

With the level firmly set on a tripod, the final adjustments are to focus the crosshairs and stadia

lines to suit your individual eyesight and then to focus the object in your view. This is done by holding a light piece of paper in front of the lens while you sight through the eyepiece. Adjust the focus for the crosshairs by rotating the eyepiece adjustment (the closest adjustment lens to your eye). Object focus is achieved by lining up the 'sighting pointer' on top of the level with your staff and then viewing through the level to the staff while adjusting the object focus screw, which is usually found on the side of the level.

If you are transferring heights only from point to point, then a stick with a clear mark on it set on the first point is probably the least confusing way to do this. However, if you need to measure the height difference between various positions, you will need a staff with measurements on it that can be read through the instrument. The process here is to set up the instrument in a place where you can read as many of the desired positions as possible before having to move the instrument to take subsequent readings.

LEARNING TASK 11.1 USING LEVELLING DEVICES

Give short answers to the following:
1 What safety precaution is needed when using a staff?
2 When using boning rods, how can you curve the surface in the middle of a road?
3 When using a spirit level, what determines the distance you can transfer a height?
4 What effect may incorrect storage have on a spirit level?
5 Hydrostatic levelling works well for two different applications. What are they?
6 What is an advantage of laser and optical levels over other levelling methods?
7 Describe the best way to adjust the base screws of a laser level when first setting it up.
8 Why is it advisable not to set a laser level at eye level?
9 What is a quick and easy way to check a wall for plumb?
10 Where is the best place to position a laser level and tripod?

Levelling staff

The **levelling staff (or stave)** is a measuring tool that may be extendable to anything from 3 m to 5 m and which is marked with large numbers (see Figure 11.19). The tool is held on a point that you want to measure the height of. An optical level is used to sight to the staff. Readings are taken through the crosshairs in the level to the staff or the sensor on the laser level. The staff is then moved to a new point and sighted with the optical or laser instrument and the measurement differences noted.

FIGURE 11.19 'E'-type staff

Using a staff requires some care to get an accurate reading and to ensure your safety.

Always check for overhead power lines.

Here are some basic rules to follow:

■ Obviously, the slogan 'Look out! Look out! There's power lines about!' applies when using a staff. A staff may be 5 m long when extended, and if it is metal it could spell disaster when working near overhead lines.

■ Keep the bottom of the staff clean of mud, so that your readings are accurate.

■ Remember to hold the staff plumb when being read. (Try leaning it over during use and see how much your readings change!) Using a spirit level with the staff will assist the novice to get the staff plumb and will minimise errors when taking readings. Some types of staff have bubble level indicators attached at the back, which makes it easier to keep the staff plumb.

■ Familiarise yourself with the numbering system. The staff is designed to be read with accuracy to the nearest 10 mm from long distances and by estimation to the millimetre between the 10 mm marks. The best way to familiarise yourself with how the measurements work on your staff is to hook your tape measure (which you are already familiar with) on the bottom of the staff and run it alongside the markings on the staff while learning what the staff measurements mean. Look for a staff with an 'E'-type pattern (because it is easier to read at greater distances), where the bars represent a 10 mm distance (see Figure 11.20). Figure 11.21 shows an example of reading the measurements of the staff.

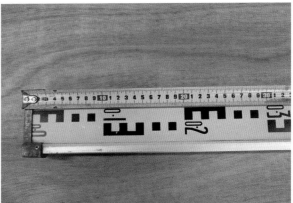

FIGURE 11.20 Using a tape to help understand the markings on a staff

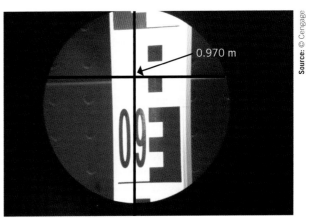

0.970 m

FIGURE 11.21 Staff showing the measurements for the points shown as seen through the optical instrument lens

1.5 Team roles and verbal and non-verbal communication signals are confirmed, as required

When you perform levelling activities on-site, you will often work with another person. If you are levelling with a laser level or over small distances with a spirit level, you may be able to work alone; otherwise, you will need to communicate with another person, and sometimes this will be over a long distance.

Your communication may be verbal or non-verbal depending on the environment at the time. Make sure you have discussed your preferred and workable communication method before you start work.

Some tools that may be needed for communicating could be signals, mobile phones or two-way radios.

2. Set up and use levelling device

2.1 Required heights or levels are identified from work instructions

The plans for the job contain one plan drawing called a 'site plan', and this drawing shows the boundaries for the site, the building layout, and a very important indication of the main point of height for referring all other height readings for the job. The datum or benchmark for the job will be the height to start for referring all the other heights on the job. Make sure you have this information on hand as you measure heights. In these plans there are various symbols and construction terminology that will be detailed. Take the time to review the chapter *CPCCCA3025 Read and interpret plans, specifications and drawings for carpentry work* to familarise yourself with this information.

FIGURE 11.22 Set tripod foot in a position where it won't slip

FIGURE 11.23b The tripod head has a hook for holding the string line and plumb bob

moving the level, taking another two readings. These four readings are then compared and the differences are the error in the instrument.

Figures 11.24 and 11.25 show how this works. Follow the steps in the 'How to' box with some sample readings shown in Table 11.1.

TABLE 11.1 Booking the two-peg test results

Instrument position	First reading (to peg A)	Second reading (to peg B)	Difference
Position 1	0.995 m	0.905 m	0.090 m
Position 2	0.967 m	0.875 m	0.092 m
Amount that the instrument is out in accuracy:			0.002 m

FIGURE 11.23a The tripod head has a screw for holding the instrument on the stage

2.2 Set up levelling device and check according to manufacturer's tolerances

At this point, we will discuss how to check if the laser level or optical level is accurate. We do this by using the two-peg test.

Two-peg test

The two-peg test involves checking the accuracy of a level in the field by taking two readings from the instrument while it is in one position and then, after

HOW TO

1 Choose some level ground and set up two pegs about 80 m apart (or a similar distance that will fit on your site).
2 Centre the instrument between the two pegs and correctly fasten the tripod and instrument.
3 Measure and record the staff sightings for each peg.
4 Move the instrument so it is close to, but beyond, one of the pegs and record the sightings under the 'Second reading' column.

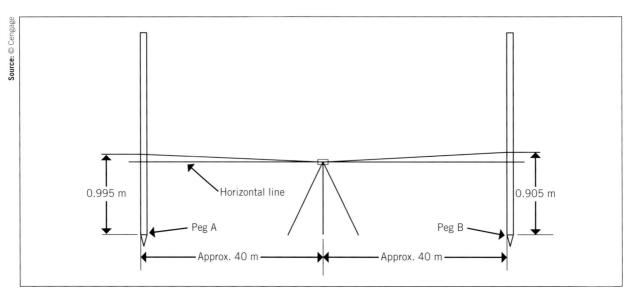

FIGURE 11.24 Performing a two-peg test for level accuracy

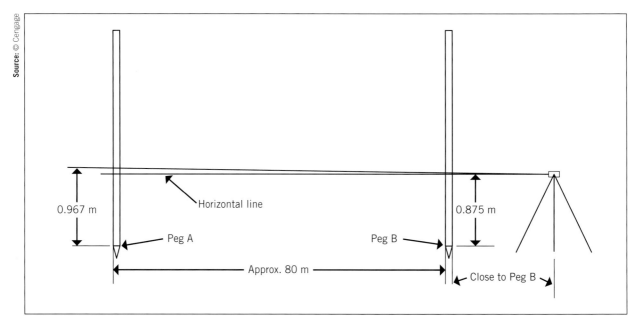

FIGURE 11.25 Checking the height difference between the different readings in a two-peg test

The difference between the two sets of readings from the two-peg test activity is the amount that the instrument is out of accuracy. In this example, there is a 2 mm error, which is not enough to worry about because it is hard to accurately read the staff to 2 mm accuracy anyway. (A surveyor would perform other calculations to eliminate any error.) By simply placing the instrument midway between the two staff positions when levelling, you will minimise errors further.

2.3 Read levels and heights and transfer and mark in required location according to job requirements

Recording the field readings accurately using a recording method is important. The method focused on

below is for field recording of heights read through an instrument and the subsequent office work to establish the heights at important points in relation to a site datum height. To do this, you may choose to use these booking sheets for normal heights and upside-down (or inverted) levels using the rise and fall booking method.

Booking sheets

Booking sheets are sheets of paper drawn up into columns and rows that help you to record the field readings accurately and enable you to calculate the reduced level of each point. (A reduced level is a level that is relative and measurable in relation to all the other points on the site.)

It is very important to place the building in the correct position on the correct site, and now we will look at some ways to ensure you are able to have a height measurement that relates to other measurements and is as free of errors as possible. To do this, you can use either the *rise and fall* or *height of collimation* booking method. As you examine these methods, remember that, in the end, all you want is a height for each point on the site that is able to be directly related to all other heights. You need to be able to measure height differences on a site so that you can calculate excavations, footings, brickwork, floor structure, etc. You can also have accurate plans and drawings prepared for the site when you know the vertical heights you are going to be working with.

Most times there is a datum or benchmark that you can use as the starting point (or the recovery point) for the job. But as you move around most sites using instruments to measure heights, you often have to move the instruments and this prevents the preceding measurements from being in relation to the subsequent ones. So, you record heights on standardised booking sheets and calculate (or reduce) the heights for all the positions that you are interested in. In the end, though, all you need is the name of the position where the height is measured and its reduced level (RL) reading. All sites should have a datum point that is away from the work so that it doesn't get damaged. If there is not one, make one. It is a good idea to make the datum point relate to the registered benchmark if there is one nearby. Allocate a height for the datum that will ensure that you won't go into negative RL numbers as you book the readings. Make the datum a rounded number, such as 10.000 m or 100.000 m.

An easy way to record simple height details is to sketch up a plan view of the area you are working on and write the height reading on that plan as you sight them (see Figure 11.26). For simple work, this can be a quick and easy way to determine differences between different points. For more detailed work and work that requires repositioning the level, you will need a more comprehensive method, such as the rise and fall method.

Inverted levels

An inverted staff reading can be used to measure the underside of items that are above the line of sight of the instrument – for example, a ceiling. The staff is placed upside down and read upside down. The measurement is recorded as a negative number in the booking sheet. It is good practice to put a 'minus' sign in front of the negative number and put a note in the comments box for the position so you don't forget to enter a negative number in the calculator when processing the numbers. Also, try to avoid a change point on an inverted staff reading as it makes it a bit more complicated.

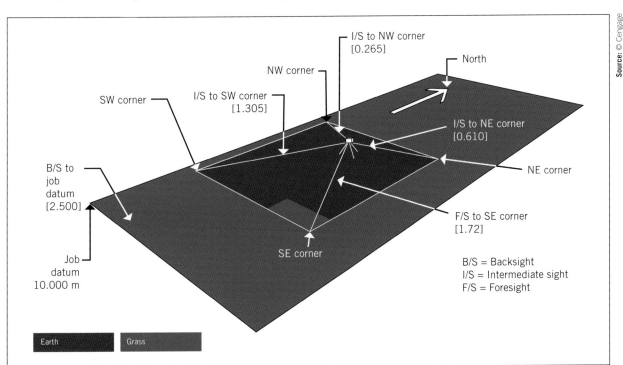

FIGURE 11.26 Set-out with the sights taken

Closed bookings for both rise and fall and height of collimation

Have an assistant help you to use an optical level, tripod and staff to perform sightings of approximately eight stations, recording the sightings as either backsights, intermediate sights or foresights. Assign a reduced level to the first reading of 100.000 m and work out the booking and calculations to get reduced levels for each station using the rise and fall and height of collimation methods. For this exercise, make sure you start and end at the same station. This is called a 'closed exercise'. Your finished point after performing the booking sheet exercises should have an RL of 100.000. Wherever possible, perform this exercise around buildings or objects that require change points. Do this until your check gets to within 5 mm.

Open bookings for both rise and fall and height of collimation

Have an assistant use an optical level, tripod and staff to perform sightings of approximately 12 stations, recording the sightings as either backsights, intermediate sights or foresights. Assign a reduced level to the first reading of 50.000 m and work out the booking and calculations to get reduced levels for each station using the rise and fall and height of collimation methods. For this exercise, make sure you start and end at different stations. This is called an 'open traverse'.

Booking using simple inverted levelling exercise

Using an assistant, perform readings of a number of positions, including approximately two inverted readings. Close off the exercise by sighting back to the start point. Book the results of this fieldwork and make sure the inverted readings are entered into the calculator as negative numbers. The resultant RLs should show very small (if any) error for the start and finish RLs.

Give short answers to the following:

1   What is the advantage of a staff with a bubble level indicator attached at the back?
2   Describe how best to familiarise yourself with the system of measurement on your staff.
3   Explain what a 'reduced level' is.
4   Name the two types of booking method.
5   What is the result of having to move instruments around the site while measuring heights?
6   When recording heights on a standard sheet, how should the items be ordered?
7   In the three-step test for a standard booking sheet, which answers should be the same?
8   What must be done in the booking sheet when it becomes necessary to measure heights around obstacles?
9   Why is it necessary to measure height differences on a site?
10 What is the name for the last staff reading that is sighted before a level is moved or packed up?

2.4 Results of levelling activities are documented according to organisational requirements

The rise and fall booking method is a table format for recording the various readings that are taken. Each reading is recorded in the order that it is sighted. The end result of this method is that you will have a column of information with a height reading (reduced level) that relates to other points on the job, so the difference between each point can easily be seen. Before proceeding, make sure you know what the terms 'backsight',

'intermediate sight', 'foresight' and 'reduced level' mean. This will assist you in using this booking method.

Completing a simple booking sheet

We will look at a practical example of a site plan and complete a simple booking sheet using the rise and fall method. Figure 11.26 shows a site plan for an L-shaped house with front patio. The building setback line is 8 m from the front boundary. In this example, we will assume that the site is level and the instrument can read all sight positions without you having to move it.

HOW TO

RISE AND FALL BOOKING SHEETS

The process of establishing the exact heights at the important points of the site is as follows:

1   Identify the points needing to be measured. (In our example, we will measure the heights of the four house corners in relation to the job datum on the front southern boundary.)
2   Label the points of the four corners. (In this exercise, we will call them according to their

orientation – for example, SW, NW, NE and SE. Multiple readings can be referred to as A, B, C, etc.) Using a standard booking sheet enables us to simplify the process of gathering and calculating the information.

3   Set up the level according to the previous discussion and then sight to the staff, which will be held on the job datum.

>>

4 Record the sighting as it appears through the scope.

5 Now move the staff to the SW corner, recording this sighting as well.

6 Continue until you have taken a sighting at each **station** and entered a record on the row in the booking sheet for that station.

7 Conclude by entering the last reading in the 'Foresight' column.

You are now ready for the calculating work (see steps 8 to 10 below). (Some values have been inserted in the booking sheet shown in Table 11.2 for the example.)

8 Record the reduced levels using this simple calculation process: on your calculator, enter the first sighting and subtract the very next sighting; the resulting answer is shown as either negative or positive. If it is negative, the answer goes in the 'R or –F' (R/F) column as a negative number (i.e. with a minus sign in front of the number); if it is positive, enter the result in the 'R or –F' column.

Note: You may see **rise** and **fall** in the same column (fall is shown as a negative number) or you may see them represented in their own column.

Either way is correct. For the purpose of this resource, the one-column ('R or –F') method has been used with a minus symbol in front of any falls. You must ensure you mark negative numbers with a clear '–' sign so you don't get confused and think it is a positive number. Alternatively, you can use a separate 'rise' and 'fall' column. I always recommend doing these bookings on a spreadsheet that requires one column.

9 Clear the answers in the calculator and re-enter the last figure entered, then subtract the next sighting and press the equals function. If the resultant answer is a negative, this means it is a fall and should be presented as a negative number in the 'R/F' column; if it is a positive number, present it as such in the 'R/F' column. (The important thing to remember is to enter the items in order of their sightings.)

10 Calculate the reduced level from the rises and falls by entering the first RL into the calculator and either adding or subtracting the rise or fall (e.g. 10.000 + 1.195 = 11.195). Leaving the answer on the screen, either add or subtract the next rise or fall and so on to the end (see Table 11.3).

TABLE 11.2 Standard booking sheet with fieldwork recording completed

Backsight	Intermediate sight	Foresight	R or –F	RL	Position
2.500				10.000	Job datum
	1.305				SW
	0.265				NW
	0.610				NE
		1.720			SE

TABLE 11.3 Calculations completed except for the three-part check

Backsight	Intermediate sight	Foresight	R or –F	RL	Position
2.500				10.000	Job datum
	1.305		1.195	11.195	SW
	0.265		1.040	12.235	NW
	0.610		−0.345	11.890	NE
		1.720	−1.110	10.780	SE

You now have a booking sheet complete except for one important thing – checking the answer.

You can check the answer using a three-step test. This test is vital: the set-out person determines where the building will go on a site, and a simple error at this stage could result in the house being placed either too high or too low. In addition, quantities of materials required are often determined based on the RLs calculated at this stage.

CHECKING FOR CALCULATION ACCURACY

The three-step test is carried out as follows:

1 Add up all the **backsights (B/S)** and record the total at the bottom, then do the same for the **foresights (F/S)**. Now, add up the rises and record the total at the bottom, then do the same for the falls.

2 Find the difference between the backsights and the foresights and the rises and the falls. These two answers should be the same.

3 Find the difference between the first RL from the last RL. The answer should be the same as the previous checked answers. If the sheet balances, the completed booking sheet will be accurate to work from (see Table 11.4).

TABLE 11.4 Completed booking sheet

Backsight	Intermediate sight	Foresight	R or –F	RL	Position
2.500				10.000	Job datum
	1.305		1.195	11.195	SW
	0.265		1.040	12.235	NW
	0.610		−0.345	11.890	NE
		1.720	−1.110	10.780	SE
2.500		1.720	**0.780**	**0.780**	
	0.780				

You have now completed a booking sheet. Having checked that it is correct, you can now use the RLs for height control during set-out.

When possible, close off the readings with a sight back to the first reading sight so that all the measurements will form a closed run. You can see from Table 11.5 how this could work on our first example, because it is a simple booking that did not involve a change point. (The 'RL' and 'Position' columns are shaded because they are the only pieces of information you need on-site to set up a job. All the other information is useful only for calculating these two columns.)

Using a spreadsheet for simple rise and fall bookings

Many of these repetitive calculation tasks can be set up to work easily on a spreadsheet program on either a computer or a smart phone. A quick search will reveal some good spreadsheet apps that can be set to give you the reduced levels quickly and easily. Another advantage of a spreadsheet is that it can eliminate error in repetitive calculations. Some basic understanding of how spreadsheets work will be required. Table 11.6 shows a simple spreadsheet showing formulas, suitable for the example booking exercise in Table 11.5.

An advantage of this type of spreadsheet, where formulas are shown, is that once a formula is entered, for example, in cell E3, then it can be copied and pasted down to all the following cells. The cells referenced by the formula should update to the new cells needed for the calculation.

Tip: To transfer a formula down to many cells, select and highlight the formula you want to transfer (say, cell E3). Then put the cursor over the bottom right-hand corner of the cell and the cursor will change to a cross. When it is a cross, simply click and drag the cross down over the cells that you want to transfer the formula into. The new cells will automatically change to the correct formula and the correct calculation answers from the formula.

TABLE 11.5 Completed booking sheet with readings closed back to the start point

Backsight	Intermediate sight	Foresight	R or –F	RL	Position
2.500				10.000	Job datum
	1.305		1.195	11.195	SW
	0.265		1.040	12.235	NW
	0.610		−0.345	11.890	NE
	1.720		−1.110	10.780	SE
		2.500	−0.780	10.000	Job datum
2.500		2.500	**0.000**	**0.000**	
	0.000				

TABLE 11.6 Spreadsheet showing formulas for a simple rise and fall booking sheet

	A	B	C	D	E	F
1	**Backsight**	**Intermediate sight**	**Foresight**	**R or –F**	**RL**	**Position**
2	2.5				10	Job datum
3		1.305		=A2–B3	=E2+D3	SW
4		0.265		=B3–B4	=E3+D4	NW
5		0.61		=B4–B5	=E4+D5	NE
6			1.72	=B5–C6	=E5+D6	SE
7	=SUM(A2:A6)		=SUM(C2:C6)	**=SUM(D3:D6)**	**=E6–E2**	
8		**=A7–C7**				

This is a very useful tool when doing a large booking sheet. However, beware of change points; these will mean the formula needs adjusting, so that it will collect the data from a different cell.

In using a spreadsheet program for rise and fall bookings, it is best to have the one column for both rise and fall. In that column, show a rise as a positive number and a fall as a negative number. (A spreadsheet automatically makes positive and negative numbers.)

Completing a complex booking sheet

If the site has a lot of slope or you need to measure heights around obstacles, then the level instrument may need to be moved. This is where we have a change point appearing in the booking sheet. A **change point** is the position where the staff is sighted to for the last reading of the instrument before it is moved and to which the staff is resighted as soon

as the instrument is set up somewhere else. A position called a change point has two readings allocated to its row on a booking sheet. Change points are necessary when the staff cannot be seen with the instrument because of corners or steep slopes, etc. For example, referring to our sample site plan, the level needs to be moved because the height difference from the front of the site to the rear of the house is too great to be measured with the staff length you have available. In this case, you will sight as much of the area as you can and then move the level. The best way to plan this is to sketch out the site or use a copy of the site plan and mark the stations and approximate level instrument positions on the plan (see **Figure 11.27** and **Table 11.7**).

Once you have completed the calculation of the reduced levels and performed the three-part check, the booking sheet will look like **Table 11.8**. It is important to follow the sighting procedure when doing the

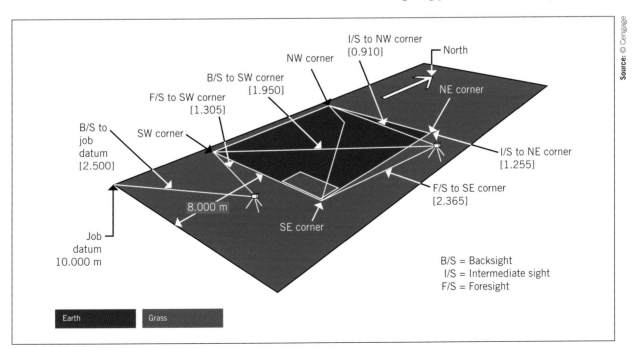

FIGURE 11.27 Site readings with change point shown

TABLE 11.7 The completed booking sheet from the field sightings, without RLs calculated or checks done

Backsight	Intermediate sight	Foresight	R or –F	RL	Position
2.500				10.000	Job datum
1.950		1.305			SW – change point
	0.910				NW
	1.255				NE
		2.365			SE

TABLE 11.8 Sighting procedure and order of calculations for a complex booking sheet after RLs calculated and three-part check completed

Backsight	Intermediate sight	Foresight	R or –F	RL	Position
2.500				10.000	Job datum
1.950		1.305	1.195	11.195	SW – change point
	0.910		1.040	12.235	NW
	1.255		−0.345	11.890	NE
		2.365	−1.110	10.780	SE
4.450		3.670			
	0.780		0.780	0.780	

calculations in a change point booking sheet. The calculations are performed as follows:

1 Enter 2.5 into the calculator and subtract 1.305. Enter the answer as either a positive or negative in the 'R or − F' column.
2 Enter 1.95 into the calculator and subtract 0.91. (This is because this is the order of sighting in the field. The arrows in Table 11.8 indicate the order of sighting and order of calculations.)
3 Continue as per spreadsheet in Table 11.8.
   You will notice in Table 11.8 that there are two entries on the row for the station SW. The rule with booking sheets is that there is a row for every station irrespective of from where the level instrument takes a sighting. The process described here applies to small, level bookings as well as large, multi-change-point bookings.
   After doing the three-part check, you can see that everything balances and that the RL agrees at the start

and end of the booking. Notice also that the RLs are the same for the simple exercise in Table 11.5 and the complex bookings with a change point in Table 11.8. This is how a booking sheet works: it is designed to give accurate RLs no matter where the instrument sights from or how many times it is moved.

Using a spreadsheet for complex rise and fall bookings

Table 11.9 shows how to complete the change point booking exercise, with one column representing both rise and fall. Remember that falls are a negative number. The columns are shown labelled with letters and the rows with numbers. The SW row has two readings (foresight and backsight) because it is a change point. Table 11.10 shows how to present the formulas for booking a complex field exercise of this type.

TABLE 11.9 Spreadsheet showing columns and rows labelled with letters and numbers

	A	B	C	D	E	F
1	b/s	i/s	f/s	r or –f	RL	Position
2	2.500				10.000	Job datum
3	1.950		1.305	1.195	11.195	SW
4		0.910		1.040	12.235	NW
5		1.255		−0.345	11.890	NE
6			2.365	−1.110	10.780	SE
7	4.450		3.670	**0.780**	**0.780**	
8		**0.780**				

TABLE 11.10 Spreadsheet showing formulas

	A	B	C	D	E	F
1	b/s	i/s	f/s	r or –f	RL	Position
2	2.500				10	Job datum
3	1.950		1.305	=A2–C3	=E2+D3	SW
4		0.910		=A3–B4	=E3+D4	NW
5		1.255		=B4–B5	=E4+D5	NE
6			2.365	=B5–C6	=E5+D6	SE
7	=SUM(A2:A6)		=SUM(C2:C6)	**=SUM(D3:D6)**	**=E6–E2**	
8		**=A7–C7**				

How to record and establish reduced levels from inverted readings

As already mentioned, inverted readings can be used to measure heights above the sight line of the instrument. An example of this could be under a beam or ceiling as in Figure 11.28. The main thing to remember is to note the inverted reading as a negative number on the booking sheet and to hold the staff upside down when sighting. Table 11.11a shows an example for rise and fall and height of

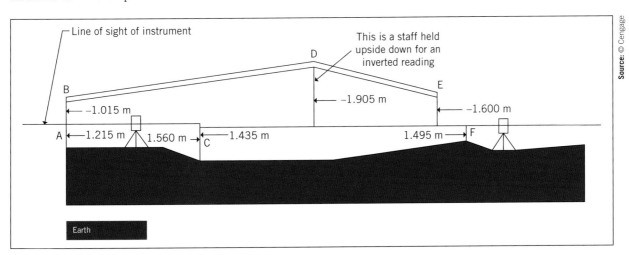

FIGURE 11.28 Example of inverted readings

TABLE 11.11a Table showing booking readings for rise and fall and height of collimation

Rise and fall method					
Backsight	Intermediate sight	Foresight	Rise/Fall	Reduced level	Comments
1.215				20.794	A.
	–1.015		2.230	23.024	B. (inverted reading)
1.435		1.560	–2.575	20.449	C.
	–1.905		3.340	23.789	D. (inverted reading)
	–1.600		–0.305	23.484	E. (inverted reading)
		1.495	–3.095	20.389	F.
Height of collimation method					
Backsight	Intermediate sight	Foresight	Height of collimation	Reduced level	Comments
1.215			22.009	20.794	A.
	–1.015			23.024	B. (inverted reading)
1.435		1.560	21.884	20.449	C.
	–1.905			23.789	D. (inverted reading)
	–1.600			23.484	E. (inverted reading)
		1.495		20.389	F.

Source: © Cengage

TABLE 11.11b Table showing formulas and calculation method

Rise and fall method					
Backsight	**Intermediate sight**	**Foresight**	**Rise/Fall**	**Reduced level**	**Comments**
1.215				20.794	A.
	−1.015		=A3−B4	=E3+D4	B. (inverted reading)
1.435		1.560	=B4−C5	=E4+D5	C.
	−1.905		=A5−B6	=E5+D6	D. (inverted reading)
	−1.600		=B6−B7	=E6+D7	E. (inverted reading)
		1.495	=B7−C8	=E7+D8	F.

Height of collimation method					
Backsight	**Intermediate sight**	**Foresight**	**Height of collimation**	**Reduced level**	**Comments**
1.215			=E14+A14	20.794	A.
	−1.015			=D14−B15	B. (inverted reading)
1.435		1.560	=E16+A16	=D14−C16	C.
	−1.905			=D16−B17	D. (inverted reading)
	−1.600			=D16−B18	E. (inverted reading)
		1.495		=D16−C19	F.

collimation and Table 11.11b shows the formulas for working it all out.

3. Clean up

3.1 Work area is cleared, materials sorted and removed or recycled according to regulations

As with all work on-site, it is important to spend time cleaning up after the job has been completed. In the case of levelling, there is not much to clean up unless you are working on a muddy site. Naturally, you will sort any materials used and stack them for later use or disposal. If part of your levelling involves marking locations on the site with marking spray or lime, then the safety data sheet for those materials will need to be read to make sure you are not doing any harm to yourself, others or the environment. Ensure left over materials from any works completed are disposed of correctly, through recycling or disposal. Materials that can be reused should be stored accordingly.

3.2 Tools and equipment are cleaned, checked, maintained and stored according to manufacturer specifications

Clean and inspect the instruments and tools and pack them away for the next use. If laser levels are being used, batteries will need to be recharged and/or replaced. Do this at the end of a job, not before the next job. Ensure you are complying with your state or territory WHS rules regarding power tool testing and tagging, and make sure this regulation work is done before you need the tools and within the time frames required. All cleaning, inspection and maintenance must be done according to the user manual for that particular tool.

 COMPLETE WORKSHEET 1

SUMMARY

In this chapter, you have learnt about the various tools and materials that can be used to measure levels on a construction site. You have also learnt how to minimise error, book field readings neatly and maintain the various tools needed for this type of work. You have been taken through the stages of levelling as listed below:

1 Plan and prepare
2 Set up and use levelling device
3 Clean up.

The importance of accurately measuring heights and distances on a building site cannot be overstated. The position and height of building structures depends on your accuracy, and it can be costly to make a mistake at this stage.

REFERENCES AND FURTHER READING

Texts

Australian Building Codes Board, *National Construction Code*, Australian Government, Canberra, **https://ncc.abcb.gov.au/**.

South Western Sydney Institute of TAFE NSW 2020, *Basic Building and Construction Skills*, 6th edn, Cengage Learning Australia, Melbourne, VIC.

Web-based resources

Some levelling content can be found by searching 'levelling'. Some of the content to look at is:

- Search for the online tutorials resource from the UK **http://www.levelling.uhi.ac.uk/**.

✓ Relevant Australian Standards
Traffic Management: Guide for Construction Work July 2014, Safe Work Australia
Work Near Overhead Power Lines, Code of Practice 2006, SafeWork NSW
AS 2397 Safe use of lasers in the building and construction industry
Safe Work Method Statement for High Risk Construction Work Information Sheet December 2014, Safe Work Australia
Excavation Work Code of Practice March 2015, Safe Work Australia
Surveying and Spatial Information Act 2002
Surveying and Spatial Information Regulation 2017

GET IT RIGHT

The photo below shows an incorrect practice that can be used when performing levelling.

Identify the incorrect method and provide reasoning for your answer.

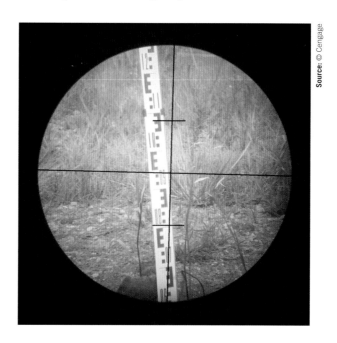

Source: © Cengage

WORKSHEET 1

Student name: _____

Enrolment year: _____

Class code: _____

Competency name/Number: _____

Task: Give short answers to the following questions on levelling.

Planning and general

1   List three different tools that could be used to obtain a level mark on the pegs of profiles. Discuss the advantages and disadvantages of each method.

a   _____

_____

_____

b   _____

_____

_____

c   _____

_____

_____

2   What are the advantages and disadvantages of steel and fibreglass long tapes?

_____

_____

_____

_____

_____

_____

3   How can a spirit level be checked for accuracy?

_____

_____

_____

4   Explain what is meant by 'levelling'.

_____

_____

_____

5   What is a 'datum'?

_____

_____

_____

6   What is a 'benchmark'?

_____

_____

_____

7   Where are the crosshair adjustment knob and the focus knob found on most optical instruments?

_____

_____

8   If the water in the water level tubes is not the same height when the tubes are checked by putting them beside each other, what would be the most likely cause?

_____

_____

_____

9   What are 'boning rods' and what are they used for?

_____

_____

10 What is the formula for the use of stadia lines?

_____

_____

11 What is an advantage and a disadvantage of a hydrostatic level?

_____

_____

_____

_____

12 What is an advantage of a laser level over an optical level?

_____

_____

13 List three important points to remember when using a levelling staff.

a _____

b _____

c _____

14 List the equipment that is needed for field levelling work with a laser or optical level.

_____

_____

_____

_____

15 List two things that a worker could do that would affect the environment while they are performing levelling.

a _____

b _____

16 Give four examples of what can be done to levelling tools when packing up so they are ready for the next job.

a _____

b _____

c _____

d _____

17 After levelling exercises, how should the work area be left when finished?

_____

_____

_____

Using levelling devices

1   What is a two-peg test used for?

_____

_____

_____

_____

2   Describe the steps involved in setting up an optical level on a tripod.

_____

_____

_____

_____

_____

_____

3   What are some safety precautions to be observed when using a laser level and staff?

_____

_____

_____

_____

_____

4   What does the adjustment do at the eyepiece end of an optical level?

_____

_____

5   What are some reasons why water levels may read incorrectly?

_____

_____

_____

_____

_____

6 Describe the steps needed to set up a tripod and optical level ready for use.

_____

_____

_____

_____

_____

_____

7 Does a staff need to be used whenever an optical level is used; and if not, what else can be used?

_____

_____

_____

_____

8 How can you be sure you have interpreted the readings on a staff when you first become familiar with it?

_____

_____

_____

_____

9 With the aid of a sketch, show how height difference could be measured between two points using a straight edge, tape and spirit level.

10 Sketch how a water level can be used to determine heights.

11 Sketch how boning rods can be used.

12 Describe two methods that could be used to mark a line on the outside of a wall that is 12 m long for a pitching plate for a verandah.

a _____

   _____

   _____

b _____

   _____

   _____

13 Write in the correct measurements indicated on the following staff readings. Note any problems that may be shown in the readings.

Source: © Cengage

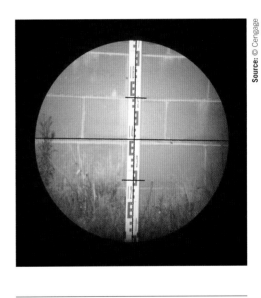

Source: © Cengage

_____

_____

_____

_____

_____

_____

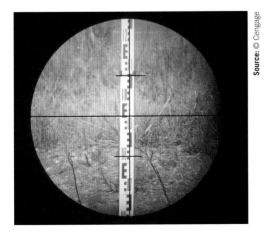

Source: © Cengage

_____

_____

_____

14 What are the main dangers of laser levels?

_____

_____

_____

_____

15 What signage may be useful for the area where levelling is being conducted?

_____

_____

_____

_____

Booking of levels

1 Why would you use a table to write up the results of your fieldwork sightings?

_____

_____

_____

_____

2 Why do you do a three-part test after completing rise and fall booking sheets?

_____

_____

_____

_____

3 List two safety precautions to be observed when using a laser level and staff.

_____

_____

_____

_____

4   Describe how to check the accuracy of booking levels on a booking sheet.

_____

_____

_____

_____

5   Explain the following terms:

   a   backsight

_____

_____

   b   foresight

_____

_____

   c   intermediate sight

_____

_____

   d   reduced level

_____

_____

   e   datum

_____

_____

   f   change point

_____

_____

   g   contour line

_____

_____

Finding the reduced levels on the booking sheet

1   Complete the following booking sheet using a rise and fall method or a height method, and check your calculations.

Backsight	Intermediate sight	Foresight	R/–F	RL	Position
1.810				100.000	Benchmark
	0.620				a
	0.850				b
3.200		0.540			c (change point)
	2.500				d
	1.950				e
	0.200				f
	– 2.100				g (note inverted reading)
	0.810				h
		0.110			Benchmark
	check		check	check	

PERFORMANCE OF SKILLS EVIDENCE

To be completed by teachers

Student competent ☐

Student not yet
competent ☐

Student name: _____

Enrolment year: _____

Class code: _____

Competency name/Number: _____

Task: Prepare for, and practise doing all of the following activities in preparation for your final assessment of your skills.

A person who demonstrates competency in this unit must satisfy all of the elements, performance criteria and foundation skills of this unit. The person must also transfer levels and record differences in height for three different projects as required by job specifications, using at least three of the following levelling devices:

- a spirit level and straight edge

- automatic/optical levelling device

- levelling with water technique

- laser levelling device.

In doing the above work, the person must:

- conduct a two-peg test with an automatic/optical level to confirm that the instrument meets manufacturer tolerances

- locate, interpret and apply relevant information in job specifications to the levelling task

- comply with site safety plan, and health and safety regulations applicable to workplace operations

- comply with organisational policies and procedures, including quality requirements

- safely and effectively use tools and equipment

- communicate and work effectively and safely with others, including using agreed communication signals

- confirm accuracy of the readings taken, including set-up and movement of device in two locations

- accurately record results of each levelling procedure according to organisational requirements.

GLOSSARY

A

accident An undesirable or unfortunate happening that occurs unintentionally and usually results in harm, injury, damage, or loss.

agenda A formal list, plan or outline of things to be done in a specific order, especially a list of things to be discussed at a meeting.

area The measurement of surface: the *area* of a flat, or plane, figure is the number of square metres the figure contains. Area can be found by multiplying the length of a figure by its width or breadth.

Australian Height Datum (AHD) This is a vertical datum in Australia. According to Geoscience Australia, in 1971 the mean sea level for 1966–1968 was assigned the value 0.000 m on the AHD at 30 tide gauges around the coast of Australia.

B

barricade A defensive barrier constructed in order to limit or prevent entry, and signify that a hazard or danger exists.

benchmark A point of reference by which something can be measured.

body language Communication by means other than by using words, e.g. through facial expressions, bodily mannerisms, postures, and hand gestures that can be interpreted as unconsciously communicating somebody's feelings or psychological state. Also described as *non-verbal communication*.

C

checklist A list of tasks, items, or points for consideration, verification, action or checking purposes.

collective agreement A workplace agreement involving the employer and a group of employees.

committee A person or group of persons elected or appointed to perform some service or function, such as to investigate, report on, or act upon a particular matter.

communication The exchange of information between people, e.g. by means of speaking, writing, or using a common system of signs or behaviour.

conflict A disagreement or clash between ideas, principles or people.

crosshairs These are two thin black lines set at 90 degrees to one another when looking through a telescope used for levelling. They provide the horizontal and vertical lines of sight.

D

dangerous goods Substances that have the potential to cause immediate harm.

datum Any known point, line or level from which a level line may be transferred to another position. Its elevation, or height, may be recorded and used as a permanent or temporary reference while carrying out a job.

dimensions Measurement in length, width, and depth (thickness).

E

efficient Performing or functioning in the best possible manner with the least waste of time and effort.

elevation A scale drawing of any side of a building or other structure, in accordance with the compass point direction it faces, that provides information relating to vertical measurements and external finishes.

experiential learning The process of learning through experience, more specifically defined as 'learning through reflection on doing'.

F

fire rating The duration for which a passive fire protection system can withstand a standard fire resistance test.

formwork An assembly or temporary construction to support and shape freshly mixed concrete until it sets and hardens.

fuel can be any combustible material, i.e. any solid, liquid or gas, that can burn. Flammable materials are any substances that can be easily ignited and will burn rapidly.

H

hazard Any situation, substance, activity, event, or environment that could potentially cause injury or ill health.

heat that may start a fire can come from many sources, e.g. flames, welding operations, grinding sparks, heat-causing friction, electrical equipment or hot exhausts.

hoardings Solid structures that prevent pedestrians from entering into building sites. Hoardings are divided into two categories: Type A hoardings are solid fences that surround the building site; Type B hoardings are similar to Type A hoardings with the addition of solid overhead protection for pedestrians walking past the building site.

I

independent contractor agreement A special form of individual contract covering independent contractors who are engaged by a business to do work that otherwise would be done by employees/workers.

individual agreement A contract between an employer and an employee covering specific employment matters on an individual basis.

induction training Compulsory training for all workers in the construction industry designed to familiarise workers with the site risks and hazards and to assist them in developing a general understanding of site safety.

industrial relations The field that looks at the relationship between management and workers, particularly groups of workers represented by a union.

injury Physical damage to the body or a body part.

instruction A spoken or written statement of what must be done, especially delivered formally, with official authority, or as an order or direction.

K

kerf The cut made by a saw or an axe into a piece of timber.

L

legislation The process of writing and passing laws, especially by a governmental assembly or official body.

level line Any horizontal line that is parallel to the surface of still water.

levelling The determination and representation of the elevation of points on the surface of the earth from a known datum, using a surveyor's level of any kind to measure the differences in elevation by direct or trigonometric methods.

linear metre A unit of measurement, measuring the distance between two points in a straight line.

M

manual handling An activity requiring a person to use force to lift, lower, push, pull, carry, move, or hold any type of object.

material The substance or substances of which a thing is made or composed.

meeting An occasion, in either a formal or informal setting, where people gather together to discuss something.

message A communication in speech, writing, or signals, containing some information, news, advice, request, opinion, fact, emotion, knowledge, warning, or any one of the many things people need to impart to others.

millimetre A unit of length equal to one thousandth of a metre.

minutes An official record of what is said or done during a meeting.

N

non-verbal communication The act and means of sending and receiving information without using spoken language. Written instructions, plans, sketches and signage are all non-verbal forms of communication.

O

orthographic projection Representing a three-dimensional object in two-dimensions.

oxygen comes mainly from the air. It may also be generated by chemical reactions.

P

parallax error When an object or point of measurement appears to shift or change position as a result of a change in position of the observer.

percentage A proportion stated in terms of one-hundredths that is calculated by multiplying a fraction by 100.

personal protective equipment (PPE) Safety clothing and equipment, designed to protect a worker's head, eyes/face, hearing, airways/lungs, hands, feet and body when exposed to harmful substances or environments or specific hazards.

person/s conducting a business or undertaking (PCBU) A PCBU conducts a business or undertaking alone or with others. The business or undertaking can operate for profit or not-for-profit. The definition of a PCBU focuses on the work arrangements and the relationships to carry out the work. In addition to employers, a PCBU can be a corporation, an association, a partnership or sole trader. A volunteer organisation which employs any person to carry out work is considered a PCBU.

pictorial representation Drawings representing the visual appearance of the completed project or construction.

plans A term used to represent all drawings including sections and details; and any supplemental drawings for complete execution of a specific project.

plant The equipment and machinery necessary for carrying on an industrial or engineering activity.

plantation timber Intensively managed stands of trees, of either native or exotic species, created by the regular placement of seedlings or seeds.

plumb Any vertical line or surface that if extended would line up with the centre of the Earth.

pneumatic tool A tool that is powered by means of compressed air. This means that the tool must be connected to an air compressor by air hoses.

power tool A tool or machine that uses a power source other than human force to power or drive the device; usual power sources include: 240-volt electricity, electric batteries, compressed air or gas or explosive gas or gun powder cartridges.

procedure An established or correct method of doing something.

Q

quadrilateral A four-sided figure.

quality The level of excellence that goes into a product or service.

quality assurance Preserving or keeping a set level of quality in the overall finished product. This is achieved by considering every stage of the construction.

quality control The system that is used to test a product by sampling and then checking it against set specifications.

quantity An exact or specified amount or measure.

R

record A document that shows what kinds of activities are being performed or what kinds of results are being achieved. It always documents and provides evidence about the past.

Reduced Level (RL) Refers to equating elevations of survey points with reference to a common assumed datum. It is a vertical distance between survey point and adopted datum plane.

resource efficient A practice in which the primary consideration of material use begins with the concept of 'Reduce-Reuse-Recycle-Repair' stated in descending order of priority.

recycle The process by which materials that would otherwise become solid waste are collected, separated or processed and returned to the economic mainstream to be reused in the form of raw materials or finished goods.

risk assessment Considers the effectiveness of existing WHS controls and then evaluates the probability and the potential severity of specific hazardous events and exposures. On the basis of such an assessment, organisations decide whether or not the risk is acceptable.

risk matrix This is traditionally presented as a table where the *hazard* is matched with the risk level; that is, the likelihood and severity of the hazard causing damage or injury. Using the table the rank of severity is gauged; for example, 'Extreme' or 'Low' and from this an appropriate level of control measures can be put into place to reduce the possibility and severity of harm.

S

safe work method statement (SWMS) Describes how work is to be carried out; identifies the work activities assessed as having risks; describes the control measures that will be applied to those work activities; and, includes a description of the equipment used in the work; the standards or codes to be complied with, and the qualifications of the personnel doing the work.

safety data sheet (SDS) A document prepared by the supplier or manufacturer of a chemical product, clearly stating its hazardous nature, ingredients, precautions to follow, health effects, safe handling/storage information and how to respond effectively in an emergency exposure situation.

scale drawing The reduction of a full-sized object to a suitable scale to enable its reproduction on drawing sheets.

sequence The order in which things are arranged, actions are carried out, or events happen.

SI units The *Système International d'Unités* (SI), or International System of Units for measurement.

signage Pictorial and written signs that give instructions or information vital for the safety of people entering or working on a construction site. All signage must comply with the relevant Australian Standards.

spalding Flaking off of the finished surface of concrete, as well as rust flakes, can also be caused by repeated freezing and thawing.

stadia lines These are short horizontal lines equally spaced above and below the horizontal cross hair in a telescope used for levelling. They are used for calculating distance from the level to the object being viewed.

suspended slab A reinforced concrete floor suspended above the ground and supported on brick walls.

sustainability Sustainability seeks to provide the best outcomes for the human and natural environments both now and into the indefinite future.

symbol A letter, figure, or other character or mark, or a combination of letters or the like, used to designate something.

T

teamwork Cooperative or coordinated effort on the part of a group of persons acting together as a team or in the interests of a common cause.

traverse A survey consisting of a continuous series of connected straight survey lines, whose lengths and bearings are measured at each survey station or nominated point. A 'closed traverse' is when the lines form a complete circuit between two known points.

V

verbal communication The act and means of sending and receiving information using spoken language.

volume The amount of space, measured in cubic metres that an object occupies. *Volume* can be found by multiplying the length by the width by the depth (thickness).

W

waste Any materials unused and rejected as worthless or unwanted.

waste management The processes and activities involved in dealing with waste, before and after it is produced, including minimisation, handling, processing, storage, recycling, transport and final disposal.

waste minimisation The use of practices and processes which reduce, as much as possible, the amount of waste generated, or the amount which requires subsequent treatment, storage and disposal.

work health and safety (WHS) Refers to all of the factors and conditions that *affect* health and safety in the workplace, or *could* affect health and safety in the workplace. These factors affect employees (permanent and temporary), contractors, visitors, and anyone else who is in the workplace.

working drawings Drawings that consist of three related views – plan, elevation and section – and give a complete understanding of the building. Working drawings show the layout of the building, the setting-out dimensions, the spaces and parts of the building, and give specific information about the junctions between the parts of the building.

INDEX

Note: Page numbers in **bold** represent defined terms.

tradesperson, 70–7
training package, 80
training pathway, 80
training qualifications, 80–1
traverse, **339**
treat, 90
trestles, 118, 356–8
trimmer, 343
trimming, tool, 338–9
tripods, **425**–6
trolleys, safety and maintenance of, 307
trowels, 335
trundle wheel, 200
trussed roof, 252
try square, 327
tube, 118
turps, 294
two-peg test, **431**–2
two-way radio, 139–40
Type A cement, 290

U

understanding safety documents, 163
unfair dismissal laws, 72
unionism, 73
union meetings, 146
unions, 73
units of competency, 82
units of roofing material, calculate
　dimensions and quantity of, 397–8
utility knife, 331–2

V

vegetated filter strips, 93–4
verbal communication, **137**
verbal report, delivering, 140
vessels, 278
vibration, 8
video conferencing, 139
vocational education and training system
　(VET system), 80
voice inflection, 139
volatile organic compound (VOC), 88
volume measurement, 209
volume of slab, **202**

W

walkie talkie, 139
wall framing, 399
wall framing materials, calculate quantity of,
　388–91
wall structures, 250
warden, 33
Warrington hammers, 328
waste, **87–8**
waste disposal, 119
　meeting all legislative and workplace
　　requirements for, 361–2

waste minimisation, **87**
　planning for, 91
water-based paints, 294
water, 291
water level, **422**–4
water systems, 88
weight, 203
wheelbarrows, 354–5
wood, 278
woodworking tools, 327
　bench grinders, 330
　clamps, 332
　diamond stones, 330
　hammers, 327–8
　hand planes, 329–30
　hand saws, 330
　marking gauge, 331
　nail punches, 328–9
　oil stones, 330
　pincers, 330–1
　pinch or wrecking bars, 330
　planks, 332
　plasterboard hammers, 328
　saw stools, 332
　squares, 327
　tapes and rules, 331
　utility or safety trimming knife, 331–2
　wood chisels, 329
word-only signs, 18
work-related mental stress, 8
workers *see* employees, 13
Work Health and Safety (*National Uniform
　Legislation*) *Act 2011*, 4
work health and safety (WHS), **3**, 278–301,
　378–9
　legislation, 3–4
　meetings, 146
　requirements, equipment and hand,
　　power and pneumatic tools, 358–60
Work Health and Safety Act 2011 (WHS
　Act), 3–4, 9, 66, 163, 167, 169, 180,
　325
Work Health and Safety Act 2012, 3–4
Work Health and Safety Act 2020, 4, 180
Work health and safety committees, 66
work health and safety legislation
　for preparation of JSA using template,
　　169
　for preparation of SWMS, 180
work health and safety requirements, 75
　emergency procedures, 76
　offences and penalties, 75
　PCBU/employer responsibilities under
　　Act, 75
　site amenities, 75–6
　workers/employees' legal
　　responsibilities, 75
working drawings, **236**
working safely, 109

work injuries, 109
work instructions, required heights or
　levels are identified from, 430–1
work method statements (WMS), 110,
　140
workplace bullying, 74
workplace committees and meetings, 65
workplace communication
　access, interpret and present
　　information, 143–4
　basic verbal report, 144–5
　basic workplace records and
　　documents, 145–6
　basic written report, 144
　convey and receive information and
　　instructions, 135–43
　participate in simple meeting process,
　　146–9
workplace, extra risks in, 163
workplace hazards, 7
*Workplace Injury Rehabilitation and
　Compensation Act 2013*, 163
workplace meetings, 66
workplace organisations, 73
workplace procedures, 27
　inspect work site for hazards and
　　report, 419
workplace requirements, 278–301
　for construction project, 378
　store tools and equipment in
　　accordance with, 307
workplace signage, 140–1
work plan, 116–21
works committees, 65–6
work site, 163–4
　area, 33
　conditions, 180
　immediately before starting work and
　　discuss JSA, 179
　inspect, 166–7
　reviewing, 163
　safety signs and symbols, 32–3
work task
　organise performance of, 116–21
work team, 68–9
work/working
　efficiently and safely, **68**
　read and interpret work instructions
　　and planned sequence of, 277–8
　to comply with laws and regulations,
　　278–301
wrecking bars, 330
written communication
　examples, 140
　gather, convey and receive through,
　　140
written documents, 142–3